An Introduction to FORMAL

LANGUAGES

AND

MACHINE

COMPUTATION

An Introduction to **FORMAL**

LANGUAGES

AND

MACHINE

COMPUTATION

Song Y Yan
University of York, UK

World Scientific
Singapore • New Jersey • London • Hong Kong

Published by

World Scientific Publishing Co. Pte. Ltd.

P O Box 128, Farrer Road, Singapore 912805

USA office: Suite 1B, 1060 Main Street, River Edge, NJ 07661

UK office: 57 Shelton Street, Covent Garden, London WC2H 9HE

Library of Congress Cataloging-in-Publication Data
Yan, Song Y.
    An introduction to formal languages and machine computation / S. Y. Yan.
      p.   cm.
    Includes bibliographical references and index.
    ISBN 9810221673 -- ISBN 9810234228 (pbk)
    1. Formal languages.   2. Machine theory.   I. Title.
    QA267.3.Y36   1998
    004'.01'51--dc21   97-44021
                                          CIP

British Library Cataloguing-in-Publication Data
A catalogue record for this book is available from the British Library.

This book is printed on acid-free paper.

Printed in Singapore by Uto-Print

Preface

Computer Science: The study of the use and sometimes construction of digital computers. It is a fashionable, interesting, difficult, and perhaps useful activity.

– CHRISTOPHER STRACHEY (1916–1975)

This book provides a *concise* and a *modern* introduction to *Formal Languages and Machine Computation*, a group of disparate topics in the *Theory of Computation*, including formal languages, automata theory, Turing machines, computability, complexity, number-theoretic computation and cryptography.

The theory of computation is essentially a mathematical subject. Indeed, many theoretical computer scientists are mathematicians, and some of them are even *pure* mathematicians. On reviewing the history of computer science, one can easily see that the key figures in the development of modern computer science, such as John von Neumann, Alan Turing, Alonzo Church, Stephen Kleene, Donald Knuth, Stephen Cook, to name a few, are mathematicians. However, as pointed out by Prof. Richard Hamlet at the University of Maryland in his book *Introduction to Computation Theory*, 1974: "The disappointing fact is that while computer science welcomes some mathematical experts, and recommends that students of the subject have a dose of theory as 'preventative mathematics', not very many practical computer scientists attempt to understand and apply the deep results of theory". The reason for this, we think, is partly because mathematics, particularly certain branches of mathematics such as number theory and topology, are very hard to understand for most practical computer scientists. But the problem is that in many areas of computer science, mathematics is a very useful and sometimes even indispensable tool. In fact, although computer science is a very practical subject, it has its *root* in mathematics; for example, logic programming has its root in first-order logic, compilers in automata theory, programming languages in formal languages, computer architecture in Boolean algebra, algorithm analysis in combinatorial mathematics, and so on. In this book we shall try to bridge the gap between mathematics and computer science by presenting some basic mathematical concepts and results which are useful in computer science, but are not too mathematically involved. In fact, the book was written essentially for students in "practical" computer science, rather than "theoretical" computer science, or mathematics. To achieve such a goal, we do not follow the traditional mathematics textbook style: "Definition-Theorem-Proof", rather, we use the more practical style: "Definition-Theorem/Algorithm-Example/Exercise", because we believe that most practical computer scientists, systems designers or computer programmers are generally more interested in the applicability and usefulness of the theorems in computer science/mathematics, rather than how to prove the theorems. However, for those who really want to know the nature of the theorems, it is a good exercise for them to try to give all the missing proofs.

The book is organized as follows. Chapter 1 provides an introduction to topics in computation-related mathematics, including mathematical logic, set theory, functions and relations, graph theory, number theory, and abstract algebra, which provide a necessary mathematical foundation for the rest of the book. Although this chapter is rather lengthy in contrast to other similar books on the market, it proves to be an invaluable source of information because practical computer scientists usually lack sufficient mathematical background to understand the results in the *theory of computation*.

Chapter 2 is about automata and formal languages. Finite-state automata (FA) are the simplest possible model for computation. In fact, the human brain is a type of finite-state automaton with some 10^{25} states. We shall first study the basic concepts and results of finite automata and their corresponding regular languages L_{REG}. There are actually two different types of finite automata: deterministic finite automata (DFA) and non-deterministic finite automata (NFA). We shall show that these two types of automata are equivalent, because they can simulate each other, and more importantly, they accept exactly the same languages. Then we shall study push-down automata (PDA) and their corresponding languages, context-free languages L_{CFL}. Unlike DFA and NFA, non-deterministic push-down automata (NPDA) can accept a larger class of languages than deterministic push-down automata (DPDA), i.e., DPDA accept only deterministic context-free languages L_{DCFL}, whereas NPDA accept non-deterministic context-free languages L_{NCFL}. Since $L_{\text{DCFL}} \subset L_{\text{NCFL}}$, NPDA have more power than DPDA. Finally, we shall study Turing machines and their corresponding languages, called recursively enumerable languages L_{RE}. We shall show that L_{RE} is acceptable but not decidable by Turing machines.

We shall also show that there are languages, called recursive languages L_{REC}, that are recursive, but not enumerable, i.e., $L_{REC} \subset L_{RE}$. This class of recursive languages L_{REC} is decidable by Turing machines.

The machines we study in Chapter 2 act as language accepters. This does not mean that automata, or more generally Turing machines, can only be used to accept languages. Automata, particularly Turing machines, the highest-level automata, are general-purpose computing machines for general computation. So in Chapter 3, we shall study the computability/decidability and computational complexity within the theoretical framework of Turing machines. More specifically, we shall study what Turing machines (or general computers) can do and what Turing machines cannot do, what problems are feasible by a Turing machine (or a general computer) and what problems are infeasible by a Turing machine. We shall also provide a brief account of the asymptotic behaviours of some common complexity functions and the analysis of algorithms.

Number theory has played an important role in the development of modern computer science and engineering, noting the titles of the seminal papers by Alan Turing "On Computable Numbers, with an Application to the Entscheidungsproblem" (1936) and by Alonzo Church "An Unsolved Problem of Elementary Number Theory" (1936). So in Chapter 4, we shall study some modern mathematical methods for large number-theoretic computations, particularly those methods for primality testing and integer factorization of large numbers. We shall also discuss some applications of number-theoretic computations in computer science, including public-key cryptography, database security, random number generation, and high-speed computation.

Turing machines were proposed by Alan Turing in 1936. Since then, many new computation models, particularly parallel computation models (e.g., neural networks, cellular automata, quantum computation and molecular biological computation) have been proposed. In the last chapter of this book, we shall present a brief introduction to the two recently developed new models of computation, namely quantum computation and molecular biological computation. We shall show that from a computability point of view, both quantum and molecular biological models will have no more computational power than Turing machines. However, from a complexity point of view, both quantum and molecular biological computers do offer some more computational power. For example, the traveling salesman problem (TSP) and the directed Hamiltonian path problem (HPP) have been proven to be $\mathcal{NP}$-complete on Turing machines, but they can be solved in polynomial biological steps by biological computers; whereas integer factorization and discrete logarithms are believed to be intractable by Turing machines, but they can be solved in polynomial time on a quantum computer. Of course, the key issue concerning quantum and molecular biological models is how to actually build a truly quantum or biological computer! At present, no quantum and molecular biological computers are available on the market, but there is no reason why we cannot build such new generations of computers. Compared with many other books in the theory of computation on the market, this book contains substantially more materials in number-theoretic computations and their applications in computer science and secure communications. As pointed out by Prof. Ian Stewart (*Nature*, Vol. 325, 15 January 1987, page 199): "computer scientists working on algorithms for factorization would be well advised to brush up on their number theory", we hope these materials will not only help computer scientists to brush up on their number theory, but also encourage them to apply results in number theory to computer science in such diverse areas as computational complexity, public-key cryptography, network and information security, computer arithmetic and architectures, graphic design, image and signal processing, and digital communications.

Acknowledgements: This book grew out from my lecture notes for the following three courses: *Mathematics for Computing, Formal Languages and Machines*, and *Algorithms and Complexity*, taught at La Trobe University, Australia (1990–1992), and The Buckinghamshire College of Brunel University, England (1993–1995). It was completely rewritten while I was in the Departments of Mathematics and Computer Science at the University of York, England, over the period of 1993 and 1997. No book is created in a vacuum, and this one is no exception; I could not have written this book without the help of many friends, colleagues, students and my family. I would like to take this opportunity to thank James Austin, Thram Dillon, William Freeman, Terence Jackson, Milton Munroe, Ian Robinson and John Zeleznikow for their encouragement and help during the writing of this book. I would also like to thank the editors at World Scientific; Tony Moore at the London Office and particularly Stephen Patt and Joy Marie Tan at the Singapore Office for their help and encouragement. Special thanks must be given to William Bloodworth, Gavin Doherty, William Freeman, Milton Munroe, and Robert Pargeter, Peter Pleasants who read the various earlier versions of the book and suggested many improvements. Finally, I would like to thank all the authors cited in this book; they are the giants on whose shoulders I can stand on.

YORK, ENGLAND
DECEMBER 1997

Contents

ix

Notations

Notation	Explanation
$\mathbb{N}$	set of natural numbers: $\mathbb{N} = \{0, 1, 2, 3, \cdots\}$
$\mathbb{Z}$	set of integers: $\mathbb{Z} = \{0, \pm 1, \pm 2, \pm 3, \cdots\}$
$\mathbb{Z}^+$	set of positive integers: $\mathbb{Z}^+ = \{1, 2, 3, \cdots\}$
$\mathbb{Q}$	set of rational numbers: $\mathbb{Q} = \left\{ \frac{a}{b} : a, b \in \mathbb{Z}, \text{ and } b \neq 0 \right\}$
$\mathbb{R}$	set of real numbers: $\mathbb{R} = \{n + 0.d_1 d_2 d_3 \cdots : n \in \mathbb{Z},$ $\forall d_i \in [0, 1, \cdots, 9]$ and no infinite sequence of 9' appears$\}$
$\mathbb{C}$	set of complex numbers: $\mathbb{C} = \{a + bi : a, b \in \mathbb{R} \text{ and } i = \sqrt{-1}\}$
$\mathbb{Z}/n\mathbb{Z}$	residue classes (ring of integers mod n; it's a field if n is prime)
$(\mathbb{Z}/n\mathbb{Z})^*$	multiplicative group of $\mathbb{Z}/n\mathbb{Z}$
$\#(\mathbb{Z}/n\mathbb{Z})^*$	order of the multiplicative group $(\mathbb{Z}/n\mathbb{Z})^*$
$\mathbb{F}_p$	finite field with p elements, where p is a prime number
$\mathbb{F}_q$	finite field with $q = p^k$ a prime power
$\mathbb{K}$	field
$\blacksquare$	end of proof
$\neg$	negation
$\wedge$	conjunction
$\vee$	disjunction
$\Rightarrow$	implication
$\Leftrightarrow$	equivalence (if and only if)
$\vdash G$	formula G is provable, derivable or deducible
$P \vdash G$	G is provable from the set of formulas P
$\models G$	formula G is valid

$G \models P$	G is logic consequence of P
$\square$	blank symbol; empty clause
$\forall$	for all, e.g., $\forall n \in \mathbb{N}$
$\exists$	there exists, e.g., $\exists c \in \mathbb{R}$
$\cup$	union
$\cap$	intersection
$\star, \oplus, \odot$	binary operations
$\emptyset$	empty set
$\overline{A}$	complement of set A
$x \in A$	element x belongs to set A; e.g., $4 \in \mathbb{N}$
$x \notin A$	element x does not belong to set A; e.g., $0.4 \notin \mathbb{N}$
$A \subseteq B$	A is a subset of B
2^A	power set of A: the set of all subsets of set A
$a \, R \, b$	a is related to b, i.e., $(a, b) \in R$
$a \, \cancel{R} \, b$	a is not related to b, i.e., $(a, b) \notin R$
$\displaystyle\sum_{i=1}^{n} x_i$	summation: $x_1 + x_2 + \cdots + x_n$
$\displaystyle\prod_{i=1}^{n} x_i$	product: $x_1 x_2 \cdots x_n$
$n!$	factorial: $n(n-1)(n-2) \cdot 3 \cdot 2 \cdot 1$
x^k	x to the k-th power: $\left(k \geq 0 \implies \displaystyle\prod_{0 \leq j < k} x; \quad \dfrac{1}{x^{-k}} \right)$
kP	$kP = \underbrace{P + P + \cdots + P}_{k \text{ times}}$, where P is a point (x, y) on an elliptic curve $E : y^2 = x^3 + ax + b$
$\mathcal{O}_E$	the point at infinity on an elliptic curve E over a field
$\log_b x$	logarithm of x to the base b ($b \neq 1$): $x = b^{\log_b x}$
$\log x$	binary logarithm: $\log_2 x$
$\ln x$	natural logarithm: $\log_e x$, where $e = \displaystyle\sum_{n \geq 0} \frac{1}{n!} \approx 2.7182818$
$\exp(x)$	exponential of x: $e^x = \displaystyle\sum_{n \geq 0} \frac{x^n}{n!}$
$a \mid b$	a divides b

$a \nmid b$	a does not divide b
$\gcd(a, b)$	greatest common divisor of a, b
$\text{lcm}(a, b)$	least common multiple of a, b
$\lfloor x \rfloor$	floor: $\max \{n : n \leq x, n \in \mathbb{Z}\}$
$\lceil x \rceil$	ceiling: $\min \{n : n \geq x, n \in \mathbb{Z}\}$
$x \pmod{n}$	remainder: $x - n\lfloor x/n \rfloor$
$x \equiv y \pmod{n}$	x congruent to y modulo n
$x \not\equiv y \pmod{n}$	x not congruent to y modulo n
$\pi(x)$	number of primes $p \leq x$
$\sigma(n)$	sum of all positive divisors of n: $\sigma(n) = \sum_{d\mid n} d$
$\phi(n)$	Euler's totient function: $\phi(n) = \sum_{\substack{0 \leq k < n \\ \gcd(k,n)=1}} 1$
$[q_0, q_1, q_2, \cdots, q_n]$	finite simple continued fraction
$C_k = \dfrac{P_k}{Q_k}$	k-th convergent of a continued fraction
$[q_0, q_1, q_2, \cdots]$	infinite simple continued fraction
$[q_0, q_1, \cdots, q_k, \overline{q_{k+1}, q_{k+2}, \cdots, q_{k+m}}]$	periodic simple continued fraction
$\left(\dfrac{a}{n}\right)$	Jacobi symbol, where n is composite
$\left(\dfrac{a}{p}\right)$	Legendre symbol, where p is prime
Σ	alphabet: a finite nonempty set of symbols
$|w|$	length of string (or word) w
λ	empty string: $|\lambda| = 0$, $\lambda w = w\lambda = w$ with w a string; Carmichael's function
Σ^*	set of strings obtained by concatenating zero or more symbols from Σ
Σ^+	$\Sigma^+ = \Sigma^* - \{\lambda\}$
$\overline{L}$	complement of language L: $\overline{L} = \Sigma^* - L$
L^*	Kleene closure of language L: $L^* = L^0 \cup L^1 \cup L^2 \cup \cdots$
L^+	positive closure of language L: $L^+ = L^1 \cup L^2 \cup L^3 \cup \cdots$
$x \rightarrow y$	form of production rules in grammar G
$w \Longrightarrow z$	w derives z: $w = uxv$ and $z = uyv$

$w \overset{*}{\Longrightarrow} v$	w derives v by an arbitrary number of steps
$w \overset{+}{\Longrightarrow} v$	w derives v by at least one step
$\vdash$	a move from one instantaneous description (ID) of a machine to another, e.g., $q_0 aabbcc \vdash xq_1 abbcc$
$\overset{*}{\vdash}$	moves from one ID to another involving an arbitrary numbers of steps, e.g., $q_0 aabbcc \overset{*}{\vdash} xxyyzz\square q_4 \square$
$\overset{+}{\vdash}$	moves from one ID to another involving at least one step
DFA/NFA	deterministic/non-deterministic finite automata
DPDA/NPDA	deterministic/non-deterministic push-down automata
LBA	linear-bounded automata
RAM/PRAM	(parallel) random-access machine
TM	Turing machine
DTM	deterministic Turing machine
NDTM	non-deterministic Turing machine
PTM	probabilistic Turing machine
L_{REG}	regular languages
L_{CFL}	context-free languages
L_{CSL}	context-sensitive languages
L_{REC}	recursive languages
L_{RE}	recursively enumerable languages
$\mathcal{L}$	class of languages decidable in logarithmic space on a DTM
$\mathcal{NL}$	class of languages decidable in logarithmic space on a NDTM
$\mathcal{P}$	class of languages decidable in polynomial time on a DTM
$\mathcal{NP}$	class of languages decidable in polynomial time on a NDTM
$\mathcal{RP}$	class of languages decidable in expected polynomial time with one-sided error on a PTM
$\mathcal{ZPP}$	class of languages decidable in expected polynomial time with zero error on a PTM
$\mathcal{BPP}$	class of languages decidable in expected polynomial time with two-sided error on a PTM.

$\mathcal{NC}$

class of languages decidable on a parallel computer
e.g., a PRAM with a polynomial number of
processors in polylogarithmic time

co-$\mathcal{RP}$, co-$\mathcal{NP}$

complementary problems of $\mathcal{RP}$, and $\mathcal{NP}$

$\mathcal{O}(\cdot)$

upper bound: $f(n) = \mathcal{O}(g(n))$ if there exists *some*
constant $c > 0$ such that $f(n) \leq c \cdot g(n)$

$o(\cdot)$

upper bound that is not asymptotically tight:
$f(n) = \mathcal{O}(g(n))$, $\forall c > 0$ such that $f(n) < c \cdot g(n)$

$\Omega(\cdot)$

low bound: $f(n) = \Omega(g(n))$ if there exists a
constant c such that $f(n) \geq \frac{1}{c} \cdot g(n)$

$\Theta(\cdot)$

tight bound: $f(n) = \Theta(n)$ if $f(n) = \mathcal{O}(g(n))$
and $f(n) = \Omega(g(n))$

$\mathcal{O}(N^k)$

polynomial-time complexity measured in terms of arithmetic
operations, denoted by $\mathcal{O}_A\left(N^k\right)$, where k is a constant

$\mathcal{O}\left((\log N)^k\right)$

polynomial-time complexity measured in terms of bit operations,
denoted by $\mathcal{O}_B\left((\log N)^k\right)$, where k is a constant

$\mathcal{O}\left((\log N)^{c\log N}\right)$

superpolynomial complexity, where c is a constant

$\mathcal{O}\left(\exp\left(c\sqrt{\log N \log\log N}\,\right)\right)$

subexponential complexity, where c is a constant;
$\mathcal{O}\left(\exp\left(c\sqrt{\log N \log\log N}\,\right)\right) = \mathcal{O}\left(N^{c\sqrt{\log\log N/\log N}}\right)$

$\mathcal{O}\left(\exp(x)\right)$

exponential complexity, sometimes denoted by $\mathcal{O}\left(e^x\right)$

$\mathcal{O}\left(N^\epsilon\right)$

exponential complexity measured in terms of bit operations;
$\mathcal{O}\left(N^\epsilon\right) = \mathcal{O}\left(2^{\epsilon\log N}\right)$, where $\epsilon > 0$ is a constant

Chapter 1

Computation-Related Mathematics

If people do not believe that mathematics is simple, it is only because they do not realize how complicated life is.

$-$ JOHN VON NEUMANN (1903–1957)

The aim of this chapter is twofold: (i) to provide mathematical preliminaries for the rest of the book, (ii) to provide a mathematical foundation for general computer science. Thus the material in this chapter is also suitable for a single course in *Mathematics for Computing*. Topics discussed in this chapter include logic, set theory, graph theory, number theory, and abstract algebraic structures. As theoretical computer science can be well expressed in the two main mathematical languages – sets and numbers, our emphasis in this chapter will be on the theory of sets and particularly the theory of numbers.

1.1   Logics and Proofs

The purpose of mathematical logic, dating back to Aristotle 2000 years ago, is to provide a model of reasoning; this is clearly explained by the contemporary mathematical logician Stephen Kleene (1909-1994) that "logic has the important function of saying what follows from what". In recent times, mathematical logic has also proved useful as a model of computation and even a programming language. There are two levels of the study of logic. The first level, called *propositional logic*, deals with relationships among whole statements. The second level, called *predicate logic*, or sometimes called first-order logic, considers the internal structure of statements. In this section, we shall provide an introduction to propositional logic, predicate logic, programming in logic and some general proof techniques.

1.1.1   Propositional Logic

Definition 1.1.1 A *proposition* is a declarative statement which is either true or false, but not both simultaneously.

Example 1.1.1 Examples of propositions are:

(i) Snow is black. (false)

(ii) Triangles have four vertices. (false)

(iii) The Sun is bigger than the Moon. (true)

(iv) 6 is less than 24. (true)

(v) $2^{2976221} - 1$ is a prime number. (true; it is the 36th known Mersenne Prime, the largest known prime at present)

(vi) Tomorrow is my birthday. (conditional true or false)

Example 1.1.2 The following examples are not propositions:

(i) Please close the door.

(ii) Long live the Queen!

Simple (*atomic*) propositions can be combined into compound propositions (called *well-formed formulas*, or wff for short), by using the following *five* logic *connectives*:

(i) negation ¬

(ii) conjunction ∧

(iii) disjunction ∨

(iv) implication $\Rightarrow$

(v) equivalence $\Leftrightarrow$ (or $\equiv$)

Definition 1.1.2 The *truth values* (defined to be either true or false) for compound propositions combined by connectives $\neg, \wedge, \vee, \Rightarrow$ and $\Leftrightarrow$ are defined in Table 1.1.

P	Q	$\neg P$	$P \wedge Q$	$P \vee Q$	$P \Rightarrow Q$	$P \Leftrightarrow Q$
T	T	F	T	T	T	T
T	F	F	F	T	F	F
F	T	T	F	T	T	F
F	F	T	F	F	T	T

Table 1.1: Definitions of Five Most Commonly Used Connectives

Table 1.1 shows how the truth values of the compound propositions $\neg P$, $P \wedge Q$, $P \vee Q$, $P \Rightarrow Q$ and $P \Leftrightarrow Q$ are obtained from the given truth values of the atomic propositions (or atoms for short) P and Q. Such a table is usually called a *truth table*. Note that we shall use the names "propositions", "statements" and "formulas" interchangeably.

Remark 1.1.1 Of the five connectives, the implication connective $\Rightarrow$ may be the one most commonly used in mathematical reasoning. A wide range of ways have been used to express $P \Rightarrow Q$; some of them are:

(i) "if P, then Q"

(ii) "P implies Q"

(iii) "P only if Q"

(iv) "Q if P"

(v) "P is sufficient for Q"

(vi) "Q is necessary for P"

Note particularly that $P \Rightarrow Q$ is false only in the case that P is true but Q is false.

P	Q	$P \Rightarrow Q$	$Q \Rightarrow P$	$(P \Rightarrow Q) \wedge (Q \Rightarrow P)$	$P \Leftrightarrow Q$
T	T	T	T	T	T
T	F	F	T	F	F
F	T	T	F	F	F
F	F	T	T	T	T

Table 1.2: The Equivalence Connective

Remark 1.1.2 The *equivalence* statement, $P \Leftrightarrow Q$, is actually the shorthand for the statement $(P \Rightarrow Q) \wedge (Q \Rightarrow P)$. This can easily be seen from the truth table 1.2. So, two statements are said to be (logically) equivalent if they have the same truth table, that is, if the set of assignments for which A is true is the same as that for B. If Γ is the set of formulas and P is a single formula, then P is called a *logic consequence* of Γ (denoted by $P \models \Gamma$) if for any assignment making all members of Γ true, P is also true. Thus, two formulas are equivalent if and only if each is a logic consequence of the other.

Definition 1.1.3 Let P and Q are propositions. Then the *converse* of the proposition $P \Rightarrow Q$ is the proposition $Q \Rightarrow P$. The *contrapositive* of $P \Rightarrow Q$ is $\neg Q \Rightarrow \neg P$. The *inverse* of $P \Rightarrow Q$ is $\neg P \Rightarrow \neg Q$.

Example 1.1.3 Let P be the proposition "a number n is divisible by 6", denoted by $6 \mid n$, Q be the proposition "n is divisible by 3", denoted by $3 \mid n$. Then the compound proposition $P \Rightarrow Q$ becomes $6 \mid n \Rightarrow 3 \mid n$, representing *if a number n is divisible by 6, then it is also divisible by 3*. The converse of the proposition is $3 \mid n \Rightarrow 6 \mid n$. The contrapositive of the proposition is $3 \nmid n \Rightarrow 6 \nmid n$. The inverse of the proposition is $6 \nmid n \Rightarrow 3 \nmid n$.

Definition 1.1.4 A proposition is called a *tautology* or is said to be *valid* if it is true for all possible assignments of truth values to its propositional variables. (If P is a tautology, we denote it by $\models P$). A proposition is called a *contradiction* if it is false for all assignments. A proposition is *satisfiable* if it is true for at least one assignment and is *unsatisfiable* if it is false for at least one assignment.

Example 1.1.4 The proposition $P \Leftrightarrow (Q \Rightarrow P \wedge Q)$ is a tautology, as can be seen from Table 1.3, since it is always true regardless of the choice of the truth values for its propositional variables. But the proposition $(P \vee \neg P) \Rightarrow (Q \wedge \neg Q)$ is a contradiction, because its truth values are always false (see Table 1.4).

P	Q	$P \wedge Q$	$Q \Rightarrow (P \wedge Q)$	$P \Leftrightarrow (Q \Rightarrow P \wedge Q)$
T	T	T	T	T
T	F	F	T	T
F	T	F	F	T
F	F	F	T	T

Table 1.3: Truth Table for a Tautology

P	Q	$P \vee \neg P$	$Q \wedge \neg Q$	$(P \vee \neg P) \Rightarrow (Q \wedge \neg Q)$
T	T	T	F	F
T	F	T	F	F
F	T	T	F	F
F	F	T	F	F

Table 1.4: Truth Table for a Contradiction

Similar to the algebraic rules of arithmetic for real numbers in high school algebra, we can have the following algebraic rules for propositional calculus, which are formulated by tautologies of the form $P \Leftrightarrow Q$, where P and Q are propositions (we also let T denote a tautology, F a contradiction).

(i) The commutative laws:

$$P \vee Q \equiv Q \vee P, \tag{1.1}$$
$$P \wedge Q \equiv Q \wedge P. \tag{1.2}$$

(ii) The associative laws:

$$P \vee (Q \vee R) \equiv (P \vee Q) \vee R, \tag{1.3}$$
$$P \wedge (Q \wedge R) \equiv (P \wedge Q) \wedge R. \tag{1.4}$$

(iii) The distributive laws:

$$P \vee (Q \wedge R) \equiv (P \vee Q) \wedge (P \vee R), \tag{1.5}$$
$$P \wedge (Q \vee R) \equiv (P \wedge Q) \vee (P \wedge R). \tag{1.6}$$

(iv) The identity laws:

$$P \vee F \equiv P, \tag{1.7}$$
$$P \wedge T \equiv P. \tag{1.8}$$

(v) The idempotent laws:

$$P \vee P \equiv P, \tag{1.9}$$
$$P \wedge P \equiv P. \tag{1.10}$$

(vi) The complementary laws:

$$P \vee \neg P \equiv T, \tag{1.11}$$
$$P \wedge \neg P \equiv F. \tag{1.12}$$

(vii) DeMorgan's[1] laws:

$$\neg(P \vee Q) \equiv \neg P \wedge \neg Q, \tag{1.13}$$
$$\neg(P \wedge Q) \equiv \neg P \vee \neg Q. \tag{1.14}$$

(viii) The law for implication and its contrapositive:

$$P \Rightarrow Q \equiv \neg Q \Rightarrow \neg P. \tag{1.15}$$

(ix) The law for converse and inverse of $P \Rightarrow Q$:

$$Q \Rightarrow P \equiv \neg P \Rightarrow \neg Q. \tag{1.16}$$

Some *basic laws* for propositional calculus:

$$\neg(\neg P) \equiv P, \tag{1.17}$$
$$P \vee T \equiv T, \tag{1.18}$$
$$P \wedge F \equiv F, \tag{1.19}$$
$$P \Rightarrow Q \equiv \neg P \vee Q, \tag{1.20}$$
$$\neg(P \Rightarrow Q) \equiv P \wedge \neg Q. \tag{1.21}$$

---

[1]In honour of the British mathematician and logician Augustus DeMorgan (1806–1871) who was responsible for developing a more symbolic approach to algebra. His name is remembered in *DeMorgan's laws*, which he formulated. In an article of 1838, he also clarified the notion of *mathematical induction*.

Example 1.1.5 Using some of the above algebraic rules and some basic facts of tautology, show that

(i) $(P \wedge (P \Rightarrow Q)) \Rightarrow Q \equiv T$

(ii) $\neg(P \Leftrightarrow Q) \equiv \neg Q \Leftrightarrow P$.

PROOF.

(i) $\begin{aligned}
\text{LHS} &\equiv (P \wedge (P \Rightarrow Q)) \Rightarrow Q \\
&\equiv \neg(P \wedge (P \Rightarrow Q)) \vee Q && \text{(by basic laws)} \\
&\equiv (\neg P \vee \neg(P \Rightarrow Q)) \vee Q && \text{(by DeMorgan laws)} \\
&\equiv (\neg P \vee (P \wedge \neg Q)) \vee Q && \text{(by basic laws)} \\
&\equiv ((\neg P \vee P) \wedge (\neg P \vee \neg Q)) \vee Q && \text{(by distributive laws)} \\
&\equiv T \wedge (\neg P \vee \neg Q)) \vee Q && \text{(by identity law)} \\
&\equiv (T \wedge \neg P) \vee (T \wedge \neg Q)) \vee Q && \text{(by distributive laws)} \\
&\equiv (\neg P \vee \neg Q) \vee Q && \text{(by basic laws)} \\
&\equiv \neg P \vee (\neg Q \vee Q) && \text{(by associative laws)} \\
&\equiv \neg P \vee T && \text{(by complementary laws)} \\
&\equiv T && \text{(by basic laws)} \\
&\equiv \text{RHS.}
\end{aligned}$

(ii) $\begin{aligned}
\text{LHS} &\equiv \neg(P \Leftrightarrow Q) \\
&\equiv \neg((P \Rightarrow Q) \wedge (Q \Rightarrow P)) && \text{(by definition of } \Leftrightarrow \text{)} \\
&\equiv \neg(P \Rightarrow Q) \vee \neg(Q \Rightarrow P) && \text{(by DeMorgan laws)} \\
&\equiv (P \wedge \neg Q) \vee (Q \wedge \neg P) && \text{(by basic laws)} \\
&\equiv ((P \wedge \neg Q) \vee Q) \wedge ((P \wedge \neg Q) \vee \neg P)) && \text{(by distributive laws)} \\
&\equiv ((P \vee Q) \wedge (\neg Q \vee Q)) \wedge ((P \vee \neg P) \\
&\qquad \wedge (\neg Q \vee \neg P)) && \text{(by distributive laws)} \\
&\equiv ((P \vee Q) \wedge T) \wedge (T \wedge (\neg Q \vee \neg P)) && \text{(by complementary laws)} \\
&\equiv (P \vee Q) \wedge (\neg Q \vee \neg P) && \text{(by basic laws)} \\
&\equiv (Q \vee P) \wedge (\neg P \vee \neg Q) && \text{(by commutative laws)} \\
&\equiv (\neg Q \Rightarrow P) \wedge (P \Rightarrow \neg Q) && \text{(by basic laws)} \\
&\equiv \neg Q \Leftrightarrow P. && \text{(by definition of } \Leftrightarrow \text{)} \\
&\equiv \text{RHS.} \quad \blacksquare
\end{aligned}$

1.1.2 Predicate Logic

A natural generalisation of propositional logic is *predicate logic*, which we obtain by adding the universal quantifier $\forall$ and existential quantifier $\exists$ to the propositions. Predicate logic is also called *first-order logic*.

Definition 1.1.5 A *propositional function*, usually also called a *predicate*, denoted by

$$P(x_1, x_2, \cdots, x_n)$$

where P is called the predicate symbol, is a statement that contains one or more variables, so that its truth value depends on the values of the variables.

Example 1.1.6 Let x be an integer, and consider the predicate

$$p(x) = (x < 4) \wedge (x > 2).$$

This predicate is true if and only if $x = 3$. When values are substituted for variables in a predicate, the predicate then becomes an ordinary proposition. For example,

$$p(2) = (2 < 4) \wedge (2 > 2)$$

is just an ordinary proposition, which is false. A predicate then stands for a collection of propositions.

Example 1.1.7 Let the predicate symbol F be "father", then the predicate $F(x, y)$ means that x is y's father. So for the predicate F, we could get a set of propositions as F(John, Mary), F(John, Bill), F(Bill, Richard), etc.

Definition 1.1.6 A *quantifier* is a rule that assigns a single proposition to a propositional function. A *existential quantifier*, denoted by $\exists$, assigns to a propositional function P the proposition "The truth set of P is not empty". The result of applying the existential quantifier $\exists$ to P is written $\exists x P(x)$, and is read "There exists x such that $P(x)$". A *universal quantifier*, denoted by $\forall$, assigns to a propositional function P the proposition "The truth set of P is equal to its domain". The result of applying the universal quantifier $\forall$ to P is written $\forall x P(x)$, and is read "For all x (or for every x), $P(x)$".

Example 1.1.8 Let P be a propositional function on $\mathbb{Z}$ ($\mathbb{Z}$ denotes the set of all integers) defined by $P(x) = x > 5$. By applying the existential quantifier $\exists$ to P we get $\exists x(x > 5)$. It is true since e.g., 6 (among other integers) belongs to the truth set of P. Thus the truth set is not empty. Similarly, by applying the universal quantifier $\forall$ to P we get $\forall x(x > 5)$. This is false because the truth set of P is $\{6, 7, 8, \cdots\}$, which is not equal to it's domain $\mathbb{Z}$.

Definition 1.1.7 Let $P(x)$ be a predicate and D the domain of x. A *universal statement* is a statement of the form $\forall x \in D, P(x)$. It is defined to be true if and only if $P(x)$ is true for every x in D. It is defined to be false if and only if $P(x)$ is false for at least one x in D. A value for x for which $Q(x)$ is false is called a *counterexample* to the universal statement.

Example 1.1.9 Consider the following two problems:

(i) Let $D = \{1, 2, 3, 4, 5\}$ and consider the statement

$$\forall x \in D, \quad x^2 \geq x$$

Show that this statement is true.

(ii). Consider the statement
$$\forall x \in \mathbb{R}, \quad x^2 \geq x$$

where $\mathbb{R}$ denotes the set of all real numbers. Find a counterexample to show that this statement is false.

Solution:

(i) Since
$$1^2 \geq 1, \ 2^2 \geq 2, \ 3^2 \geq 3, \ 4^2 \geq 4, \ 5^2 \geq 5$$

then "$\forall x \in D, \quad x^2 \geq x$" is true.

(ii) Take $x = \frac{1}{2}$. Then we have

$$\left(\frac{1}{2}\right)^2 = \frac{1}{2} \not\geq \frac{1}{2}.$$

Thus we have discovered a counterexample, which shows that "$\forall x \in \mathbb{R}, x^2 \geq x$" is false.

Definition 1.1.8 Let $P(x)$ be a predicate and D the domain of x. An *existential statement* is a statement of the form $\exists x \in D, P(x)$. It is defined to be true if and only if $P(x)$ is true for at least one x in D. It is defined to be false if and only if $P(x)$ is false for all x in D.

Example 1.1.10 Consider the following two problems:

(i) Let $E = \{5, 6, 7, 8, 9, 10\}$ and consider the statement

$$\exists m \in E \text{ such that } m^2 = m$$

Show that this statement is false.

(ii) Consider the statement

$$\exists m \in \mathbb{Z} \text{ such that } m^2 = m$$

Show that this statement is true.

Solution:

(i) Since
$$5^2 \neq 5, \ 6^2 \neq 6, \ 7^2 \neq 7, \ 8^2 \neq 8, \ 9^2 \neq 9, \ 10^2 \neq 10$$
then "$\exists m \in E, x^2 \geq x$" is false. the "$\exists m \in E$ such that $m^2 = m$" is true.

(ii) Take $m = 1$ we have $1^2 = 1$. So "$\exists m \in \mathbb{Z}$ such that $m^2 = m$" is true.

The quantifier $\forall$ and $\exists$ just discussed are closely related. For example, consider the proposition $\exists x P(x)$ about a predicate P. If $\exists x P(x)$ is false, the truth set of P is $\emptyset$. But if the truth set of P is $\emptyset$, then truth set of $\neg P$ is all of the domain of P. That is, $\forall \neg P(x)$ is true. Conversely, $\forall \neg P(x)$ is true, the truth set of P is $\emptyset$, so $\exists x P(x)$ is false. Thus we see that for any predicate P, the propositions $\exists P(x)$ is equivalent to $\neg \forall \neg P(x)$. Thus, we have two more equivalence laws for predicate logic:

$$\neg \exists x P(x) \equiv \forall \neg P(x), \tag{1.22}$$
$$\neg \forall x P(x) \equiv \exists \neg P(x). \tag{1.23}$$

Plus some tautologies from propositional logic:

$$\neg(P \vee Q) \equiv \neg P \wedge \neg Q, \tag{1.24}$$
$$\neg(P \wedge Q) \equiv \neg P \vee \neg Q. \tag{1.25}$$
$$\neg(P \Rightarrow Q) \equiv P \wedge \neg Q. \tag{1.26}$$

So we have five rules for transforming statements beginning with negation into more easily understood forms.

1.1.3   Programming in Logic

Mathematical logic has many successful applications in computer science; one of them is the application in computer programming, resulting in the birth of *logic programming* [72] in general and PROLOG in particular [28]. In this subsection, we shall briefly introduce the basic ideas and definitions of logic programming based on Horn clauses.

(I) Basic Definitions for Logic Programs

Let us first begin with formal definitions for the alphabet of predicate languages, and terms and formulas over the alphabet:

Definition 1.1.9 An *alphabet* for the predicate language consists of seven classes of symbols:

(i) variables,

(ii) constants,

(iii) function symbols,

(iv) predicate symbols,

(v) connectives,

(vi) quantifiers,

(vii) punctuation symbols.

Definition 1.1.10 A *term* is defined inductively as follows:

(i) a variable is a term,

(ii) a constant is a term,

(iii) if f is an n-ary function symbol and $t_1, t_2, \cdots, t_n$ are terms, then $f(t_1, t_2, \cdots, t_n)$ is a term.

Definition 1.1.11 A *well-formed formula* (or just *formula* for short) is defined inductively as follows:

(i) if p is an n-ary predicate symbol and $t_1, t_2, \cdots, t_n$ are terms, then $p(t_1, t_2, \cdots, t_n)$ is a formula,

(ii) if P and Q are formulas, then so are $\neg P$, $P \wedge Q$, $P \vee Q$, $P \Rightarrow Q$ and $P \Leftrightarrow Q$,

(iii) if P is a formula and x is a variable, then $\forall x P$ and $\exists x P$ are formulas.

If P is of the form $p(t_1, t_2, \cdots, t_n)$, then P is called an atomic formula (or just *atom* for short), and $\neg P$ a negation of the atom P. (It is customarily to include the empty clause, denoted by $\square$, as an atom). A *literal* is either an atom or a negation of the atom.

In logic programming, we are generally more interested in the following clausal formulas.

Definition 1.1.12 A *clause* is a formula of the form

$$\forall x_1 \forall x_2 \cdots \forall x_k (L_1 \vee L_2 \vee \cdots \vee L_n) \tag{1.27}$$

where each L_i is a *literal* and $x_1, x_2, \cdots, x_k$ are all the variables occuring in $L_1 \vee L_2 \vee \cdots \vee L_n$. On the assumption that all variables in $L_1 \vee L_2 \vee \cdots \vee L_n$ are universally quantified at the front of the clause, (1.27) can be rewritten briefly as follows:

$$L_1 \vee L_2 \vee \cdots \vee L_n. \tag{1.28}$$

More specifically, a clause is a formula of the form

$$A_1 \vee A_2 \vee \cdots \vee A_m \vee \neg B_1 \vee \neg B_2 \vee \cdots \vee \neg B_n \tag{1.29}$$

where we have listed the positive literals first. It is clear that all the following forms of clauses are logically equivalent:

$$
\begin{aligned}
A_1 \vee A_2 &\vee \cdots \vee A_m \vee \neg B_1 \vee \neg B_2 \vee \cdots \vee \neg B_n \\
&\Longleftrightarrow A_1 \vee A_2 \vee \cdots \vee A_m \vee \neg(B_1 \wedge B_2 \wedge \cdots \wedge B_n) \\
&\Longleftrightarrow A_1 \vee A_2 \vee \cdots \vee A_m \leftarrow B_1 \wedge B_2 \wedge \cdots \wedge B_n \\
&\Longleftrightarrow B_1 \wedge B_2 \wedge \cdots \wedge B_n \rightarrow A_1 \vee A_2 \vee \cdots \vee A_m.
\end{aligned}
$$

But in logic programming, (1.29) is written in another logically equivalent formula:

$$A_1, A_2, \cdots, A_m \leftarrow B_1, B_2, \cdots, B_n. \tag{1.30}$$

Definition 1.1.13 If a clause is of the form

$$\leftarrow \tag{1.31}$$

we call it *empty clause*, and denote it by $\square$.

Definition 1.1.14 A *Horn clause* is a clause having one of the following three forms:

$A \vee \neg B_1 \vee \neg B_2 \vee \cdots \vee \neg B_n$ (one positive and some negative literals)
A (one positive and no negative literals)
$\neg B_1 \vee \neg B_2 \vee \cdots \vee \neg B_n$ (no positive and all negative literals).

That is,

$$A \;\leftarrow\; B_1 \wedge B_2 \wedge \cdots \wedge B_n \tag{1.32}$$

$$A \;\leftarrow \tag{1.33}$$

$$\leftarrow\; B_1 \wedge B_2 \wedge \cdots \wedge B_n. \tag{1.34}$$

Definition 1.1.15 A (definite) logic *program clause* is a Horn clause of the form:

$$A \leftarrow B_1, B_2, \cdots, B_n \tag{1.35}$$

where A is called the *head* and $B_1, B_2, \cdots, B_n$ the *body* of the clause, respectively. The body $B_1, B_2, \cdots, B_n$ may be absent, in which case, (1.35) becomes $A \leftarrow$. The procedure interpretation for (1.35) is that "to solve A, solve $B_1, B_2, \cdots, B_n$"; whereas the semantical interpretation for (1.35) is just the same as the predicate formula $B_1 \vee B_2 \vee \cdots \vee B_n \to A$, that is, "if $B_1, B_2, \cdots, B_n$ are true, then A is true". A (definite) logic program is a finite set of such program clauses, whereas a defined *goal* for a logic program is a Horn clause of the form:

$$\leftarrow B_1, B_2, \cdots, B_n. \tag{1.36}$$

Example 1.1.11 For example, The following is a simple logic program:

father(Fred, Bill) $\leftarrow$
father(Bill, John) $\leftarrow$
grandfather(X, Z) $\leftarrow$ father(X, Y), father(Y, Z)

To run the program, we give it a goal G such as

$$\leftarrow \text{grandfather(Fred, John)}$$

which is understood as a request to find "if Fred is John's grandfather". More formally, it is a request to show if the goal G is the logical consequence of the program P, or if G is provable from P, or if $P \cup \{G\}$ is unsatisfiable.

In what follows, we shall show how answers in logic programming can be inferred from the resolution principle.

(II) Resolution and Unification

Resolution is a method of mechanical theorem proving, first introduced by J. A. Robinson in 1965 [121]. The basic idea of *resolution* is as follows. It takes two clauses and constructs a new clause from them by deleting e.g., all the occurrences of p in the first clause and then all the occurrences of $\neg p$ in the second clause:

$$\frac{\begin{array}{c} P \\ \neg P \end{array}}{\square} \qquad\qquad \frac{\begin{array}{c} P \vee Q \\ \neg P \vee R \end{array}}{Q \vee R}$$

The new clause is called the *resolvent* of the previous two clauses. For simplicity, we shall first consider the resolution rule for propositional clauses (i.e., clauses containing no variables):

$$\frac{\begin{array}{c} A_1, A_2, \cdots, A_m \leftarrow B_1, B_2, \cdots, B_n, P \\ P, C_1, C_2, \cdots, C_j \leftarrow D_1, D_2, \cdots, D_k \end{array}}{A_1, A_2, \cdots, A_m, C_1, C_2, \cdots, C_j \leftarrow B_1, B_2, \cdots, B_n, D_1, D_2, \cdots, D_k}$$

A proof by resolution is called a *refutation* that uses only the resolution rule (no axioms will be needed).

Example 1.1.12 Let the program P be

$$\begin{array}{l} \text{T} \leftarrow \text{Q, R} \\ \text{Q} \leftarrow \\ \text{R} \leftarrow \end{array}$$

and the goal G be $\leftarrow$ T. Then the computation for the answer to G wrt P is a refutation, which tries to prove $P \cup \{G\}$ is unsatisfiable:

1. T ← Q, R	P
2. Q ←	P
3. R ←	P
4. ← P	G
5. ← Q, R	Resolution on 4 and 1
6. ← R	Resolution on 5 and 2
7. □	Resolution on 6 and 3.

Since the proof terminates at an empty clause, then $P \cup \{G\}$ is unsatisfiable, and hence G is provable from P.

Remark 1.1.3 Note that the set of clauses of a predicate language (e.g., $P \cup \{G\}$) is contradictory (i.e., unsatisfiable) if and only if the empty clause can be derived from it by meanings of resolution. Note also that in mathematical logic, we usually denote $P \vdash G$ if G is provable, or derivable, or deducible from P; if $\vdash G$, the G is said to be provable or to be a theorem.

When resolution is used to predicate logic, the situation becomes more complicated, and some extra operations such as *substitutions* and *unifications* will be needed. Let us first provide some definitions and examples of expressions and substitutions.

Definition 1.1.16 An *expression* is either a term, a literal, a conjunction, or a disjunction of literals. A simple expression is either a term or an atom. A *substitution* θ is a finite set of the form $\{v_1/t_1, v_2/t_2, \cdots, v_n/t_n, \}$ where each v_i is a variable and each t_i is a term distinct from v_i. Each v_i/t_i is called a *binding*. If θ and σ are substitutions then the *composition* $\theta\sigma$ of θ and σ satisfies $(t\theta)\sigma = t(\theta\sigma)$ for any term t.

The following is an example of applying substitutions to expressions.

Example 1.1.13 Let

$$\theta = \{x/b, y/x\} \quad \text{and}$$
$$E = p(x, y, f(a)).$$

Then $E\theta = p(b, x, f(a))$.

Definition 1.1.17 Let

$$\theta = \{u_1/s_1, u_2/s_2, \cdots, u_n/s_m, \} \text{ and}$$
$$\sigma = \{v_1/t_1, v_2/t_2, \cdots, v_n/t_n, \}$$

be substitutions. Then the *composition* $\theta\sigma$ of θ and σ is the substitution obtained from the set

$$\{u_1/s_1\sigma, u_2/s_2\sigma, \cdots, u_n/s_m\sigma, v_1/t_1, v_2/t_2, \cdots, v_n/t_n\}$$

by deleting any binding $u_i/s_i\sigma$ for which $u_i = s_i\sigma$ and deleting any binding v_j/t_j for which $j \in \{u_1, u_2, \cdots, u_m\}$.

It is clear that

(i) $\theta\epsilon = \epsilon\theta = \theta$, where ϵ is an identity substitution.

(ii) $(E\theta)\sigma = E(\theta\sigma)$ for any expression E,

(iii) $(\theta\sigma)\gamma = \theta(\sigma\gamma)$.

Example 1.1.14 Let $\theta = \{x/f(y), y/z\}$ and $\sigma = \{x/a, y/b, z/y\}$. Then $\theta\sigma = \{x/f(b), z/y\}$.

In logic programming, we will be particularly interested in substitutions that unify a set of expressions, that is, make each expression in the set syntactically identical.

Definition 1.1.18 Let S be a finite set of simple expressions. A substitution θ is called a *unifier* for S if $S\theta$ is a singleton. A unifier is called a *most general unifier* (mgu) for S if for each unifier σ of S there exists a substitution γ such that $\sigma = \theta\gamma$.

Example 1.1.15 $\{p(f(x), a), \ p(y, f(w))\}$ is not unifiable, since the second arguments cannot be unified, that is, they have no unifier. But $\{p(f(x), z), \ p(y, a)\}$ is unifiable, since $\sigma = \{y/f(x), x/a, z/a\}$ is a unifier; a most general unifier is $\theta = \{y/f(x), z/a\}$ (note that $\sigma = \theta\{x/a\}$).

The resolution rule for predicate clauses (i.e., clauses containing variables) is an elaboration, using unification, of that for propositional clauses:

$$\frac{\begin{array}{c} A \vee \neg B_1 \vee \neg B_2 \vee \cdots \vee \neg B_n \\ \vee \neg C_1 \vee \cdots \vee \neg C_j \vee \cdots \vee \neg C_k \end{array}}{(\neg C_1 \vee \cdots \vee \neg C_{j-1} \vee \neg B_1 \vee \neg B_2 \vee \cdots \vee \neg B_n \vee \neg C_{j+1} \vee \cdots \vee \neg C_k)\,\sigma}$$

that is,

$$\frac{\begin{array}{c} A \leftarrow B_1, B_2, \cdots, B_n \\ \leftarrow C_1, \cdots, C_j, \cdots, C_k \end{array}}{\leftarrow (C_1, \cdots, C_{j-1}, B_1, B_2, \cdots, B_n, C_{j+1}, \cdots, C_k)\,\sigma}$$

where σ is the mgu of A and C_j, and the set of variables occuring in $A \leftarrow B_1, B_2, \cdots, B_n$ is disjoint from the set of those occuring in $\leftarrow C_1, C_2, \cdots, C_k$; if these set are not already disjoint then they can be made so by *renaming substitution*, which can be defined as follows. Let E be an expression, and V the set of variables occuring in E. A renaming substitution of E is just a substitution $\{x_1/y_1, x_2/y_2, \cdots, x_n/y_n\}$ such that $\{x_1, x_2, \cdots, x_n\} \subseteq V$, the y_i are distinct, and $(V \backslash \{x_1, x_2, \cdots, x_n\}) \cap \{y_1, y_2, \cdots, y_n\} = \emptyset$. The following is an example of the more general resolution rule for predicate clauses:

$$\frac{\begin{array}{c} P(x, f(x)) \leftarrow Q(x,y), R(y) \\ \leftarrow P(a,z), R(z) \end{array}}{\leftarrow Q(a,y), R(y), R(f(a)).}$$

(III) SLD-Resolution for Logic Programs

To evaluate a Horn clause logic program, we use the so-called SLD-resolution: SLD stands for Selective Linear resolution for Definite (i.e., Horn) clauses. Let us begin with some formal definitions related to SLD-resolution.

Definition 1.1.19 Let G be $\leftarrow C_1, \cdots, C_j, \cdots, C_n$ and P be $A \leftarrow B_1, B_2, \cdots, B_m$. The G' is derived from G and P using mgu θ if the following conditions satisfy:

(i) C_j is an atom, called the *selected atom*, in G;

(ii) θ is an mgu of C_j and A;

(iii) G' is the new goal $\leftarrow C_1, \cdots, C_{j-1}, B_1, B_2, \cdots, B_m, C_{j+1}, \cdots, C_n$.

Definition 1.1.20 Let P be a program and G a goal. A SLD-derivation of $P \cup \{G\}$ consists of a sequence $G_0 = G, G_1, G_2, \cdots$ of goals, a sequence of variants of program clauses of P and a sequence $\theta_1, \theta_2, \cdots$ of mgu's such that each G_{i+1} is derived from G_i and C_{i+1} using θ_{i+1}. A SLD-refutation of $P \cup \{G\}$ is a finite SLD-derivation of $P \cup \{G\}$ which has the empty clause $\square$ as the last goal in the derivation. If $G_n = \square$, then the refutation has length n.

Definition 1.1.21 Let P be a program and G a goal $\leftarrow B_1, b_2, \cdots, B_n$. An *answer* for $P \cup \{G\}$ is a substitution θ for variables in G. θ is called a *correct answer* for $P \cup \{G\}$ if $\forall((B_1, b_2, \cdots, B_n)\theta)$ is a logical consequence of P.

Theorem 1.1.1 Let P be a program and G a goal. If there exists an SLD-refutation of $P \cup \{G\}$, then $P \cup \{G\}$ is unsatisfiable.

Informally, SLD-resolution for (definite) logic program can be described as follows: Resolve the goal $\leftarrow C_1, \cdots, C_j, \cdots, C_n$ with the program clause $A \leftarrow B_1, B_2, \cdots, B_n$ by first unifying A via mgu θ. Then replace C_j in the goal by the body $B_1, B_2, \cdots, B_n$ and apply θ to the resulting goal to obtain the resolvent $\leftarrow (C_1, \cdots, C_{j-1}, B_1, B_2, \cdots, B_n, C_{j+1}, \cdots, C_k)\theta$. To construct a logic program proof, we start by listing each program clause and the goal as a premise. Then we use the SLD-resolution repeatedly to add new resolvent to the proof, each new resolvent being constructed from the goal on the previous line together with some program clauses. We shall introduce the use of the SLD-resolution in the following example.

Example 1.1.16 Let P be the following logic program:

$$\text{father(William, Bill)} \leftarrow$$
$$\text{father(Bill, John)} \leftarrow$$
$$\text{father(Bill, Mary)} \leftarrow$$
$$\text{grandfather(X, Z)} \leftarrow \text{father(X, Y), father(Y, Z)}$$

and the goal G be

$$\leftarrow \text{grandfather(R, S)}.$$

For convenience, we briefly denote the program clauses in P and the goal G as follows:

$$f(w, b) \leftarrow$$
$$f(b, j) \leftarrow$$
$$f(b, m) \leftarrow$$
$$g(X, Z) \leftarrow f(X, Y), f(Y, Z)$$
$$\leftarrow g(R, S).$$

To compute the answer for G with respect to P is equivalent to prove that $P \cup \{G\}$ is unsatisfiable. The proof by SLD-resolution is as follows:

PROOF.

1.	$f(w, b) \leftarrow$	P (program clause)
2.	$f(b, j) \leftarrow$	P (program clause)
3.	$f(b, m) \leftarrow$	P (program clause)
4.	$g(X, Z) \leftarrow f(X, Y), f(Y, Z)$	P (program clause)
5.	$\leftarrow g(R, S)$	G (initial goal)
6.	$\leftarrow f(R, Y), f(Y, S)$	Resolution on 4 and 5; $\theta_1 = \{R/X, S/Z\}$
7.	$\leftarrow f(b,S)$	Resolution on 1 and 6; $\theta_2 = \{R/w, Y/b\}$
8.	$\leftarrow f(b,j)$	Resolution on 2 and 7; $\theta_3 = \{S/j\}$
9.	$\square$	Resolution on 2 and 8; $\theta_4 = \{ \}$ ∎

So, we get $R = w$, and $S = j$. That is, William is John's grandfather. There is, in fact, another possible answer for the goal $\leftarrow g(R, S)$, namely $R = w$, and $S = m$, that is William is also Mary's grandfather. To find all possible answers for a goal G with respect to a program P is to derive the whole SLD proof tree (SLD-tree for short) for that goal G wrt the program P. For example, the SLD-tree for the goal $\leftarrow g(R, S)$ wrt the P defined above can be pictured as in Figure 1.1 (suppose we use the standard depth-first search and left-most computation strategies).

Exercise 1.1.1 Let the program P be defined as follows:

$$p(x, x) \leftarrow$$
$$q(a, b) \leftarrow$$
$$p(x, z) \leftarrow q(x, y), f(y, z)$$

and the goal G be $\leftarrow p(x, b)$. Derive the SLD-tree for the goal G, and find all the answers for the goal G with respect to the program P (suppose we use the standard depth-first search and left-most computation strategies).

In this subsection, we have just introduced the very basic idea and theory of logic programming, those who desire a more detailed exposition are advised to consult e.g., [56] and [85].

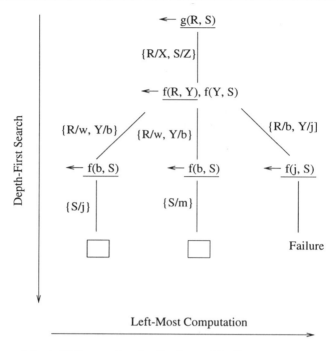

Figure 1.1: A SLD-tree for the Goal $\leftarrow g(R, S)$ wrt the Program P

1.1.4   Proof Techniques

In the previous subsection, we have introduced a special machine-oriented theorem proving method (i.e., the resolution principle) and its application in programming. In this subsection, we shall introduce some general proof techniques
widely used in mathematics and computer science.

(I) Direct Proof

Proof is not only useful in mathematics, but also in computer science. In fact,
a computer program can be regarded as a mathematical theorem. Thus, the
methods used for proving mathematics theorems can then be used to prove the
correctness of computer programs.

There are many proving techniques one can use to prove mathematical theorems. The obvious approach to proving a theorem

$$P \Rightarrow Q$$

is *direct proof* – assume the hypothesis P and deduce the conclusion Q. The best way to introduce the method of direct proof is via some examples:

Example 1.1.17 Prove that $\forall x \in \mathbb{Z}$, $6 \mid x \Rightarrow 3 \mid x$. That is, we need to prove that "if an integer is divisible by 6, then it is also divisible by 3".

PROOF.

Hypothesis: $6 \mid x$

$x = 6 \cdot k$ for some integer k	(definition of divisibility)
$x = (2 \cdot 3)k$	
$x = 3(2 \cdot k)$	
$x = 3 \cdot k'$	(k' is an integer)
Conclusion: $3 \mid x$.	(definition of divisibility) ∎

Example 1.1.18 Prove that $\forall x \in \mathbb{Z}$, $x \mid a \ \& \ x \mid b \Rightarrow x \mid (a \pm b)$.

PROOF.

Hypothesis: $x \mid a \ \& \ x \mid b$

$\exists a, b$ such that $a = xu$ and $b = xv$	(definition of divisibility)
$a \pm b = xu \pm xv = x(u \pm v)$	(substitution)
$x \mid [x(u \pm v)]$	
Conclusion: $x \mid (a \pm b)$.	∎

(II) Proof by Contraposition

Very often, it is not practical to give a direct proof of a theorem. Instead, an *indirect proof* is given. An indirect proof consists of two steps: we first prove some statement rather than the statement of the theorem, then we use logic to show that the theorem follows from the statement just proved. There are two common methods of indirect proof: *proof by contraposition* and *proof by contradiction*.

The idea of proofs by contraposition is based on the equivalence of $P \Rightarrow Q$ and $\neg Q \Rightarrow \neg P$. That is, to prove $P \Rightarrow Q$ by contraposition, we prove instead that $\neg Q \Rightarrow \neg P$, which is valid, because

$$(\neg Q \Rightarrow \neg P) \Rightarrow (P \Rightarrow Q)$$

is a tautology. The following are some example of proof by contraposition:

Example 1.1.19 Prove by contraposition that $\forall x \in \mathbb{Z}, \ 6 \mid x \Rightarrow 3 \mid x$.

PROOF.   First notice that the contraposition of "$6 \mid x \Rightarrow 3 \mid x$" is "$3 \nmid n \Rightarrow 6 \nmid n$".

Hypothesis: $3 \nmid x$
$(3k) \nmid x$ for any integer k
$(3 \cdot 2k') \nmid x$ for any integer k'
$(6 \cdot k') \nmid x$
Conclusion: $6 \nmid x$.   ∎

Example 1.1.20 A positive integer is called perfect[2] if it is the sum of all its proper divisors. For example, 6 is perfect, since $6 = 1 + 2 + 3$. Prove that "a perfect number cannot be prime".

PROOF.   We prove the assertion by contraposition, that is, we prove that "if a number is prime, then it cannot be perfect". Suppose a positive integer p is prime, then the only proper divisor of p is 1, so the sum of the proper divisors is 1, which is not equal to p. Hence, p cannot be perfect. ∎

Example 1.1.21 Let $\sigma(m)$ be the sum of all positive divisors of m (including 1 and m). A pair of positive integers (m, n) with $m < n$ is called amicable[3] if $\sigma(m) = \sigma(n) = m + n$ (where e.g., $\sigma(m)$ denotes the sum of all positive divisors of m, including 1 and m itself). For example, $(220, 284)$ is an amicable pair, because $\sigma(220) = \sigma(280) = 220 + 280 = 504$. Prove by contraposition that "an amicable pair cannot be both primes".

PROOF.   The contraposition of the statement is that "if a pair of positive integers are both primes, then it cannot be amicable". Suppose the pair of

---

[2]The famous Euclid-Euler's theorem states that n is perfect if and only if $n = 2^{p-1}(2^p - 1)$, where $2^p - 1$ is Mersenne prime (the number $2^p - 1$ is called Mersenne prime whenever p is prime $2^p - 1$ is also prime; e.g., $2^{13} - 1 = 8191$ is Mersenne prime, but $2^{11} - 1 = 2047$ is not Mersenne prime since $2047 = 23 \times 89$). The first four perfect numbers 6, 28, 496 and 8128 were found in Euclid's *Elements* 2000 years ago, but the 36th perfect number $2^{2976220}(2^{2976221} - 1)$, the largest known perfect number at present was not found until 24 August 1997. Note that $2^{2976221} - 1$ is also the largest known prime at the present time!

[3]The first amicable pair $(220, 284)$ was known to the legendary Pythagoras 2000 years ago, but the second smallest amicable pair $(1184, 1210)$ was not found until 1866 when it was discovered by a 16-year-old Italian school-boy, Nicolo Paganini. Two large amicable pairs of record size (1739 resp. 1923 decimal digits) were found by Mariano Garcia in New York and sent to Herman te Riele in Amsterdam on 10 February 1997. Readers who are interested in perfect and amicable numbers as well as Mersenne primes are advised to consult [147].

positive integers (p_1, p_2) is prime. Then $\sigma(p_1) = p_1 + 1$ and $\sigma(p_2) = p_2 + 1$. So $\sigma(p_1) \neq \sigma(p_2) \neq p_1 + p_2$. Thus, the pair (p_1, p_2) cannot be an amicable pair. ∎

(III) Proof by Contradiction

Proofs by contradiction are also used quite often in mathematics and computer science. This technique is powerful in some situations where other attacks fail. The idea of proof by contradiction is that let F stand for any contradiction, i.e., any proposition whose truth value is always false ($\neg A \wedge A$ would be such a proposition). Suppose we are trying to prove $P \Rightarrow Q$. By constructing a truth table, we see that

$$(P \wedge \neg Q \Rightarrow F) \Rightarrow (P \Rightarrow Q)$$

is a tautology, so to prove the theorem $P \Rightarrow Q$, it is sufficient to prove $P \wedge \neg Q \Rightarrow F$. Therefore, in a proof by contradiction, we assume both the assumption and the negation to be true and then try to deduce some contradiction by using these assumptions.

Example 1.1.22 Prove by contradiction that $\sqrt{2}$ is an irrational number.

PROOF. Suppose that $\sqrt{2}$ is a rational number. Then

$$\sqrt{2} = \frac{m}{n}, \quad \text{for some integers } m \text{ and } n.$$

By cancelling any common factors, we may assume that *m and n have no common factor*. Then

$$2 = \frac{m^2}{n^2}, \quad \text{or} \quad m^2 = 2n^2.$$

This implies that m^2 is even, and thus m is even. So,

$$m = 2k \quad \text{for some integer } k.$$

Thus,

$$(2k)^2 = 2n^2,$$
$$4k^2 = 2n^2,$$
$$2k^2 = n^2.$$

Therefore, n^2 is even, and so is n. This contradicts the assumption that m and n have no common factor. Thus $\sqrt{2}$ is not in the form m/n, so $\sqrt{2}$ is irrational.
∎

Example 1.1.23 Prove by contradiction that the product of two odd integers is odd.

PROOF. Let the two odd integers be

$$x = 2m + 1$$
$$y = 2n + 1$$

and assume the product xy is even. Then we have

$$\begin{aligned}
xy &= (2m + 1)(2n + 1)\\
&= (2m + 1)(2n + 1)\\
&= 4mn + 2m + 2n + 1\\
&= 2(2mn + m + n) + 1.
\end{aligned}$$

Since $2(2mn + m + n)$ is an even integer, then $2(2mn + m + n) + 1$ is odd. This contradicts the assumption that xy is even. So the product of two odd integers is odd. ∎

(IV) Proof by Induction

In proof by induction, we have a sequence of statements

$$P_1, P_2, \cdots, P_n, P_{n+1}$$

about which we want to make some claim. Suppose that we know that the claim holds for all statements $P_1, P_2, \ldots$ up to P_n. We then try to argue that this implies that the claim also holds for P_{n+1}. If we can carry out this inductive step for all positive n, and if we have some starting point for induction, we can then say the claim holds for all statements in the sequence.

   The starting point for induction is called the *basis*. The assumption that the claim holds for the statements $P_1, P_2, \ldots P_n$ is the *inductive assumption*, and the argument connecting the inductive assumption to P_{n+1} is the *inductive step*.

Example 1.1.24 Prove by induction that

$$1 + 3 + 6 + \cdots + \frac{n(n + 1)}{2} = \frac{n(n + 1)(n + 2)}{6}.$$

PROOF. We first prove that P_1 is correct (we use LHS and RHS to represent the left-hand side and right-hand side of the formula, respectively):

$$\begin{aligned} \text{RHS} &= \frac{1(1+1)(1+2)}{6} \\ &= 1 \\ &= \text{LHS}. \end{aligned}$$

Now we assume that P_k:

$$1 + 3 + 6 + \cdots + \frac{k(k+1)}{2} = \frac{k(k+1)(k+2)}{6}$$

is correct and want to prove that P_{k+1}:

$$1 + 3 + 6 + \cdots + \frac{k(k+1)}{2} + \frac{(k+1)(k+2)}{2} = \frac{(k+1)(k+2)(k+3)}{6}$$

is also correct. Now

$$\begin{aligned} \text{LHS} &= 1 + 3 + 6 + \cdots + \frac{k(k+1)}{2} + \frac{(k+1)(k+2)}{2} \\ &= \frac{k(k+1)(k+2)}{6} + \frac{(k+1)(k+2)}{2} \qquad \text{(by the inductive hypothesis)} \\ &= \frac{k(k+1)(k+2)}{6} + \frac{3(k+1)(k+2)}{6} \\ &= \frac{(k+1)(k+2)(k+3)}{6} \\ &= \text{RHS} \end{aligned}$$

So by the inductive hypothesis, the formula holds for all positive integers n. ∎

Example 1.1.25 Prove by induction that for any positive integer n, the number $2^{2n} - 1$ is divisible by 3.

PROOF. We first prove the base case P_1:

$$2^{2 \cdot 1} - 1 = 2^2 - 1 = 3$$

which is divisible by 3.

Now we assume $2^{2k} - 1$ is divisible by 3, which means $2^{2k} - 1 = 3m$ for some integer m, or $2^{2k} = 3m + 1$. Next we want to show that $2^{2(k+1)} - 1$ is also divisible by 3.

$$\begin{aligned} 2^{2(k+1)} - 1 &= 2^{2k+2} - 1 \\ &= 2^{2k} \cdot 2^2 - 1 \\ &= (3m+1) \cdot 2^2 - 1 \qquad \text{(by the inductive hypothesis)} \\ &= 12m + 4 - 1 \\ &= 12m + 3 \\ &= 3(4m+1) \qquad \qquad \text{($4m+1$ is an integer)} \end{aligned}$$

Thus, $2^{2(k+1)} - 1$ is divisible by 3. Hence $2^{2n} - 1$ is divisible by 3 for any positive integer n. ∎

Proof by induction has an application in verifying *loop invariants* in computer programming.

Example 1.1.26 Consider the following pseudocode program, which is supposed to compute xy for nonnegative integers x and y.

> 1. given $x, y \geq 0$
> 2. $i \leftarrow 0$ (assign 0 to i)
> 3. $j \leftarrow 0$
> 4. while $x \neq y$ do
>   begin
> $j \leftarrow j + y$
> $i \leftarrow i + 1$
>   end
> 5. write j

There is a loop at step 4. The quantities x and y are given and remain unchanged throughout the program; values of i and j change. Let i_n and j_n denote the values if i and j , respectively, after n iterations of the loop. If $P(n)$ is the statement $j_n = i_n \cdot y$, then $P(n)$ is a loop invariant. We prove in the following by induction that $P(n)$ holds for all $n \geq 0$.

PROOF. We first prove the base case $P(0)$, which is the statement

$$j_0 = i_0 \cdot y.$$

As at this stage, $j_0 = 0$ and $i_0 \cdot y = 0 \cdot y = 0$, $P(0)$ is correct.

Now we assume $P(k) : j_k = i_k \cdot y$ is correct.

We now want to show $P(k+1)$: $j_{k+1} = i_{k+1} \cdot y$ is correct. Since

$$j_{k+1} = j_k + y$$
$$i_{k+1} = i_k + 1.$$

Then

$$\begin{aligned} j_{k+1} &= j_k + y \\ &= i_k \cdot y + y \qquad \text{(by the inductive hypothesis)} \\ &= y(i_k + 1) \\ &= y \cdot i_{k+1}. \end{aligned}$$

At the loop termination, $i = x$ and $j = y \cdot i = y \cdot x$. ∎

1.2   Sets, Functions and Graphs

In this section, we provide an introduction to the basic concepts and results of sets, functions, relations, graphs and matrices.

1.2.1   Sets and Operations on Sets

A *set* is a collection of elements, without any structure other than membership. The purpose of *set theory* is to study the properties of sets.

Example 1.2.1 The following are examples of sets about numbers, which we shall use throughout the book (we have, in fact, already met some of the sets in the previous sections):

(i) All the *natural numbers* (or *nonnegative integers*) form the set $\mathbb{N}$:

$$\mathbb{N} = \{0, 1, 2, 3, 4, \cdots\}. \qquad (1.37)$$

Note that some authors do not consider the number "0" to be a natural number. It is really only a matter of definition whether or not "0" is regarded as a natural number, and it is true that "0" was invented long after all others. Sometimes, we also use $\mathbb{N}_{>1}$ to represent all the positive integers greater than 1.

(ii) All the integers form the set $\mathbb{Z}$ ("$\mathbb{Z}$" stands for "Zehlen"; the German word for "numbers"):

$$\mathbb{Z} = \{\cdots, -4, -3, -2, -1, 0, 1, 2, 3, 4, \cdots\}. \qquad (1.38)$$

All the *positive integers* form the set $\mathbb{Z}^+$:

$$\mathbb{Z}^+ = \{1, 2, 3, 4, \cdots\}, \qquad (1.39)$$

and all the *negative integers* the set $\mathbb{Z}^-$:

$$\mathbb{Z}^- = \{-1, -2, -3, -4, \cdots\}. \qquad (1.40)$$

(iii) All the *rational numbers* form the set $\mathbb{Q}$ ("$\mathbb{Q}$" stands for "quotient"):

$$\mathbb{Q} = \left\{\frac{p}{q} : p, q \in \mathbb{Z}, \ q \neq 0\right\}. \qquad (1.41)$$

(iv) All the *real numbers* form the set $\mathbb{R}$; real numbers are defined to be converging sequences of rational numbers or as decimals that might or might not repeat; for example, $\sqrt{2}$ and π are real numbers.

(v) All the *complex numbers* form the set $\mathbb{C}$:

$$\mathbb{C} = \{a + bi : a, b \in \mathbb{R}, \ i = \sqrt{-1}\}. \tag{1.42}$$

Definition 1.2.1 The size of a set S, often call the cardinality of a set S, denoted by $|S|$, is the number of elements in S.

Definition 1.2.2 Suppose A and B are two sets. The set operations *union* ($\cup$), *intersection* ($\cap$), *difference* ($-$) on sets of A and B and *complement* ($\overline{A}$) of set A are defined as:

$$A \cup B = \{x \in U : \ x \in A \text{ or } x \in B\} \tag{1.43}$$
$$A \cap B = \{x \in U : \ x \in A \text{ and } x \in B\} \tag{1.44}$$
$$A - B = \{x \in U : \ x \in A \text{ and } x \notin B\} \tag{1.45}$$
$$\overline{A} = \{x \in U : \ x \notin A\} \tag{1.46}$$

where U is the *universal set*, a set to which all the sets considered are subsets of U.

It is often useful to picture sets using *Venn diagrams*[4] . The Venn diagrams for $A \cup B$, $A \cap B$, $A - B$ and A^c are shown in Figure 1.2.

Definition 1.2.3 The set with no elements, denoted by $\emptyset$, is called the *empty set* or *null set*. A set with only one element is called a *singleton*.

Definition 1.2.4 A set A is called a *subset*, denoted by $A \subseteq B$, of B if every element of A is also an element of B: Two sets A and B are said to be equal if $A \subset B$ and $B \subset A$. If A is a subset of B but not equal to B, then A is a *proper subset* of B, denoted by $A \subset B$. The set of all subsets of a set A is called the *powerset* of A and is denoted by 2^A. Two sets A and B are call *disjoint* if

$$A \cap B = \emptyset. \tag{1.47}$$

The Venn Diagrams for subset and disjoin are shown in Figure 1.3.

---

[4]After the British logician John Venn (1834–1923) who introduced the diagrams in his work *symbolic Logic* of 1881.

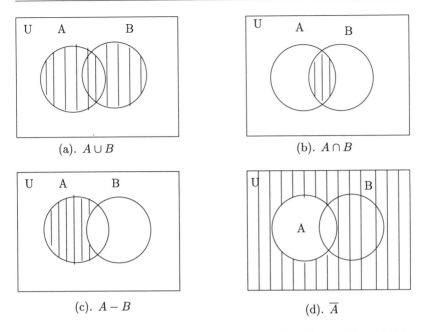

(a). $A \cup B$

(b). $A \cap B$

(c). $A - B$

(d). $\overline{A}$

Figure 1.2: Venn diagrams for $A \cup B$, $A \cap B$, $A - B$, and A^c

Definition 1.2.5 The *Cartesian product* S of two sets S_1 and S_2 are defined by

$$S = S_1 \times S_2 = \{(x, y) : x \in S_1 \text{ and } y \in S_2\}. \tag{1.48}$$

Set theory has important applications in computer science. We will see in the next chapter that the regular language generated by a regular grammar is a set, and the context-free language generated by the context-free grammar is also a set.

Similar to the algebraic rules for propositions, we can have the following algebraic rules for sets (let A, B, C be any sets):

(i) The commutative laws:

$$A \cup B = B \cup A, \tag{1.49}$$

$$A \cap B = B \cap A. \tag{1.50}$$

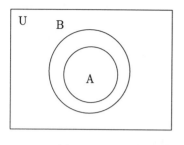

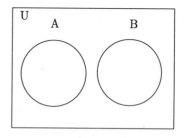

(a). $A \subset B$

(b). $A \cap B = \emptyset$

Figure 1.3: Venn Diagrams for $A \subset B$ and $A \cap B = \emptyset$

(ii) The associative laws:

$$A \cup (B \cup C) = (A \cup B) \cup C, \tag{1.51}$$
$$A \cap (B \cap C) = (A \cap B) \cap C. \tag{1.52}$$

(iii) The distributive laws:

$$A \cup (B \cap C) = (A \cup B) \cap (A \cup C), \tag{1.53}$$
$$A \cap (B \cup C) = (A \cap B) \cup (A \cap C). \tag{1.54}$$

(iv) The idempotent laws:

$$A \cup A = A, \tag{1.55}$$
$$A \cap A = A. \tag{1.56}$$

(v) The identity laws:

$$A \cup \emptyset = A, \tag{1.57}$$
$$A \cap U = A. \tag{1.58}$$

(vi) The complementary laws:

$$A \cup \overline{A} = U, \tag{1.59}$$
$$A \cap \overline{A} = \emptyset, \tag{1.60}$$
$$\overline{\overline{A}} = A. \tag{1.61}$$

(vii) DeMorgan's laws:

$$\overline{P \cup Q} = \overline{P} \cap \overline{Q}, \tag{1.62}$$
$$\overline{P \cap Q} = \overline{P} \cup \overline{Q}. \tag{1.63}$$

We now study a very important structure in computer science – Boolean algebra[5] using the concept of sets.

A *Boolean algebra* is a set S together with two operations, denoted by $\oplus$ and $\odot$, such that $\forall a, b \in S$, both $a \oplus b$ and $a \odot b$ are in S and such that the following axioms hold ($\forall a, b, c \in S$):

(i) The commutative laws:

$$a \oplus b = b \oplus a, \tag{1.64}$$
$$a \odot b = b \odot a. \tag{1.65}$$

(ii) The associative laws:

$$a \oplus (b \oplus c) = (a \oplus b) \oplus c, \tag{1.66}$$
$$a \odot (b \odot c) = (a \odot b) \odot c. \tag{1.67}$$

(iii) The distributive laws:

$$a \oplus (b \cdot c) = (a \oplus b) \cdot (a \oplus c), \tag{1.68}$$
$$a \odot (b \oplus c) = (a \odot b) \oplus (a \odot c). \tag{1.69}$$

(iv) The identity laws:

$$a \oplus 0 = a, \tag{1.70}$$
$$a \odot 1 = a. \tag{1.71}$$

(v) The complementary laws:

$$a \oplus \overline{a} = 1, \tag{1.72}$$
$$a \odot \overline{a} = 0. \tag{1.73}$$

---

[5] After the British mathematician George Boole (1815–1864) who introduced in 1855 what is now called Boolean algebra in his honour. Boole was also one of the founding fathers of mathematical logic.

Example 1.2.2 Suppose S is a Boolean algebra. Show that:

(i) $\forall a \in S$:

    (a) $a \oplus a = a$,

    (b) $a \odot a = a$.

(ii) $\forall a, b \in S$, $a \odot b = 1 \Rightarrow a = b = 1$.

(iii) $\forall a \in S$:

    (a) $a \oplus 1 = 1$,

    (b) $a \odot 0 = 0$.

PROOF. Let a, b, c be any elements of S.

(i) $\forall a \in S$. Then

    (a)

$$\begin{aligned}
a \oplus a &= (a \oplus a) \odot 1 && \text{(by identity law)} \\
&= (a \oplus a) \odot (a \oplus \bar{a}) && \text{(by complementary law)} \\
&= a \oplus (a \odot \bar{a}) && \text{(by distributive law)} \\
&= a \oplus 0 && \text{(by complementary law)} \\
&= a. && \text{(by identity law)}
\end{aligned}$$

    (b)

$$\begin{aligned}
a \odot a &= (a \odot a) \oplus 0 && \text{(by identity law)} \\
&= (a \odot a) \oplus (a \odot \bar{a}) && \text{(by complementary law)} \\
&= a \odot (a \oplus \bar{a}) && \text{(by distributive law)} \\
&= a \odot 1 && \text{(by complementary law)} \\
&= a. && \text{(by identity law)}
\end{aligned}$$

(ii) Let a and b be any elements of S, and $a \odot b = 1$. Then

    (a)

$$\begin{aligned}
1 &= a \odot b, && \text{(by assumption)} \\
&= (a \odot a) \odot b, && \text{(as } a = a \odot a\text{)} \\
&= a \odot (a \odot b), && \text{(by associative law)} \\
&= a \odot 1, && \text{(by assumption)} \\
&= a. && \text{(by identity law)}
\end{aligned}$$

(b)

$$
\begin{aligned}
1 &= a \odot b, & &\text{(by assumption)} \\
&= a \odot (b \odot b), & &\text{(as } a = a \odot a) \\
&= (a \odot b) \odot b, & &\text{(by associative law)} \\
&= 1 \odot b, & &\text{(by assumption)} \\
&= b. & &\text{(by identity law)}
\end{aligned}
$$

(iii) $\forall a \in S.$

(a)

$$
\begin{aligned}
1 &= a \oplus \bar{a} & &\text{(by complementary law)} \\
&= (a \oplus a) \oplus \bar{a} & &\text{(as } a = a \oplus a) \\
&= a \oplus (a \oplus \bar{a}) & &\text{(by associative law)} \\
&= a \oplus 1. & &\text{(by complementary law)}
\end{aligned}
$$

(b)

$$
\begin{aligned}
0 &= a \odot \bar{a} & &\text{(by complementary law)} \\
&= (a \odot a) \odot \bar{a} & &\text{(as } a = a \odot a) \\
&= a \odot (a \odot \bar{a}) & &\text{(by associative law)} \\
&= a \odot 0. & &\text{(by complementary law)} \quad \blacksquare
\end{aligned}
$$

1.2.2 Functions and Relations

In this section, we introduce some basic concepts of functions and relations.

Definition 1.2.6 A *function* or a *map* f, denoted by

$$f : A \to B. \tag{1.74}$$

is a rule that assigns to each element, say a, of set A a unique element of another set B. A is called the *domain* of f, and B the *codomain* of f. We may use the notation $a \longmapsto f(a)$ to indicate that $f(a)$ in B is the value assigned to a; we call $f(a)$ the *image* of a. The image of f, denoted by

$$f(A) = \{ b : b \in B \text{ and } b = f(a) \text{ for some } a \in A \} \tag{1.75}$$

is the subset of B comprising the f-images of elements of A. This is the part of B that f actually "ranges over". If the domain is all of A, then f is a *total function*; otherwise, f is said to be a *partial function*.

Example 1.2.3 If $A = \{1, 2, 4\}$ and $B = \{1, 2, 6, 8\}$, then the rule that assigns

$$
\begin{array}{ll}
1 \longmapsto 6 & \text{(i.e., } f(1) = 6\text{); or} \\
2 \longmapsto 2 & \text{(i.e., } f(2) = 2\text{); or} \\
4 \longmapsto 8 & \text{(i.e., } f(4) = 8\text{).}
\end{array}
$$

is a function.

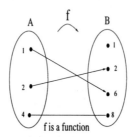

f is a function

The domain of f is $\{1, 2, 4\}$, the codomain of f is $\{1, 2, 6, 8\}$, and the range of f is $\{2, 6, 8\}$.

Definition 1.2.7 Let f be a function from set A to set B, $f : A \to B$.

(i) f is said to be *surjective* (or *onto*), if every $b \in B$ is the image of some $a \in A$. That is,

$$f : A \to B \text{ is surjective} \iff \forall b \in B, \exists a \in A \text{ such that } f(a) = b.$$
$$(1.76)$$

(ii) A function $f : A \to B$ is *injective* (or *one-to-one*), if all elements in A have distinct images. That is,

$$f : A \to B \text{ is injective} \iff \forall a, a' \in A, \ f(a) \neq f(a') \Longrightarrow a \neq a'.$$
$$(1.77)$$

(iii) A function $f : A \to B$ is a *bijection* (or *one-to-one correspondence*), if it is both one-to-one and onto.

Functions defined on finite sets A and B can be represented by so-called *arrow diagrams* as follows:

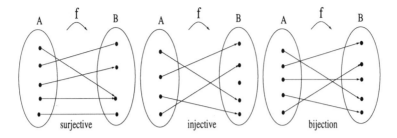

Example 1.2.4 The following arrow diagrams show three functions where the first is injective but not surjective, the second is surjective but not injective, the third is not a function, since it maps one element $a \in A$ to two different elements $2, 3 \in B$.

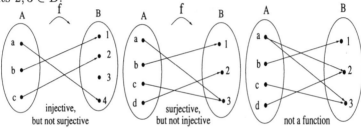

Definition 1.2.8 Let f and g be functions from set A to set B. Then f equals g, written $f = g$, if is the function $i_A : A \to A$ defined by

$$f(x) = g(x), \quad \forall x \in A. \tag{1.78}$$

Definition 1.2.9 Let A be a set. Then the *identity function* is the function $i_A : A \to A$ defined by

$$i_A(a) = a, \quad \forall a \in A. \tag{1.79}$$

That is, the identity function on A maps each element of A to itself.

Definition 1.2.10 Let $f : A \to B$ be a bijection. Then the *inverse function*, denoted by $f^{-1} : B \to A$, is the function that assigns to $b \in B$ the unique element $a \in A$ such that $f(a) = b$. That is,

$$f^{-1}(b) = x \iff b = f(a). \tag{1.80}$$

Example 1.2.5 Let $f : \mathbb{Z} \to \mathbb{Z}$ such that $f(x) = x + 12$. Then by the basic properties of integer addition, f is a bijective function. To invert the bijection, suppose that y is the image of x, so that $y = x + 11$. Then $x = y - 12$. This means that $y - 12$ is the unique element of $\mathbb{Z}$ that is assigned to y by f. So $f^{-1} = y - 12$. But if $f : \mathbb{Z} \to \mathbb{Z}$ is defined by $f(x) = x^2$, then f is not a bijective function, since $f(-1) = f(1) = 1$. Therefore, f is not invertible, since if f were invertible, it would have to assign two elements to 1.

Definition 1.2.11 Let $g : A \to B$ and $f : B \to C$. The *composition* of the functions f and g, denoted by $f \circ g$, is defined by

$$(f \circ g)(a) = f(g(a)). \tag{1.81}$$

The following diagrams illustrate the concepts of an inverse function and a composition of two functions:

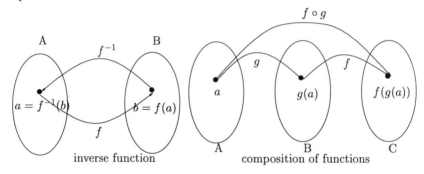

inverse function   composition of functions

Example 1.2.6 Let $f, g : \mathbb{Z} \to \mathbb{Z}$ defined by $f(x) = 2x - 3$ and $g(x) = 3x - 2$. Then

$$
\begin{aligned}
(f \circ g)(x) &= f(g(x)) \\
&= f(3x - 2) \\
&= 2(3x - 2) - 3 \\
&= 6x - 7
\end{aligned}
$$

$$
\begin{aligned}
(g \circ f)(x) &= g(f(x)) \\
&= g(2x - 3) \\
&= 3(2x - 3) - 2 \\
&= 6x - 11.
\end{aligned}
$$

Definition 1.2.12 A function $f : A \to B$ is an *isomorphism* if it has an inverse, i.e., a function $g : B \to A$ such that

$$g \cdot f = i_A \quad \text{and} \quad f \cdot g = i_B. \tag{1.82}$$

Theorem 1.2.1 A function $f : A \to B$ is an isomorphism if and only if it is a bijection.

PROOF. It is easy to verify that if f is an isomorphism, then it is surjective and injective. ■

We now introduce some important functions widely used in computer science.

Definition 1.2.13 The *floor function* maps the real number x to the largest integer n that is less than or equal to x. That is,

$$\lfloor x \rfloor = \text{ the unique integer } n \text{ such that } n \leq x < n + 1. \tag{1.83}$$

The *ceiling function* maps the real number x to the smallest integer that is greater than or equal to x. That is,

$$\lceil x \rceil = \text{ the unique integer } n \text{ such that } n - 1 < x \leq n. \tag{1.84}$$

Example 1.2.7 $\lfloor 0.5 \rfloor = 0$, $\lfloor -0.5 \rfloor = -1$; $\lceil 0.5 \rceil = 1$, $\lceil -0.5 \rceil = 0$.

Definition 1.2.14 Let b be a positive real number. For each positive real number x, the *logarithm* with base b of x, written $\log_b x$, is defined by

$$\log_b x = y \iff b^y = x. \tag{1.85}$$

The *logarithm function* with base b is the function from $\mathbb{R}^+$ to $\mathbb{R}$ that takes each positive real number x to $\log_b x$.

Example 1.2.8 $\log_3 27 = 3$ since $3^3 = 27$, and $\log_2 128 = 7$ since $2^7 = 128$.

Definition 1.2.15 A *sequence* is a function whose domain is a set (possibly infinite) of consecutive integers.

Example 1.2.9 Let the sequence $s : \mathbb{N} \to \mathbb{N}$ be

$$\begin{cases} s_0 = 0 \\ s_1 = 1 \\ s_n = s_{n-1} + s_{n-2} \quad \text{for } n > 1 \end{cases} \tag{1.86}$$

Then the first few elements of the sequence are

$$0, 1, 1, 2, 3, 5, 8, 13, 21, 34, 55, \cdots$$

This sequence of numbers is called *Fibonacci Sequence*.

Definition 1.2.16 Let $\Sigma = \{0, 1\}$. The Σ^n is the set of all strings consisting of 1's and 0's of length n. The Hamming[6] distance function

$$H : \Sigma^n \times \Sigma^n \to \mathbb{N} \tag{1.87}$$

is defined as follows: for each pair $(s, t) \in \Sigma^n \times \Sigma^n$, $H(s, t) =$ the number of positions in which s and t have different values.

Example 1.2.10 It is easy to verify that $H(1010000001, 1010110010) = 4$, and $H(0111101010011, 0100101101101) = 7$.

Another useful function in computer science is the *hash function* h that assigns memory location $h(k)$ to the record that has k as its key:

$$h : h(k) \to k, \quad k \in \mathbb{N} \tag{1.88}$$

A typical hash function is the remainder modulo p, where p is a prime number:

$$h(k) \equiv k \pmod{p}. \tag{1.89}$$

For example, in a database, the record of the student with social security number 629731485 may be assigned to memory location 75, since

$$629731485 \equiv 75 \pmod{131}.$$

Now we move on to the discussion of the basic concepts of *relations*.

---

[6]Richard W. Hamming (1915–) received his PhD in mathematics from the University of Illinois in 1942 and had been staff member in Bell Telephone Laboratories and the Naval Postgraduate School. He is perhaps best known for his invention of the error-detecting/correcting codes which bear his name. Hamming received the 1968 ACM Turing Award and the 1980 IEEE Computer Society Pioneer Award.

Definition 1.2.17 Let A and B be sets. A (binary) *relation* $R : A \rightarrow B$ is a subset of $A \times B$. Given an ordered pair (a, b) in $A \times B$, a is related to b, written $a\,R\,b$, if $(a, b) \in R$; a is not related to b, written $a\cancel{R}\,b$, if $(a, b) \notin R$. That is;

$$a\,R\,b \Longleftrightarrow (a, b) \in R, \quad \text{and} \quad a\cancel{R}\,b \Longleftrightarrow (a, b) \notin R. \tag{1.90}$$

Definition 1.2.18 A relation on the set A is a subset of $A \times A$.

Example 1.2.11 Let $A = \{1, 2, 3, 4\}$. Then

$$R = \{(a, b) : (a, b) \in A \times A \text{ and } a \mid b\}$$

is a relation on A. It is clear That

$$R = \{(1,1),\ (1,2),\ (1,3),\ (1,4),\ (2,2),\ (2,4),\ (3,3),\ (4,4)\}.$$

The pairs $(a, b) \in R$ are shown in the following directed graph (the formal definition of *directed graph* will be given in the next subsection):

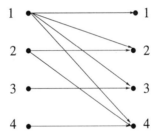

Note that a function can also be defined as a special type of binary relations:

Definition 1.2.19 A function $F : A \to B$ is a relation $R : A \rightarrow B$ that satisfies the following two properties:

(i) $\forall a \in A$, $\exists b \in B$ such that $(a, b) \in F$.

(ii) $\forall a \in A$, $\forall b, c \in B$, $(a, b) \in F$ & $(a, c) \in F \Longrightarrow b = c$.

If F is a function from A to B, we write

$$y = F(x) \Longleftrightarrow (a, b) \in F.$$

Just as with functions, we can define an inverse relation as follows:

Definition 1.2.20 Let R be a relation from A to B. Then the inverse relation $R^{-1} : B \to A$ is defined by

$$R^{-1} = \{(a, b) \in B \times A : (a, b) \in R\}. \tag{1.91}$$

Definition 1.2.21 Given sets $A_1, A_2, \cdots, A_n$, an *n-ary relation* R on $A_1 \times A_2 \times \cdots \times A_n$ is a subset of $A_1 \times A_2 \times \cdots \times A_n$.

Clearly, a binary relation is a special kind of n-ary relation. N-ary relations form the mathematical foundation of relational database theory. Say, for example, we define a 6-ary relation R about university student academic records on $A_1 \times A_2 \times A_3 \times A_4 \times A_5 \times A_6$ as follows:

$$(a_1, a_2, a_3, a_4, a_5) \stackrel{\text{def}}{=} (\text{ID No, Name, Maths, Physics, PASCAL, OS/UNIX}).$$

A typical database in this relation might be:

(976601, John Smith, 75, 88, 69, 59)
(976601, Michael Lewis, 70, 66, 81, 72)
(976601, Mary Johnson, 66, 82, 79, 60)
(976601, William Lloyd, 60, 73, 59, 63)

Definition 1.2.22 Let R be a relation on a set X, then

(i) R is *reflexive* if xRx, $\forall x \in X$.

(ii) R is *symmetric* if whenever xRy, then yRx.

(iii) R is *transitive* if whenever xRy and yRz, then xRz.

(iv) R is *antisymmetric* if whenever xRy and yRx, then $x = y$.

(v) R is said to be an *equivalence relation* if R is reflexive, symmetric, and transitive.

(vi) R is said to be an *order relation* (or partial order relation) if R reflexive, antisymmetric, and transitive. A set A together with a partial order relation R, denoted by (A, R), is called a *partially ordered set* (or *poset* for short).

Example 1.2.12 Let $A = \{0, 1, 2, 3\}$ and define R as follows:

$$R = \{(0,0),\ (0,1),\ (0,3),\ (1,0),\ (1,1),\ (2,2),\ (3,0),\ (3,3)\}.$$

We first draw the direct graph of R, so that we would be able to investigate the properties of R easily:

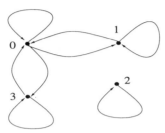

Then

(i) R is *reflexive*, since there is a loop at each node of the directed graph. This means that each element in A is related to itself.

(ii) R is *symmetric*, since once there is an arrow going from one point of the graph to a second, there is an arrow going back from the second to the first. This means that whenever one element of A is related by R to a second, then the second is also related to the first.

(iii) R is *not transitive*, since there is an arrow from 1 to 0 and an arrow from 0 to 3, but no arrow from 1 to 3. This means that $\exists 0, 1, 3 \in A$, such that $1\,R\,0$ and $0\,R\,3$, but $1\,R\!\!\!/\,3$.

Example 1.2.13 The divisibility relation " $\mid$ " is a partial order relation, since it is reflexive, antisymmetric, and transitive. The set of all positive integers $\mathbb{N}$ together with the divisibility relation, $(\mathbb{N},\ \mid\ )$, forms a poset.

Definition 1.2.23 Let R be an equivalence relation on a set A. The set of all elements that are related to an element $a \in A$ is called the *equivalence class* of a, denoted by $[a]_R$. (When only one relation is under consideration, we will delete the subscript R and write $[a]$ for the equivalence class). That is,

$$[a]_R = \{s : (a, s) \in R\}. \tag{1.92}$$

If $b \in [a]_R$, b is called a *representative* of this equivalence class.

Theorem 1.2.2 Let R be an equivalence relation on a set A. Then the equivalence classes of R form a partition of A. Conversely, given a partition $\{A_i : i \in I\}$ of the set A, there is an equivalence relation R that has the sets A_i, $i \in I$, as its equivalence classes.

Example 1.2.14 The set of integers $\mathbb{Z}$ modulo 2 is an equivalence relation that partitions $\mathbb{Z}$ into two equivalence classes, namely;

(i) the equivalence class of 0, that contains all even numbers:

$$[0]_2 = \{\cdots, -6, -4, -2, 0, 2, 4, 6, 8, \cdots\},$$

(ii) the equivalence class of 1, that contains all odd numbers:

$$[1]_2 = \{\cdots, -5, -3, -1, 1, 3, 5, 7, 9, \cdots\}.$$

1.2.3   Directed and Undirected Graphs

Graph theory has a strong connection with computer science, for example, an automaton can be represented by a directed graph. In this section, We shall survey some basic concepts and results in graph theory.

Definition 1.2.24 A *directed graph* (or *digraph* for short) G, is a pair of (V, E), where V is a finite set and E is a binary relation on V. The set V is called the *vertex set* of G, and its elements are called *vertices* or nodes. The set E is called the *edge set* of G, and its elements are called *edges*. An *undirected graph* $G = (V, E)$ is a graph in which the edge set E consists of *unordered* pairs of vertices, rather than ordered pairs as in a digraph.

Definition 1.2.25 If (u, v) is an edge in a digraph $G = (V, E)$, we say that (u, v) is *incident* from or *leaves* vertex u and is *incident to* or *enters* vertex v. If (u, v) is an edge in an undirected $G = (V, E)$, we say that (u, v) is *incident on* vertices u and v. If (u, v) is an edge in a graph $G = (V, E)$, we say that the vertex v is *adjacent* to vertex u, or u and v are adjacent. An edge with just one vertex is called a *loop*, and two distinct edges with the same set of vertices are said to be *parallel*. A *simple graph* is a graph with no parallel edges and no loops. A vertex on which no edges are incident is called *isolated*. A graph with no vertices is called *empty*, and one with at least one vertex is called *nonempty*.

Definition 1.2.26 A graph $G' = (V', E')$ is called a (proper) *subgraph* of a graph $G = (V, E)$, if $V' \subset V$ and $E' \subset E$. The *union* of two simple graphs $G_1 = (V_1, E_1)$ and $G_2 = (V_2, E_2)$, denoted by $G_1 \cup G_2$, is the simple graph $G = (V, E)$ with vertex set $V = V_1 \cup V_2$ and edge set $E = E_1 \cup E_2$.

Definition 1.2.27 The *degree* of a vertex v, denoted by $deg(v)$, in an undirected graph is the number of edges incident on the vertex, with each loop at the vertex being counted twice. In a digraph, the *out-degree* of a vertex is the number of edges leaving it, and the *in-degree* of a vertex is the number of edges entering it. The degree of a vertex in a digraph is its in-degree plus its out-degree.

Definition 1.2.28 A *path* of *length* k from a vertex u to a vertex v in a graph $G = (V, E)$ is a sequence $\langle v_0, v_1, \cdots, v_k \rangle$ of vertices such that $u = v_0$, $v = v_k$, and $(v_{i-1}, v_i) \in E$ for $i = 1, 2, \cdots, k$. The length of the path is the number of edges in the graph. The path contains the vertices $v_0, v_1, \cdots, v_k$ and the edges (v_0, v_1), (v_1, v_2), $\cdots$, (v_{k-1}, v_k). If there exists a path p from u to v, we say that v is *reachable* from u via p. A path is *simple*, if all vertices in the path are distinct. In a digraph, a path $\langle v_0, v_1, \cdots, v_k \rangle$ forms a *cycle* if $v_0 = v_k$, and the path contains at least one edge. The cycle is *simple* if, in addition, $\langle v_0, v_1, \cdots, v_k \rangle$ are distinct.

Definition 1.2.29 An (undirected) graph is called a *complete graph*, if it is a simple graph and every vertex is adjacent to every other vertex. The number of vertices is called the *order* of the complete graph. A complete graph is usually denoted by K_m, where m is the order of the complete graph K.

Example 1.2.15 The following graphs (from left to right) are complete graphs K_2, K_3, K_4 and K_5:

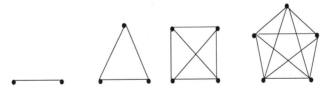

Definition 1.2.30 An (undirected) graph G with vertices V is called *bipartite* if it is possible to write V as a union of two nonempty disjoint subsets V_1 and V_2, such that each edge of G connects a vertex of V_1 with a vertex of V_2.

Definition 1.2.31 The *complete bipartite graph* on (m, n) vertices, denoted by $K_{m,n}$, is a graph with vertices $V = V_1 \cup V_2$, where V_1 has m elements, V_2 has n elements, $V_1 \cap V_2 = \emptyset$, and each vertex of V_1 is connected to each vertex of V_2 by exactly one edge. The pair (m, n) is called the order of the complete bipartite graph.

Example 1.2.16 The following graphs (from left to right) are bipartite graphs K_{23}, K_{33}, and K_{34}:

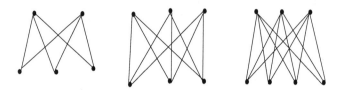

Definition 1.2.32 Let $G_1 = (V_1, E_1)$ and $G_2 = (V_2, E_2)$ be simple graphs. Then G_1 is *isomorphic* to G_2, if there exists a bijection function $f : V_1 \to V_2$ with the property that u and v are adjacent in G_1 if $f(u)$ and $f(v)$ are are adjacent in G_2, for all u and v in V_1.

Example 1.2.17 In the following four graphs (from left to right), the first two are isomorphic, whereas the last two are not:

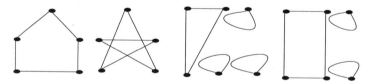

There are several useful ways to represent graphs inside a computer. One of the most powerful and convenient ways is using matrices. In what follows, we shall introduce a standard method called the *adjacency matrix representation* of graphs.

Definition 1.2.33 Let G be a graph $G = (V, E)$. We assume that the vertices are numbered $1, 2, \cdots, |V|$ in some arbitrary manner. The *adjacency matrix* representation of a graph G is the $|V| \times |V|$ matrix $A = (a_{ij})$ such that

$$a_{ij} = \begin{cases} 1 & \text{if } (i, j) \in E \\ 0 & \text{otherwsie.} \end{cases} \tag{1.93}$$

Example 1.2.18 The following two matrices (from left to right):

	1	2	3	4	5	6
1	0	1	0	0	0	1
2	0	1	0	1	1	0
3	0	0	1	1	1	0
4	0	0	0	1	0	0
5	1	0	0	0	0	1
6	0	1	0	0	0	0

	1	2	3	4
1	0	3	2	0
2	3	1	1	1
3	2	1	0	2
4	0	1	2	1

are the two adjacency matrix representations of the following two graphs (again from left to right):

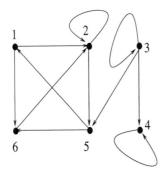 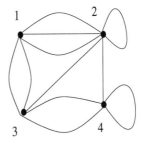

It should be noted that once a graph G is given, it is always possible to get its adjacency matrix representation; similarly, once an adjacency matrix representation of G is given, it is always possible to get the graph representation of G.

1.3 Divisibility, Continued Fractions and Congruences

The theory of numbers is the study of the properties of integers, particularly positive integers. In this section, we provide an introduction to some basic concepts and results of elementary theory of numbers, which are very useful in computer science.

1.3.1   Divisibility

Divisibility has been studied for at least three thousand years. From before the time of Pythagoras, the Greeks considered questions about even and odd numbers, perfect and amicable numbers, and the primes, among many others; even today a few of these questions are still unanswered.

Definition 1.3.1 Let a and b be integers with $a \neq 0$. We say a divides b, denoted by $a \mid b$, if there exists an integer c such that $b = ac$. When a divides b, we say that a is a *divisor* (or *factor*) of b, and b is a *multiple* of a. If a does not divide b, we write $a \nmid b$

Example 1.3.1 The integer 12 has divisors

$$-12, -6, -4, -3, -2, -1, \ 1, \ 2, \ 3, \ 4, \ 6, \ 12,$$

so we can write:

$$-12 \mid 12, \ -6 \mid 12, \ 3 \mid 12, \ 4 \mid 12, \cdots.$$

But $5 \nmid 12$.

Note that it is usually sufficient to consider only the positive divisors.

Definition 1.3.2 A divisor of n is called a *trivial divisor* of n if it is either 1 or n itself. A divisor of n is called a *nontrivial divisor* if it is a divisor of n but is neither 1 nor n.

Example 1.3.2 1 and 18 are the trivial divisors of 18, but 2, 3, 6 and 9 are nontrivial divisors of 18.

Some basic properties of divisibility are given in the following theorem:

Theorem 1.3.1 Let a, b and c be integers. Then

(i) if $a \mid b$ and $a \mid c$, then $a \mid (b + c)$.

(ii) if $a \mid b$, then $a \mid bc$, for all integers c.

(iii) if $a \mid b$ and $b \mid c$, then $a \mid c$.

Definition 1.3.3 A positive integer n greater than 1 is called *prime* if its only divisors are n and 1. A positive integer n that is greater than 1 and is not prime is called *composite*.

Example 1.3.3 The integer 23 is prime since its only divisors are 1 and 23, whereas 22 is composite since it is divisible by 2.

Theorem 1.3.2 Every composite has a prime factor.

PROOF. Let n be a composite number. Then

$$n = n_1 n_2$$

where n_1 and n_2 are positive integers with $n_1, n_2 < n$. If either n_1 or n_2 is a prime, then the theorem is proved. If n_1 is not a prime, then

$$n_1 = n_3 n_4$$

where n_3 and n_4 are positive integers with $n_3, n_4 < n_1$. Again if n_3 or n_4 is a prime, then the theorem is proved. If n_3 is not a prime, then we can write

$$n_3 = n_5 n_6$$

where n_5 and n_6 are positive integers with $n_5, n_6 < n_3$. In general, after k steps we write

$$n_{2k-1} = n_{2k+1} n_{2k+2}$$

where n_{2k+1} and n_{2k+2} are positive integers with $n_{2k+1}, n_{2k+1} < n_{2k-1}$. Since

$$n > n_1 > n_3 > n_5 > \cdots n_{2k-1} > 0$$

for any value k, the process must terminate. So there must exist an n_{2k-1} for some value of k, that is prime. Hence, every composite has a prime factor. ∎

The process of deciding whether or not a number is prime is called *primality testing*. Primality testing of large numbers is a hot topic nowadays. The simplest test for primality is the trial division method, which is based on the following observation.

Theorem 1.3.3 If n is a composite, then n has a prime divisor less than or equal to $\sqrt{n}$.

Corollary 1.3.1 If n is not divisible by any primes less than or equal to $\sqrt{n}$, then n is prime.

The most important result about primes is the prime number theory (PNT):

Theorem 1.3.4 (Prime Number Theorem): The number of primes less than or equal to x is asymptotic to $\frac{x}{\ln x}$. That is, let $\pi(x)$ denote the number of primes less than or equal to x, then

$$\lim_{x \to \infty} \frac{\pi(x)}{x/\ln x} = 1 \tag{1.94}$$

where $\pi(x) \sim \frac{x}{\ln x}$ is the prime distribution function.

The Prime Number Theorem (PNT for short) was postulated by both Legendre and Gauss on numerical evidence (it was known that Gauss constructed by hand a table of all primes up to three million, and investigated the number of primes occuring in each group of 1000). But both Legendre and Gauss were unable to prove it. It was through the collective genius and hard work of a few mathematicians in the 19th century – notably Dirichlet, Chebyshev, Riemann, Hadamard and de la Vallée-Poussin – that the theorem was finally established independently in 1896 by a proof given by Hadamard[7] and de la Vallée-Poussin[8], using very complicated analytical methods. A *technically elementary* proof (no use of integral calculus) was also given independently by Atle Selberg[9] and Paul Erdős[10] in 1948 and 1949. These elementary proofs were considered so important that Selberg got a Fields medal in 1950 and Erdős received the American Mathematical Society's Cole Prize in 1951 and the Wolf prize in 1984. (The

---

[7] Jacques Hadamard (1865–1963), French mathematician, best known for his proof in 1896 of the *Prime Number Theorem*. He also made important contributions to complex analysis, functional analysis and mathematical physics.

[8] Charles-Jean de la Vallée-Poussin (1866–1962), Belgian mathematician who in 1896 proved the *Prime Number Theorem* independently of Hadamard. Notice that both Hadamard and Vallée-Poussin lived well into their 90's (Hadamard 98, Vallée-Poussin 96); it is a common belief among mathematicians that anyone who produces a proof of the *Prime Number Theorem* is guaranteed longevity!

[9] Atle Selberg (1917–), Norwegian number theorist, analyst, and Fields Medal recipient (1950). Selberg contributed significantly to the Riemann zeta function, but proved the PNT without its use. Selberg is currently Professor Emeritus in the Institute of Advanced Study at Princeton.

[10] The legendary Paul Erdős was born in Budapest, Hungary on 26 March 1913 and died on 20 September 1996 while attending a minisemester at the Banach Center in Warsaw, Poland. Erdős was one of the most versatile and prolific mathematicians of our time, and indeed probably of all times. He made many significant contributions to algebra, analysis, combinatorics, geometry, topology and number theory, and wrote about 1500 papers, about five times as many as other prolific mathematicians, and had close to 500 collaborators. Through his prodigious output and many collaborators, he greatly influenced many branches of mathematics.

Fields medal and the Wolf prize are regarded as the most prestigious in mathematics, the equivalent of a Nobel prize in the physical sciences). Since Euclid discovered 2000 years ago a theorem that *"there are infinitely many prime numbers"*, thousands of theorems about prime numbers have been discovered; many are significant, some are beautiful, but only *this* serious theorem is called the *Prime Number Theorem*.

Primes are the building blocks of positive integers, as the following theorem shows:

Theorem 1.3.5 (Fundamental Theorem of Arithmetic) Every positive integer n greater than 1 can be written uniquely as the product of primes:

$$n = p_1^{\alpha_1} p_2^{\alpha_2} \cdots p_k^{\alpha_k} = \prod_{i=1}^{k} p_i^{\alpha_i} \tag{1.95}$$

where $p_1, p_2, \cdots, p_k$ are distinct primes, and $\alpha_1, \alpha_2, \cdots, \alpha_k$ are natural numbers.

PROOF. We shall first show that a factorization exists. Starting from $n > 1$, if n is a prime, then it stands as a *product* with a single factor. Otherwise, n can be factored into, say, ab, where $a > 1$ and $b > 1$. Apply the same argument to a and b: each is either a prime or a product of two numbers both > 1. The numbers other than primes involved in the expression for n are greater than 1 and decrease at every step; hence eventually all the numbers must be prime.

Now we come to uniqueness. Suppose that the theorem is false and let $n > 1$ be the smallest number having more than one expression as the product of primes, say

$$n = p_1 p_2 \cdots p_r = q_1 q_2 \cdots q_s$$

where each p_i $(i = 1, 2, \cdots, r)$ and each q_j $(j = 1, 2, \cdots, s)$ is prime. Clearly both r and s must be greater than 1 (otherwise n is prime, or a prime is equal to a composite). If for example p_1 were one of the q_j $(j = 1, 2, \cdots, s)$, then n/p_1 would have two expressions as a product of primes, but $n/p_1 < n$ so this would contradict the definition of n. Hence p_1 is not equal to any of the q_j $(j = 1, 2, \cdots, s)$, and similarly none of the p_i $(i = 1, 2, \cdots, r)$ equals any of the q_j $(j = 1, 2, \cdots, s)$. Next, there is no loss of generality in presuming that $p_1 < q_1$, and we define the positive integer N as

$$N = (q_1 - p_1)q_2 q_3 \cdots q_s = p_1(p_2 p_3 \cdots p_r - q_2 q_3 \cdots q_s).$$

Certainly $1 < N < n$, so N is uniquely factorable into primes. However, $p_1 \nmid (q_1 - p_1)$, since $p_1 < q_1$ and q_1 is prime. Hence one of the above expressions for N contains p_1 and the other does not. This contradiction proves the result: there cannot be any exceptions to the theorem. ∎

Note that if n is prime, then the product is, of course, n itself.

Example 1.3.4 The following are some sample prime factorizations:

$$643 = 643$$
$$644 = 2^2 \cdot 7 \cdot 23$$
$$645 = 3 \cdot 5 \cdot 43$$
$$646 = 2 \cdot 17 \cdot 19$$
$$647 = 647$$

$$2^{31} - 1 = 2147483647$$
$$2^{31} + 1 = 3 \cdot 715827883$$
$$2^{32} - 1 = 3 \cdot 5 \cdot 17 \cdot 257 \cdot 65537$$
$$2^{32} + 1 = 641 \cdot 6700417$$
$$2^{32} + 2 = 2 \cdot 5^2 \cdot 13 \cdot 41 \cdot 61 \cdot 1321$$

Definition 1.3.4 Let a and b be integers, not both zero. The largest divisor d such that $d \mid a$ and $d \mid b$ is called the *greatest common divisor* of a and b. The greatest common divisor of a and b is denoted by $\gcd(a, b)$.

Example 1.3.5 The sets of positive divisors of 111 and 333 are as follows:

$$1, 3, 37, 111,$$
$$1, 3, 9, 37, 111, 333,$$

so $\gcd(111, 333) = 111$. But $\gcd(91, 111) = 1$, since 91 and 111 have no common divisors other than 1.

Definition 1.3.5 The integers a and b are called *relatively prime* if $\gcd(a, b) = 1$. We say that integers $n_1, n_2, ..., n_k$ are *pairwise relatively prime* if, whenever $i \neq j$, we have $\gcd(n_i, n_j) = 1$.

Example 1.3.6 91 and 111 are relatively prime, since $\gcd(91, 111) = 1$.

Theorem 1.3.6 If $a \mid bc$ and $\gcd(a, b) = 1$, then $a \mid c$.

Definition 1.3.6 If d is a multiple of a and also a multiple of b, then d is a common multiple of a and b. The least common multiple (lcm) of two integers a and b, not 1, is the smallest of the common multiples of a and b. The least common multiple of a and b is denoted by $\text{lcm}(a, b)$.

Theorem 1.3.7 Suppose a and b are not both zero (i.e., one of the a and b can be zero, but not both zero), and that $m = \text{lcm}(a, b)$. If x is a common multiple of a and b, then $m \mid x$. That is, every common multiple of a and b is a multiple of the least common multiple.

The following theorem asserts that lcm can be calculated from gcd:

Theorem 1.3.8 Suppose a and b are positive integers, then

$$\text{lcm}(a, b) = \frac{ab}{\gcd(a, b)}. \tag{1.96}$$

One way to calculate $\gcd(a, b)$ or $\text{lcm}(a, b)$ is to use the standard prime factorizations of a and b. That is:

Theorem 1.3.9 If

$$a = p_1^{\alpha_1} p_2^{\alpha_2} \cdots p_k^{\alpha_k}$$

and

$$b = p_1^{\beta_1} p_2^{\beta_2} \cdots p_k^{\beta_k},$$

then

$$\gcd(a, b) = p_1^{\min(\alpha_1, \beta_1)} p_2^{\min(\alpha_2, \beta_2)} \cdots p_k^{\min(\alpha_k, \beta_k)}, \tag{1.97}$$

$$\text{lcm}(a, b) = p_1^{\max(\alpha_1, \beta_1)} p_2^{\max(\alpha_2, \beta_2)} \cdots p_k^{\max(\alpha_k, \beta_k)}. \tag{1.98}$$

Example 1.3.7 Since the prime factorization of 240 and 560 are

$$240 = 2^4 \cdot 3 \cdot 5$$
$$560 = 2^4 \cdot 5 \cdot 7,$$

then

$$\gcd(240, 560) = 2^{\min(4,4)} \cdot 3^{\min(1,0)} \cdot 5^{\min(1,1)} \cdot 7^{\min(0,1)}$$
$$= 2^4 \cdot 3^0 \cdot 5^1 \cdot 7^0$$
$$= 80.$$
$$\text{lcm}(240, 560) = 2^{\max(4,4)} \cdot 3^{\max(1,0)} \cdot 5^{\max(1,1)} \cdot 7^{\max(0,1)}$$
$$= 2^4 \cdot 3^1 \cdot 5^1 \cdot 7^1$$
$$= 1680.$$

Since there is no known efficient algorithm for integer factorization, the most common method to find the gcd or lcm is still Euclid's algorithm, attributed to Euclid[11] at least 2000 years ago, which we shall introduce in the following.

Euclid's algorithm is perhaps the oldest nontrivial algorithm that has survived to the present day. It is based on the following so-called *division algorithm* (The division algorithm is actually not an algorithm, it is just, as we shall see, an algebraic equation):

Theorem 1.3.10 (Division algorithm) For any integer a and any positive integer b, there are unique q and r such that

$$a = bq + r, \quad 0 \leq r < b. \tag{1.99}$$

Further more, $b \mid a$ if and only if $r = 0$.

In the equality given in the above division algorithm, b is called the divisor, a is called the *dividend*, q is called the *quotient*, and r is called the *remainder*. Apply the division algorithm to the pair of integers (a, b) recursively as follows:

$$
\begin{array}{lll}
a = bq_0 + r_1, & 0 \leq r_1 < b, & \text{(dividing } b \text{ into } a\text{)}, \\
b = r_1 q_1 + r_2, & 0 \leq r_2 < r_1, & \text{(dividing } r_1 \text{ into } b\text{)}, \\
r_1 = r_2 q_2 + r_3, & 0 \leq r_3 < r_2, & \text{(dividing } r_2 \text{ into } r_1\text{)}, \\
r_2 = r_3 q_3 + r_4, & 0 \leq r_4 < r_3, & \text{(dividing } r_3 \text{ into } r_2\text{)}, \\
\cdots\cdots & \cdots\cdots & \\
r_{n-2} = r_{n-1} q_{n-1} + r_n, & 0 \leq r_n < r_{n-1}, & \\
r_{n-1} = r_n q_n + 0 & r_{n+1} = 0. &
\end{array}
$$

That is,

---

[11]Euclid (around 350 B.C.) was the author of the most successful mathematical textbook ever written, namely his *Elements*, which has appeared in over a thousand different editions from ancient to modern times. It provides an introduction to plane and solid geometry and number theory. Euclid's algorithm for computing the greatest common divisor of two positive integers is found in Book VII of the 13 books in the *Elements*, and his proofs for the infinitude of primes and the sufficient condition for even perfect numbers are found in Book IX. The "Axiom-Definition-Theorem-Proof" style of Euclid's work has become the standard for formal mathematical writing up to the present day.

$$
\begin{array}{c|c|c}
a & & \\
\hline
-bq_1 & q_0 & b \\
\hline
r_1 & q_1 & r_1 q_2 \\
\hline
r_2 q_3 & q_2 & r_2 \\
\hline
r_3 & q_3 & r_3 q_4 \\
\vdots & \vdots & \vdots \\
r_{n-1} & q_{n-1} & r_{n-1} q_n \\
\hline
r_n q_{n+1} & q_n & \boxed{r_n} \\
\hline
r_{n+1} = 0 & &
\end{array}
$$

Then the greatest common divisor gcd of a and b is r_n. That is,

$$\gcd(a, b) = r_n. \tag{1.100}$$

Theorem 1.3.11 (Euclid's algorithm) Let a and b be positive integers with $a \geq b$. If $b \mid a$, then $\gcd(a, b) = b$. If $b \nmid a$, then apply the division algorithm repeatedly as follows:

$$
\begin{cases}
a = bq_0 + r_1, & 0 \leq r_1 < b, \\
b = r_1 q_1 + r_2, & 0 \leq r_2 < r_1, \\
r_1 = r_2 q_2 + r_3, & 0 \leq r_3 < r_2, \\
r_2 = r_3 q_3 + r_4, & 0 \leq r_4 < r_3, \\
\cdots\cdots\ \cdots\cdots & \\
r_{n-2} = r_{n-1}q_{n-1} + r_n, & 0 \leq r_n < r_{n-1}, \\
r_{n-1} = r_n q_n + 0
\end{cases}
\tag{1.101}
$$

This process ends when a remainder of 0 is obtained. This must occur after a finite number of steps; that is, $r_{n+1} = 0$ for some n. Then r_n, the last nonzero remainder, is the *greatest common divisor* of a and b. That is,

$$\gcd(a, b) = r_n. \tag{1.102}$$

Note that Euclid's algorithm is found in Book VII, Proposition 1 and 2 of his *Elements*, but it probably wasn't his own invention. Scholars believe that the method was known up to 200 years earlier. However, it first appeared in Euclid's *Elements*.

It is evident that the algorithm cannot recur indefinitely, since the second argument strictly decreases in each recursive call. Therefore, the algorithm always terminates with the correct answer. More importantly, it can be performed in polynomial time. That is, if Euclid's algorithm is applied to two positive integers a and b with $a \geq b$, then the number of divisions required to find $\gcd(a, b)$ is $\mathcal{O}(\log b)$, a polynomial-time complexity[12].

Example 1.3.8 Use Euclid's algorithm to find the gcd of 1281 and 243. Since

$$
\begin{array}{r|c|c}
1281 & & \\
-\ 1215 & 5 & 243 \\
\hline
66 & 3 & 198 \\
\hline
45 & 1 & 45 \\
\hline
21 & 2 & 42 \\
\hline
21 & 7 & \boxed{3} \\
\hline
0 & &
\end{array}
$$

Then $\gcd(1281, 243) = 3$.

The result about $\gcd(a, b)$ will be equal to unity more than 60 percent of the time for random inputs is a consequence of the following well-known result of number theory [69]:

Theorem 1.3.12 If a and b are integers chosen at random, the probability that $\gcd(a, b) = 1$ is $6/\pi^2 = 0.60793$.

1.3.2 Continued Fractions

Euclid's algorithm for computing the greatest common divisor of two integers is intimately connected with continued fractions.

---

[12]We shall introduce the $\mathcal{O}$-notation and discuss computational complexity formally in Chapter 3.

Definition 1.3.7 Let a and b be integers and let Euclid's algorithm run as in equation (1.101). Then the fraction $\frac{a}{b}$ can be expressed as a simple continued fraction:

$$\frac{a}{b} = q_0 + \cfrac{1}{q_1 + \cfrac{1}{q_2 + \cfrac{1}{\ddots\, q_{n-1} + \cfrac{1}{q_n}}}}, \tag{1.103}$$

where $q_0, q_1, \cdots, q_{n-1}, q_n$ are taken directly from Euclid's algorithm expressed in (1.101). For simplicity, the continued fraction expansion (1.103) of $\frac{a}{b}$ is usually written as

$$\frac{a}{b} = q_0 + \frac{1}{q_1+} \frac{1}{q_2+} \cdots \frac{1}{q_{n-1}+} \frac{1}{q_n} \tag{1.104}$$

or even more briefly as

$$\frac{a}{b} = [q_0, q_1, q_2, \cdots q_{n-1}, q_n]. \tag{1.105}$$

If each q_i is an integer, the continued fraction is called *simple*; a simple continued fraction can either be *finite* or *infinite*. A continued fraction formed from $[q_0, q_1, q_2, \cdots q_{n-1}, q_n]$ by neglecting all of the terms after a given term is called a *convergent* to the original continued fraction. If we denote the k-th convergent by $C_k = \frac{P_k}{Q_k}$, then

(i) $C_0 = \frac{P_0}{Q_0} = \frac{q_0}{1}$; $C_1 = \frac{P_1}{Q_1} = \frac{q_0 q_1 + 1}{q_1}$; $C_k = \frac{P_k}{Q_k} = \frac{q_k P_{k-1} + P_{k-2}}{q_k Q_{k-1} + Q_{k-2}}$, $k \geq 2$.

(ii) If $P_k = q_k Q_{k-1} + Q_{k-2}$ and $Q_k = q_k P_{k-1} + P_{k-2}$, then $\gcd(P_k, Q_k) = 1$.

(iii) $P_k Q_{k-1} - P_{k-1} Q_k = (-1)^{k-1}$, for $k \geq 1$.

In the following example, we show how to use Euclid's algorithm to express a rational number as a finite simple continued fraction.

Example 1.3.9 Expand the rational number $\frac{1281}{243}$ as a simple continued fraction. First let $a = 1281$ and $b = 243$, and then let Euclid's algorithm run as follows:

$$
\begin{array}{c|c|c}
1281 & & \\
-\,1215 & 5 & 243 \\
\hline
66 & 3 & -\,198 \\
\hline
-\,45 & 1 & 45 \\
\hline
21 & 2 & -\,42 \\
\hline
-\,21 & 7 & \boxed{3} \\
\hline
0 & &
\end{array}
$$

So $\frac{1281}{243} = [5, 3, 1, 2, 7]$. Thus

$$
\frac{1281}{243} = 5 + \cfrac{1}{3 + \cfrac{1}{1 + \cfrac{1}{2 + \cfrac{1}{7}}}}.
$$

Of course, as a by-product, we also find that $\gcd(1281, 243) = 3$.

Theorem 1.3.13 Any *finite* simple continued fraction represents a *rational* number. Conversely, any *rational* number can be expressed as a *finite* simple continued fraction in exactly two ways, one with an odd number of terms and one with an even number of terms.

Definition 1.3.8 Let $q_0, q_1, q_2, \cdots$ be a sequence of integers, all positive except possibly q_0. Then the expression $[q_0, q_1, q_2, \cdots]$ is called an *infinite* simple continued fraction and is defined to be equal to the number $\lim_{n \to \infty} [q_0, q_1, q_2, \cdots, q_{n-1}, q_n]$.

Theorem 1.3.14 Any *irrational* number can be written uniquely as an *infinite* simple continued fraction. Conversely, if α is an infinite simple continued fraction, then α is irrational.

Definition 1.3.9 A real irrational number which is the root of a quadratic equation with integer coefficients is called *quadratic irrational*.

For example, $\sqrt{3}$, $\sqrt{5}$, $\sqrt{7}$ are all quadratic irrationals. For convenience, we shall denote $\sqrt{N}$, with N not a perfect square, as a quadratic irrational. Quadratic irrationals are the simplest possible irrationals.

Definition 1.3.10 An infinite simple continued fraction is said to be *periodic* if it is of the form $[q_0, q_1, \cdots, q_k, \overline{q_{k+1}, q_{k+2}, \cdots, q_{k+m}}]$. If it is of the form $[\overline{q_0, q_1, \cdots, q_{m-1}}]$, then it is called *purely periodic*. The smallest positive integer m satisfying the above relationship is called the *period* of the expansion.

Theorem 1.3.15 Any *periodic* simple continued fraction is a quadratic irrational. Conversely, any *quadratic irrational* has a periodic expansion as a simple continued fraction.

What we are more interested in is *how* to represent a quadratic irrational as a periodic simple continued fraction. The following theorem provides us with a systematic way, based on Euclid's algorithm, of expressing a quadratic irrational number, say, e.g., $\sqrt{N}$ with N not a perfect square, as an infinite simple continued fraction; the *period* of the continued fraction expansion of $\sqrt{N}$ can be easily determined at a very earlier stage of the computation.

Theorem 1.3.16 Let $\sqrt{N}$, with N not a perfect square, be a quadratic irrational number. Then $\sqrt{N}$ can be expressed as an infinite (but periodic) continued fraction:

$$[q_0, q_1, q_2, \cdots, q_n, q_{n+1}, \cdots]$$

where

$$\begin{cases} x_0 = \sqrt{N} \\[2mm] q_i = \lfloor x_i \rfloor \\[2mm] x_{i+1} = \dfrac{1}{x_i - q_i} \end{cases}$$

for $i = 1, 2, \cdots, n, n+1, \cdots$.

By the above theorem, an efficient algorithm for representing $\sqrt{N}$ as an infinite simple continued fraction can be derived as follows:

Algorithm 1.3.1 (Algorithm for Computing Continued Fractions of $\sqrt{N}$) Let $x_0 = \sqrt{N}$ be the quadratic irrational number. This algorithm generates the infinite simple continued fraction $[q_0, q_1, q_2, \cdots, q_n, q_{n+1}, \cdots]$ for $\sqrt{N}$ as follows:

$$q_0 = \lfloor x_0 \rfloor, \qquad\qquad x_1 = \frac{1}{x_0 - q_0}$$

$$q_1 = \lfloor x_1 \rfloor, \qquad\qquad x_2 = \frac{1}{x_1 - q_1}$$

$$\cdots\cdots \qquad\qquad\qquad \cdots\cdots$$

$$\cdots\cdots \qquad\qquad\qquad \cdots\cdots$$

$$q_n = \lfloor x_n \rfloor, \qquad\qquad x_{n+1} = \frac{1}{x_n - q_n}$$

$$q_{n+1} = \lfloor x_{n+1} \rfloor, \qquad\qquad x_{n+2} = \frac{1}{x_{n+1} - q_{n+1}}$$

$$\cdots\cdots \qquad\qquad\qquad \cdots\cdots$$

$$\cdots\cdots \qquad\qquad\qquad \cdots\cdots$$

Of course, we do not need to calculate the infinitely many q_i's, since according to Theorem 1.3.15, any quadratic irrational number is periodic and can be written as an infinite simple continued fraction of the form $[q_0, q_1, q_2, \cdots, q_k, \overline{q_{k+1}, \cdots, q_{k+m}}]$.

Now we are ready to use Algorithm 1.3.1 to represent any quadratic irrational as a periodic simple continued fraction.

Example 1.3.10 Expand $\sqrt{3}$ as a periodic simple continued fraction. Let $x_0 = \sqrt{3}$. Then we have:

$$q_0 = \lfloor x_0 \rfloor = \lfloor \sqrt{3} \rfloor = 1$$

$$x_1 = \frac{1}{x_0 - q_0} = \frac{1}{\sqrt{3} - 1} = \frac{1}{\sqrt{3} - 1} = \frac{\sqrt{3} + 1}{2}$$

$$q_1 = \lfloor x_1 \rfloor = \lfloor \frac{\sqrt{3} + 1}{2} \rfloor = \lfloor 1 + \frac{\sqrt{3} - 1}{2} \rfloor = 1$$

$$x_2 = \frac{1}{x_1 - q_1} = \frac{1}{\frac{\sqrt{3} + 1}{2} - 1} = \frac{1}{\frac{\sqrt{3} - 1}{2}} = \frac{2(\sqrt{3} + 1)}{(\sqrt{3} - 1)(\sqrt{3} + 1)} = \sqrt{3} + 1$$

$$q_2 = \lfloor x_2 \rfloor = \lfloor \sqrt{3} + 1 \rfloor = 2$$

$$x_3 = \frac{1}{x_2 - q_2} = \frac{1}{\sqrt{3} + 1 - 2} = \frac{1}{\sqrt{3} - 1} = \frac{\sqrt{3} + 1}{2} = x_1$$

$$q_3 = \lfloor x_3 \rfloor = \lfloor \frac{\sqrt{3} + 1}{2} \rfloor = \lfloor 1 + \frac{\sqrt{3} - 1}{2} \rfloor = 1 = q_1$$

$$x_4 = \frac{1}{x_3 - q_3} = \frac{1}{\frac{\sqrt{3} + 1}{2} - 1} = \frac{1}{\frac{\sqrt{3} - 1}{2}} = \frac{2(\sqrt{3} + 1)}{(\sqrt{3} - 1)(\sqrt{3} + 1)} = \sqrt{3} + 1 = x_2$$

$$q_4 = \lfloor x_3 \rfloor = \lfloor \sqrt{3} + 1 \rfloor = 2 = q_2$$

$$x_5 = \frac{1}{x_4 - q_4} = \frac{1}{\sqrt{3} + 1 - 2} = \frac{1}{\sqrt{3} - 1} = \frac{\sqrt{3} + 1}{2} = x_3 = x_1$$

$$q_5 = \lfloor x_5 \rfloor = \lfloor x_3 \rfloor = 1 = q_3 = q_1$$

$$\cdots \cdots \cdots$$

$$\cdots \cdots \cdots$$

So, for $n = 1, 2, 3, \cdots$, we have $q_{2n-1} = 1$ and $q_{2n} = 2$. Thus, the *period* of the continued fraction expansion of $\sqrt{3}$ is 2. Therefore, we finally get

$$\sqrt{3} = 1 + \cfrac{1}{1 + \cfrac{1}{2 + \cfrac{1}{1 + \cfrac{1}{2 + \cfrac{1}{\ddots}}}}} = [1, \overline{1, 2}]. \qquad (1.106)$$

1.3.3 Diophantine Equations

In this subsection, we shall introduce some basic concepts of Diophantine equations. More specifically, we shall show how to use continued fractions to solve linear and quadratic Diophantine equations. Diophantine equations get their name from the ancient Greek mathematician Diophantus[13], who was the first to consider a whole class of problems having as their common feature that there were more unknowns than equations.

Definition 1.3.11 The algebraic equation with two variables

$$ax + by = c \qquad (1.107)$$

---

[13] Diophantus (around 250 A.D.) lived in the great city of Alexandria about 1700 years ago. He is perhaps best known as the writer of the book *Arithmetica*, of which ten of the thirteen books are extant. About 130 problems in Arithmetic and Algebra are considered in the book, some of them are surprisingly hard. The work of Diophantus was forgotten until a copy of the book was discovered in 1570. Italian mathematicians in the 16th century introduced his works to Europe where they were read with great interest and where they stimulated the study of Algebra, more specifically, Diophantine Analysis. Very little knowledge about his personal life has survived except his epitaph which contains the clues to his age: One sixth of his life was spent as a child; after one twelfth more he grew a beard; when one seventh more had passed, he married. Five years later a son was born; the son lived to half his father's age; four years after the son's death, he also died.

is called a *linear Diophantine equation* , for which we wish to find integer solutions in x and y.

A linear Diophantine equation is a type of algebraic equation with two linear variables. For this reason, it is sometimes also called a *bilinear Diophantine equation*. In this type of equation $ax + by = c$, we are only interested in the integer solutions in x and y.

Theorem 1.3.17 Let a, b, c be integers with not both a and b equal to 0, and let $d = \gcd(a, b)$. The linear Diophantine equation $ax + by = c$ has integer solutions in x and y if and only if $d \mid c$. If $d \nmid c$, then the equation has no integer solution.

Theorem 1.3.18 Let a, b, c be integers with not both a and b equal to 0, and let $d = \gcd(a, b)$. Suppose x_0, y_0 is any particular pair of integral solutions of the linear Diophantine equation $ax + by = c$, then all other (or the complete set of the) solutions of the equation are given by

$$\begin{cases} x = x_0 + \dfrac{b}{d} \cdot t \\ y = y_0 - \dfrac{a}{d} \cdot t \end{cases} \tag{1.108}$$

for varying integer t.

Theorem 1.3.17 only tells us whether or not the equation $ax + by = c$ is soluble in integers, but does not tell us how to find the integer solutions of x and y. There are several methods for finding such integer solutions. We shall show in the following how to compute x and y by using the continued fraction method.

Suppose that a and b are two integers whose *gcd* is d and we wish to solve

$$ax - by = d. \tag{1.109}$$

We expand a/b as a finite continued fraction with convergents

$$\left[\frac{P_0}{Q_0}, \frac{P_1}{Q_1}, \cdots, \frac{P_{n-1}}{Q_{n-1}}, \frac{P_n}{Q_n} \right] = \frac{a}{b}. \tag{1.110}$$

Since $d = \gcd(a, b)$ we must have $a = da'$, $b = db'$ and $\gcd(a', b') = 1$. Then $P_n/Q_n = a'/b'$ and both fractions are in their lowest terms, giving $P_n = a'$, $Q_n = b'$. So equation (1.109) gives

$$P_n Q_{n-1} - Q_n P_{n-1} = a' Q_{n-1} - b' P_{n-1} = (-1)^{n-1} \tag{1.111}$$

Hence

$$aQ_{n-1} - bP_{n-1} = da'Q_{n-1} - db'P_{n-1} = (-1)^{n-1}d \qquad (1.112)$$

or

$$(-1)^{n-1}aQ_{n-1} - (-1)^{n-1}bP_{n-1} = d \qquad (1.113)$$

A solution to equation $ax - by = d$ is therefore given by

$$\begin{cases} x = (-1)^{n-1}Q_{n-1}, \\ y = (-1)^{n-1}P_{n-1}. \end{cases} \qquad (1.114)$$

To conclude the above analysis, we have the following theorem for solving the linear Diophantine equation $ax - by = d$:

Theorem 1.3.19 Let the convergents of the finite continued fraction of a/b be as follows:

$$\left[\frac{P_0}{Q_0}, \frac{P_1}{Q_1}, \cdots, \frac{P_{n-1}}{Q_{n-1}}, \frac{P_n}{Q_n} \right] = \frac{a}{b}. \qquad (1.115)$$

Then the integer solutions in x and y of the equation $ax - by = d$ are

$$\begin{cases} x = (-1)^{n-1}Q_{n-1}, \\ y = (-1)^{n-1}P_{n-1}. \end{cases} \qquad (1.116)$$

Example 1.3.11 Use the continued fraction method to solve the following linear Diophantine equation:

$$364x - 227y = 1.$$

Since $364/227$ can be expanded as a finite continued fraction with convergents

$$\left[1, \ 2, \ \frac{3}{2}, \ \frac{5}{3}, \ \frac{8}{5}, \ \frac{85}{53}, \ \frac{93}{58}, \ \frac{364}{227} \right]$$

then we have

$$\begin{cases} x = (-1)^{n-1}q_{n-1} = (-1)^{7-1}58 = 58, \\ y = (-1)^{n-1}p_{n-1} = (-1)^{7-1}93 = 93. \end{cases}$$

That is,

$$364 \cdot 58 - 227 \cdot 93 = 1.$$

Example 1.3.12 Use the continued fraction method to solve the following linear Diophantine equation:

$$20719x + 13871y = 1.$$

Note first that

$$20719x + 13871y = 1 \iff 20719x - (-13871y) = 1.$$

Now since $20719/13871$ can be expanded as a finite simple continued fraction with convergents

$$\left[1, \ \frac{3}{2}, \ \frac{118}{79}, \ \frac{829}{555}, \ \frac{947}{634}, \ \frac{1776}{1189}, \ \frac{2723}{1823}, \ \frac{4499}{3012}, \ \frac{20719}{13871} \right]$$

then we have

$$\begin{cases} x = (-1)^{n-1}q_{n-1} = (-1)^{8-1}3012 = -3012, \\ y = (-1)^{n-1}p_{n-1} = (-1)^{8-1}4499 = -4499. \end{cases}$$

That is,

$$20719 \cdot (-3012) - 13871 \cdot (-4499) = 1.$$

We now move on to the solution of quadratic Diophantine equations.

Definition 1.3.12 We call the Diophantine equation of the form

$$x^2 - Ny^2 = 1 \quad \text{or} \quad x^2 - Ny^2 = -1 \tag{1.117}$$

Pell's equation, where N is a positive integer other than a perfect square.

Remark 1.3.1 Pell's equation is named after the 17th British mathematician John Pell. It probably should be called Fermat's equation since Fermat initiated its comparatively recent study of the topic. But in one of his works, Euler mistakenly attributed the equation to Pell who apparently made no contribution to this topic.

The solutions to Pell's equation can be easily obtained in terms of the continued fraction for $\sqrt{N}$.

Theorem 1.3.20 Let N be a positive integer other than a perfect square. If (x_0, y_0) is a positive integral solution of $x^2 - Ny^2 = \pm 1$, then $x_0 = P_n$ and $y_0 = Q_n$, where $\frac{P_n}{Q_n}$ is one of the convergents to $\sqrt{N}$.

Theorem 1.3.21 Let N be a positive integer other than a perfect square, and m the period of the expansion of $\sqrt{N}$ as a simple continued fraction. Then we have:

(I) m is even

(i) The positive integer solutions of $x^2 - Ny^2 = 1$ are

$$\begin{cases} x = P_{km-1} \\ y = Q_{km-1} \quad \text{for } k = 1, 2, 3, \cdots, \end{cases} \tag{1.118}$$

with

$$\begin{cases} x = P_{m-1} \\ y = Q_{m-1} \end{cases} \tag{1.119}$$

as the *smallest* positive integer solutions.

(ii) The equation $x^2 - Ny^2 = -1$ has no integer solution.

(II) m is odd

(i) The positive integer solutions of $x^2 - Ny^2 = 1$ are

$$\begin{cases} x = P_{km-1} \\ y = Q_{km-1} \quad \text{for } k = 2, 4, 6, \cdots, \end{cases} \tag{1.120}$$

with

$$\begin{cases} x = P_{2m-1} \\ y = Q_{2m-1} \end{cases} \tag{1.121}$$

as the *smallest* positive integer solutions.

(ii) The positive integer solutions of $x^2 - Ny^2 = -1$ are

$$\begin{cases} x = P_{km-1} \\ y = Q_{km-1} \quad \text{for } k = 1, 3, 5, \cdots, \end{cases} \tag{1.122}$$

with

$$\begin{cases} x = P_{m-1} \\ y = Q_{m-1} \end{cases} \tag{1.123}$$

as the *smallest* positive integer solutions.

Example 1.3.13 Find the integer solutions of $x^2 - 97y^2 = \pm 1$. Note first that

$$\sqrt{97} = [9, \overline{1, 5, 1, 1, 1, 1, 1, 1, 5, 1, 18}].$$

So the period $m = 11$ and of course m is odd. Thus, both equations are soluble.

(i) The smallest positive integral solutions of $x^2 - 97y^2 = 1$ are

$$\begin{cases} x = P_{km-1} = P_{2\cdot 11-1} = P_{21} = 62809633 \\ y = Q_{km-1} = Q_{2\cdot 11-1} = Q_{21} = 6377352 \end{cases} \tag{1.124}$$

That is, $62809633^2 - 97 \cdot 6377352^2 = 1$

(ii) The positive integer solutions of $x^2 - 97y^2 = -1$ are

$$\begin{cases} x = P_{km-1} = P_{1\cdot 11-1} = P_{10} = 5604 \\ y = Q_{km-1} = Q_{1\cdot 11-1} = Q_{10} = 569 \end{cases} \tag{1.125}$$

That is, $5604^2 - 97 \cdot 569^2 = 1$.

Remark 1.3.2 Incidentally, the continued fraction for $\sqrt{N}$ always has the form

$$\sqrt{N} = [q_0, \overline{q_1, q_2, q_3, \cdots, q_3, q_2, q_1, 2q_0}]$$

as for

$$\sqrt{97} = [9, \overline{1, 5, 1, 1, 1, 1, 1, 1, 5, 1, 18}]$$

discussed above. Moreover, the period m of $\sqrt{N}$ is always soluble if N is a prime $p \equiv 1 \pmod 4$. In fact, for each prime p, the equation

$$x^2 - py^2 = -1$$

is always soluble.

1.3.4   Congruences

The notion of congruences was first introduced by Gauss, though the ancient Greeks and Chinese had already had the idea, who gave the definition of congruences in his celebrated *Disquistiones Arithmeticae* in 1801.

Definition 1.3.13 Let a and b be integers and m a positive integer. We define "$a \bmod m$" to be the remainder r when a is divided by m.

Example 1.3.14 The following are some examples of $a \bmod m$:

$$35 \bmod 12 = 11,$$
$$-129 \bmod 7 = 4,$$
$$3210 \bmod 101 = 79,$$
$$1412^{13115} \bmod 12349 = 1275.$$

Definition 1.3.14 Let a and b be integers and m a positive integer. We say that "a is *congruent* to b modulo m" if m is a divisor of $a - b$ (i.e., $m \mid (a - b)$). This relation is written $a \equiv b \pmod{m}$. m is called the *modulus* of the congruence. If $m \nmid (a - b)$, then we say a is not congruent (or incongruent) to b modulo m and write $a \not\equiv b \pmod{m}$.

Remark 1.3.3 The above definition, introduced by Gauss[14] in *Disquistiones Arithmeticae*, does not introduce any new idea, since "$a \equiv b \pmod{m}$" and "$m \mid (a - b)$ (resp. "$a \not\equiv b \pmod{m}$" and "$m \nmid (a - b)$) have the same meaning, although each of them has its advantages. Note that Gauss did present many new ideas, such as his proof of the *Reciprocity Law* and his extensive theory of binary quadratic form, a complete treatment of primitive roots and indices, etc. What we are more interested in is not one of these new things, but his new way of looking at the old things.

Remark 1.3.4 It is evident that $a \equiv b \pmod{m}$ if and only if $a \bmod m = b \bmod m$, and $a \not\equiv b \pmod{m}$ if and only if $a \bmod m \neq b \bmod m$.

Example 1.3.15 The following are some examples of congruences or incongruences:

$$35 \equiv 11 \pmod{12} \qquad \text{since } 12 \mid (35 - 11),$$
$$\not\equiv 12 \pmod{11} \qquad \text{since } 11 \nmid (35 - 12),$$
$$\equiv 2 \pmod{11} \qquad \text{since } 11 \mid (35 - 2).$$

---

[14]Carl Friedrich Gauss (1777–1855), the greatest mathematician of all time (Prince of Mathematicians), was the son of a German bricklayer. It was quickly apparent that he was a child prodigy. In fact, at the age of three he corrected an error in his father's payroll. Gauss made fundamental contributions to astronomy including calculating the orbit of the asteroid Ceres. On the basis of this calculation, Gauss was appointed director of the Göttingen Observatory. He laid the foundations of modern number theory with his book *Disquistiones Arithmeticae* in 1801. Gauss conceived most of his discoveries before the age of 20, but spent the rest of his life polishing and refining them. Although Gauss adorned every branch of mathematics, he always held number theory in high esteem and affection. He insisted that, "Mathematics is the Queen of the Sciences, and Number Theory is the Queen of Mathematics".

The relation of congruences have many properties in common with the relation of equality. For example, we know from high-school algebra that equality is

(i) *reflexive*: $a = a$, $\forall a \in \mathbb{Z}$;

(ii) *symmetric*: if $a = b$, then $b = a$, $\forall a, b \in \mathbb{Z}$;

(iii) *transitive*: $a = b$ and $b = c$, then $a = c$, $\forall a, b, c \in \mathbb{Z}$.

We shall see in the following theorem that congruence modulo m is also reflexive, symmetric and transitive.

Theorem 1.3.22 Let m be a positive integer. Then

(i) *reflexive*: $a \equiv a \pmod{m}$, $\forall a \in \mathbb{Z}$;

(ii) *symmetric*: if $a \equiv b \pmod{m}$, then $b \equiv a \pmod{m}$, $\forall a, b \in \mathbb{Z}$;

(iii) *transitive*: $a \equiv b \pmod{m}$ and $b \equiv c \pmod{m}$, then $a \equiv c \pmod{m}$, $\forall a, b, c \in \mathbb{Z}$.

Theorem 1.3.22 shows that congruence modulo m is an equivalence relation on the set of integers $\mathbb{Z}$, which partitions $\mathbb{Z}$ into m *equivalent* classes. In number theory, we call these classes *congruent classes*, or *residue classes*. More formally, we have:

Definition 1.3.15 If $x \equiv a \pmod{m}$, then a is called a *residue of x modulo m*. denoted by $[a]$. The *residue class* or the *congruence class* of a modulo m, denoted by $[a]$, is the set of all those integers that are congruent to a modulo m. That is,

$$\begin{aligned} [a] &= \{x : x \in \mathbb{Z} \text{ and } x \equiv a \pmod{m}\} \\ &= \{a + km : k \in \mathbb{Z}\}. \end{aligned} \tag{1.126}$$

Example 1.3.16 Let $m = 5$. Then there are five residue classes, modulo 5, namely the sets:

$$\begin{aligned} [0] &= \{\cdots, -15, -10, -5, 0, 5, 10, 15, 20, \cdots\} \\ [1] &= \{\cdots, -14, -9, -4, 1, 6, 11, 16, 21, \cdots\} \\ [2] &= \{\cdots, -13, -8, -3, 2, 7, 12, 17, 22, \cdots\} \\ [3] &= \{\cdots, -12, -7, -2, 3, 8, 13, 18, 23, \cdots\} \\ [4] &= \{\cdots, -11, -6, -1, 4, 9, 14, 19, 24, \cdots\}. \end{aligned}$$

The first set containing all those integers congruent to 0 (mod 5), the second set, those congruent to 1 (mod 5), $\cdots$, the fifth (i.e., the last) set those congruent to 4 (mod 5). So, say, for example, the residue class [2] can be represented by any one of the elements in the set

$$\{\cdots, -13, \ -8, -3, 2, 7, 12, 17, 22, \cdots\}.$$

Clearly, there are infinitely many elements in the set [2].

Example 1.3.17 In residue classes modulo 2, [0] is the set of all even integers, and [1] is the set of all odd integers:

$$[0] = \{\cdots, -6, -4, -2, 0, 2, 4, 6, 8, \cdots\}$$
$$[1] = \{\cdots, -5, -3, -1, 1, 3, 5, 7, 9, \cdots\}$$

Example 1.3.18 In congruence modulo 5, we have

$$\begin{aligned}[9] &= \{9 + 5k : k \in \mathbb{Z}\} = \{9, 9 \pm 5, 9 \pm 10, 9 \pm 15, \cdots\} \\ &= \{\cdots, -11, -6, -1, 4, 9, 14, 19, 24, \cdots\}\end{aligned}$$

we can also have

$$\begin{aligned}[4] &= \{4 + 5k : k \in \mathbb{Z}\} = \{4, 4 \pm 5, 4 \pm 10, 4 \pm 15, \cdots\} \\ &= \{\cdots, -11, -6, -1, 4, 9, 14, 19, 24, \cdots\}\end{aligned}$$

So, clearly, [4] = [9] modulo 5.

Definition 1.3.16 If $x \equiv a \pmod{m}$ and $0 \le a \le m - 1$, then a is called the *least residue* of x modulo m.

Example 1.3.19 Let $m = 7$. There are seven residue classes, modulo 7. In each of these seven residue classes, there is exactly one least residue of x modulo 7. So the complete set of all least residues x modulo 7 is $\{0, 1, 2, 3, 4, 5, 6\}$.

The following theorem gives some elementary properties of residue classes:

Theorem 1.3.23 Let m be a positive integer. Then we have:

(i) $[a] = [b]$ if and only if $a \equiv b \pmod{m}$.

(ii) $[a] \neq [c]$ if and only if $[a] \cap [b] = \emptyset$.

(iii) Two residue classes modulo m are either disjoint or identical.

(iv) There are exactly m distinct residue classes modulo m, namely, $[0], [1], [2], [3], \cdots, [m-1]$, and they contain all of the integers.

Definition 1.3.17 Let m be a positive integer. A set of integers $a_1, a_2, \cdots, a_m$ is called a *complete system of residues* modulo m, if the set contains exactly one element from each residue class, modulo m.

Example 1.3.20 Let $m = 4$. Then $\{-12, 9, -6, -1\}$ is a complete system of residues modulo 4, and $\{12, -7, 18, -9\}$ is another complete system of residues modulo 4.

Definition 1.3.18 Let m be a positive integer, and $\phi(m)$ is the number of residue classes modulo m, that is relatively prime to m. The function $\phi(m)$ is usually called Euler's totient function (or just Euler's function for short). A set of integers $\{a_1, a_2, \cdots, a_{\phi(m)}\}$ is called a *reduced system of residues* modulo m, if the set contains exactly one element from each residue class modulo m that is relatively prime to m.

Example 1.3.21 In Example 1.3.16, we know that $[1]$, $[2]$, $[3]$, $[4]$ modulo 5 are residue classes that are relatively prime to 5, so by choosing -4 from $[1]$, 17 from $[2]$, 23 from $[3]$ and 19 from $[4]$ modulo 10, we get a reduced system of residues modulo 5: $\{-4, 17, 23, 19\}$. Similarly, $\{1, -13, 13, -1\}$ is another reduced system of residues modulo 5.

Theorem 1.3.24 Let $m = p_1^{\alpha_1} p_2^{\alpha_2} \cdots p_k^{\alpha_k}$. Then the Euler's function can be evaluated by

$$\phi(m) = \prod_{i=1}^{k} p_i^{\alpha_i - 1}(p_i - 1) = m \prod_{i=1}^{k} \left(1 - \frac{1}{p_i}\right). \qquad (1.127)$$

If $m = p^\alpha$ with p prime and α positive integer, then

$$\phi(p^\alpha) = p^\alpha \left(1 - \frac{1}{p}\right) = p^{\alpha-1}(p - 1). \qquad (1.128)$$

If $m = p$ is prime, then

$$\phi(p) = p - 1. \qquad (1.129)$$

Congruences have much in common with equations. In fact, linear congruence $ax \equiv b \pmod{m}$ is equivalent to the linear Diophantine equation $ax - my = b$. That is,

$$ax \equiv b \pmod{m} \Longleftrightarrow ax - my = b. \tag{1.130}$$

So linear congruences can be solved by using the continued fraction method just as that for linear Diophantine equations. More interestingly, systems of linear congruences, similar to systems of linear equations, can be solved by the Chinese Remainder Theorem (CRT), invented by the ancient Chinese mathematician Sun Tsu[15].

Theorem 1.3.25 (The Chinese Remainder Theorem CRT) If $m_1, m_2, ..., m_n$ are pairwise relatively prime and greater than 1, and $a_1, a_2, ..., a_n$ are any integers, then there is a solution x to the following simultaneous congruences:

$$\begin{cases} x \equiv a_1 \pmod{m_1}, \\ x \equiv a_2 \pmod{m_2}, \\ \qquad \cdots \cdots \\ \qquad \cdots \cdots \\ x \equiv a_n \pmod{m_n}. \end{cases} \tag{1.131}$$

If x and x' are two solutions, then $x \equiv x' \pmod{M}$, where $M = m_1 m_2 \cdots m_n$.

PROOF. (Existence) Let us first solve a special case of the simultaneous congruences (1.131), where i is some fixed subscript, $a_i = 1$, and $a_1 = a_2 = \cdots = a_{i-1} = a_{i+1} = \cdots = a_n = 0$. Let

---

[15]Sun Zi (known as Sun Tsu in the West), Chinese mathematician lived sometime between 200 B.C. and 200 A.D. He is perhaps best known for his invention of the Chinese Remainder Theorem which arose in Problem 26 in Volume 3 of his classic three-volume mathematics book *Mathematical Manual*: find a number that leaves a remainder of 2 when divided by 3, a remainder of 3 when divided by 5, and a remainder of 2 when divided by 7; in modern algebraic language, to find the smallest positive integer satisfying the following system of congruences: $\langle x \equiv 2 \pmod{3}, \ x \equiv 3 \pmod{5}, \ x \equiv 2 \pmod{7} \rangle$. Sun Zi gave a rule called "tai-yen" ("great generalisation") to find the solution. Sun Zi's rule was generalized in today's "theorem-form" by the great Chinese mathematician Qin Jiushao (1202–1261), known as Chhin Chiu Shao in the West, in his book *Shushu Jiuzhang* (Mathematical Treatise in Nine Sections) in 1247; Qin also rediscovered Euclid's Algorithm, and gave a complete procedure for solving numerically polynomial equations of any degree, which is very similar to or almost the same as what is now called the *Horner method* published by William Horner in 1819.

$k_i = m_1 m_2 \cdots m_{i-1} m_{i+1} \cdots m_n$. Then k_i and m_i are relatively prime, so we can find integers r and s such that $r k_i + s m_i = 1$. This gives the congruences:

$$\begin{cases} r k_i \equiv 0 \pmod{k_i}, \\ r k_i \equiv 1 \pmod{m_i}. \end{cases}$$

Since $m_1, m_2, \cdots, m_{i-1}, m_{i+1}, \cdots m_n$ all divide k_i, it follows that $x_i = r k_i$ satisfies the simultaneous congruences:

$$\begin{cases} x_i \equiv 0 \pmod{m_1}, \\ x_i \equiv 0 \pmod{m_2}, \\ \quad \cdots \cdots \\ x_i \equiv 0 \pmod{m_{i-1}}, \\ x_i \equiv 1 \pmod{m_i}, \\ x_i \equiv 0 \pmod{m_{i+1}}, \\ \quad \cdots \cdots \\ x_i \equiv 0 \pmod{m_n}. \end{cases}$$

For each subscript i, $1 \le i \le n$, we find such an x_i. Now to solve the system of the simultaneous congruences (1.131), set $x = a_1 x_1 + a_2 x_2 + \cdots + a_n x_n$. Then $x \equiv a_i x_i \equiv a_i \pmod{m_i}$ for each i, $1 \le i \le n$, x is the solution of the simultaneous congruences.

(Uniqueness) Let x' be another solution to the simultaneous congruences (1.131), but different from the solution x, such that $x' \equiv x \pmod{m_i}$ for each x_i. Then $x - x' \equiv 0 \pmod{m_i}$ for each i. So m_i divides $x - x'$ for each i; hence the least common multiple of all the m_j's divides $x - x'$. But since the m_i are pairwise relatively prime, this least common multiple is the product M. So $x \equiv x' \pmod{M}$. ∎

Remark 1.3.5 The proof of the existence of the CRT is constructive, which itself provides a method for finding all solutions of systems of simultaneous congruences 1.131. The Chinese Remainder Theorem has many important applications in areas such as algebra, number theory, computer arithmetic, fast computation, cryptography, computer security, and hash functions.

Now we are moving on to the introduction of some more useful results based on the theory of congruences:

Theorem 1.3.26 (Fermat's little theorem[16] , 1640) Let a be a positive integer, and p prime. If $\gcd(a,p) = 1$, then

$$a^{p-1} \equiv 1 \ (\mathrm{mod}\ p). \tag{1.132}$$

PROOF. First notice that the residues modulo p of a, $2a$, $\cdots$, $(p-1)a$ are $1, 2, \cdots, (p-1)$ in some order, because no two of them can be equal. So, if we multiply them together, we get

$$
\begin{aligned}
a \cdot 2a \cdots (p-1)a &\equiv [(a \bmod p) \cdot (2a \bmod p) \cdots (p-1)a \bmod p)] \ (\mathrm{mod}\ p) \\
&\equiv (p-1)! \ (\mathrm{mod}\ p).
\end{aligned}
$$

This means that

$$(p-1)!a^{p-1} \equiv (p-1)! \ (\mathrm{mod}\ p).$$

Now we can cancel the $(p-1)!$ since $p \nmid (p-1)!$, the result thus follows. ∎

An alternative form of Fermat's little theorem (or just *Fermat's theorem*, for short) is sometimes more convenient and more general:

$$a^p \equiv a \ (\mathrm{mod}\ p), \tag{1.133}$$

for $a \in \mathbb{N}$. The proof is easy: if $\gcd(a,p) = 1$, we simply multiply (1.132) by a. If not, then $p \mid a$. So $a^p \equiv 0 \equiv a \ (\mathrm{mod}\ p)$.

Fermat's theorem has several important consequences which are very useful in compositeness or primality testing; one of the these consequences is as follows:

Corollary 1.3.2 (Converse of Fermat's theorem) Let n be an odd positive integer. If $\gcd(a,n) = 1$ and

$$a^{n-1} \not\equiv 1 \ (\mathrm{mod}\ n), \tag{1.134}$$

then n is composite.

---

[16]In honor of the great French mathematician Pierre de Fermat (1601–1665). Fermat was born in Beaumont-de-Lomagne, a small town in southwest France. He pursued a career in local government and the judiciary, and produced high quality work in number theory and other areas of mathematics as a hobby. He published almost nothing, revealing most of his results in his extensive correspondence with friends, and generally kept his proofs to himself. Perhaps the most remarkable reference to his work is his *Last Theorem*, which asserts that if $n > 2$, the equation $x^n + y^n = z^n$ cannot be solved in integers x, y, z, with $xyz \neq 0$. He claimed in a margin of his copy of Diophantus' book *Arithmetica* that he had found a beautiful proof of this theorem, but the margin was too small to contain his proof. Later on mathematicians in the world struggled to find a proof for this theorem but without a success. The theorem remained open for more than 350 years, and was finally proved in 1994 by Andrew J. Wiles of Princeton University after a decade of concentrated effort.

Based on Fermat's theorem, Euler[17] established a more general result in 1760:

Theorem 1.3.27 (Euler's theorem) Let a and n be positive integers with $\gcd(a, n) = 1$. Then

$$a^{\phi(n)} \equiv 1 \ (\text{mod } n), \tag{1.135}$$

where $\phi(n)$ is the Euler's totient function. [Recall that $\phi(n)$ is defined to be the number of residue classes modulo n that is relatively prime to n. That is, $\phi(n) = \prod_i p_i^{\alpha_i - 1}(p_i - 1)$].

The following is a useful, but often forgotten generalization of Euler's theorem:

Theorem 1.3.28 (Carmichael's theorem[18]) Let a and n be positive integers with $\gcd(a, n) = 1$. Then

$$a^{\lambda(n)} \equiv 1 \ (\text{mod } n), \tag{1.136}$$

where $\lambda(n)$ is the *Carmichael's function* and is defined by

$$\begin{cases} \lambda(p^\alpha) = \phi(p^\alpha) & \text{for } p = 2 \text{ and } \alpha \le 2, \text{ and for } p \ge 3 \\ \lambda(2^\alpha) = \frac{1}{2}\phi(2^\alpha) & \text{for } \alpha \ge 3 \\ \lambda(n) = \text{lcm}\left(\lambda(p_1^{\alpha_1})\lambda(p_2^{\alpha_2}) \cdots \lambda(p_k^{\alpha_k})\right) & \text{if } n = \prod_{i=1}^{k} p_i^{\alpha_i}. \end{cases} \tag{1.137}$$

---

[17]In honor of the Swiss mathematician Leonhard Euler (1707–1783). A key figure in the 18th century mathematics, Euler spent most of his life in the Imperial Academy in St. Petersburg, Russia (1727–1741 and 1766–1783). Mainly known for his work in analysis, Euler wrote a calculus textbook and introduced the present-day symbols for e, ϕ and i. Among Euler's discoveries in number theory is the law of quadratic reciprocity, which connects the solvability of the congruences: $\langle x^2 \equiv p \ (\text{mod } q), \ y^2 \equiv q \ (\text{mod } p)\rangle$, where p and q are distinct primes, although it remained for Gauss to provide the first proof. Euler also gave a marvelous proof of the existence of infinitely many primes based on the divergence of the harmonic series $\sum n^{-1}$.

[18]In honor of the American mathematician Robert D. Carmichael (1879–1967). Carmichael received his PhD in 1911 from Princeton University. His thesis, written under G. D. Birkhoff, was considered the first significant American contribution to differential equations. Perhaps best known in number theory for his *Carmichael numbers*, *Carmichael's function*, and *Carmichael's theorem*, Carmichael worked in a wide range of areas, including real analysis, differential equations, mathematical physics, group theory and number theory.

Example 1.3.22 Let $n = 65520 = 2^4 \cdot 3^2 \cdot 5 \cdot 7 \cdot 13$ and $a = 11$ such that $\gcd(65520, 11) = 1$. Then we have:

$$\phi(65520) = 8 \cdot 6 \cdot 4 \cdot 6 \cdot 12 = 13824 \Longrightarrow 11^{13824} \equiv 1 \;(\text{mod } 65520)$$

$$\lambda(65520) = \text{lcm}(4, 6, 4, 6, 12) = 12 \Longrightarrow 11^{12} \equiv 1 \;(\text{mod } 65520).$$

In 1770 the British mathematician E. Waring published the following result, which is attributed to Wilson:

Theorem 1.3.29 (Wilson's theorem) If p is a prime, then

$$(p - 1)! \equiv -1 \;(\text{mod } p). \tag{1.138}$$

The congruences $ax \equiv b \;(\text{mod } m)$ we have studied so far are a special type of high-order congruences, that is, they are all linear congruences. We can, of course, also consider quadratic, cubic and higher degree congruences defined by:

Definition 1.3.19 Let m be a positive integer, and let

$$f(x) = a_0 + a_1 x + a_2 x^2 + \cdots + a_n x^n$$

be any polynomial with integer coefficients. Then a *high-order congruence* or a *polynomial congruence* is a congruence of the form

$$f(x) \equiv 0 \quad (\text{mod } m). \tag{1.139}$$

A polynomial congruence is also called a *polynomial congruential equation* .

Let us consider the polynomial congruence

$$f(x) = x^3 + 5x - 4 \equiv 0 \quad (\text{mod } 7).$$

This congruence holds when $x = 2$, since

$$f(2) = 2^3 + 5 \cdot 2 - 4 \equiv 0 \quad (\text{mod } 7).$$

Just as for algebraic equations, we say that $x = 2$ is a root or a solution of the congruence. In fact, any value of x which satisfies the following condition

$$x \equiv 2 \quad (\text{mod } 7)$$

is also a solution of the congruence. In general, as in linear congruence, when a solution x_0 has been found, all values x for which

$$x \equiv x_0 \pmod{m}$$

are also solutions. But by convention, we still consider them as a *single* solution. Thus, our problem is to find all incongruent (different) solutions of $f(x) \equiv 0 \pmod{m}$. In general, this problem is very difficult, and many techniques for solution depend partially on trial-and-error methods. For example, to find all solutions of the congruence $f(x) \equiv 0 \pmod{m}$, we could certainly try all values $0, 1, 2, \cdots, m-1$ (or the numbers in the complete residue system modulo m), and determine which of them satisfy the congruence; this gives us the total number of *incongruent* solutions modulo m.

Theorem 1.3.30 Let $M = m_1 m_2 \cdots m_n$, where $m_1, m_2, ..., m_n$ are pairwise relatively prime. Then the integer x_0 is a solution of

$$f(x) \equiv 0 \pmod{M} \tag{1.140}$$

if and only if x_0 is a solution of the system of polynomial congruences:

$$\begin{cases} f(x) \equiv 0 \pmod{m_1}, \\ f(x) \equiv 0 \pmod{m_2}, \\ \cdots\cdots \\ \cdots\cdots \\ f(x) \equiv 0 \pmod{m_n}. \end{cases} \tag{1.141}$$

If x and x' are two solutions, then $x \equiv x' \pmod{M}$, where $M = m_1 m_2 \cdots m_n$.

We now restrict ourselves to quadratic congruences, the simplest possible nonlinear polynomial congruences.

Definition 1.3.20 A quadratic congruence is a congruence of the form:

$$x^2 \equiv a \pmod{m} \tag{1.142}$$

where $\gcd(a, m) = 1$. To solve the congruence is to find an integral solution for x which satisfies the congruence.

In most cases, it is sufficient to study this congruence rather than the more general quadratic congruence

$$ax^2 + bx + c \equiv 0 \pmod{m} \tag{1.143}$$

since if $\gcd(a, m) = 1$ and b is even or m is odd, then the congruence (1.142) can be reduced to a congruence of type (1.143). The problem can even be further reduced for solving the congruence of the type:

$$x^2 \equiv a \pmod{p_1^{\alpha_1} p_2^{\alpha_2} \cdots p_k^{\alpha_k}} \tag{1.144}$$

because to solve congruence (1.144) is equivalent to solving the following system of congruences:

$$\begin{cases} x^2 \equiv a \pmod{p_1^{\alpha_1}} \\ x^2 \equiv a \pmod{p_2^{\alpha_2}} \\ \quad\ldots\ldots \\ x^2 \equiv a \pmod{p_k^{\alpha_k}} \end{cases} \tag{1.145}$$

So in this section, we shall only deal with quadratic congruences of the form

$$x^2 \equiv a \pmod{p} \tag{1.146}$$

where p is an odd prime and $a \not\equiv 0 \pmod{p}$.

Definition 1.3.21 Let a be any integer and m a natural number. Suppose that $\gcd(a, m) = 1$. Then a is called a *quadratic residue* (mod n) if the congruence

$$x^2 \equiv a \pmod{m}$$

is soluble; otherwise it is called a *quadratic non-residue* (mod m).

Example 1.3.23 Decide the quadratic residues or quadratic non-residues for moduli $5, 7, 11$, respectively.

(i) $1, 4$ are quadratic residues and $2, 3$ are quadratic non-residues (mod 5), since $1^2 \equiv 4^2 \equiv 1 \pmod 5$ and $2^2 \equiv 3^2 \equiv 4 \pmod 5$.

(ii) $1, 2, 4$ are quadratic residues and $3, 5, 6$ are quadratic non-residues (mod 7), since $1^2 \equiv 6^2 \equiv 1 \pmod 7$, $2^2 \equiv 5^2 \equiv 4 \pmod 7$ and $3^2 \equiv 4^2 \equiv 2 \pmod 7$.

(iii) $1, 3, 4, 5, 9$ are quadratic residues and $2, 6, 7, 8, 10$ are quadratic non-residues (mod 11), since

$$1^2 \equiv 10^2 \equiv 1 \pmod{11} \quad 2^2 \equiv 9^2 \equiv 4 \pmod{11} \quad 3^2 \equiv 8^2 \equiv 9 \pmod{11}$$
$$4^2 \equiv 7^2 \equiv 5 \pmod{11} \qquad 5^2 \equiv 6^2 \equiv 3 \pmod{11}$$

Definition 1.3.22 Let p be an odd prime and a an integer. Suppose that $\gcd(a, p) = 1$. Then the *Legendre symbol*, in honor of the French mathematician Adrien Marie Legendre (1752–1883), $\left(\frac{a}{p}\right)$ is defined by

$$\left(\frac{a}{p}\right) = \begin{cases} = 1, & \text{if } a \text{ is a quadratic residue (mod p)} \\ = -1, & \text{if } a \text{ is a non-quadratic residue (mod p)} \end{cases} \tag{1.147}$$

Example 1.3.24 Let $p = 7$ and

$$1^2 \equiv 1 \pmod{7}, \quad 2^2 \equiv 4 \pmod{7}, \quad 3^2 \equiv 2 \pmod{7},$$
$$4^2 \equiv 2 \pmod{7}, \quad 5^2 \equiv 4 \pmod{7}, \quad 6^2 \equiv 1 \pmod{7}.$$

Then

$$\left(\frac{1}{7}\right) = \left(\frac{2}{7}\right) = \left(\frac{4}{7}\right) = 1, \qquad \left(\frac{3}{7}\right) = \left(\frac{5}{7}\right) = \left(\frac{6}{7}\right) = -1.$$

A serious problem with the computation of the Legendre symbol is that we need to factor a. The best way to overcome the difficulty of factoring a is to introduce the following so-called Jacobi symbol, in honor of the German mathematician Carl Gustav Jacobi (1804–1851), which is a natural generalisation of the Legendre symbol:

Definition 1.3.23 Let n be a positive integer greater than 1 and

$$n = p_1^{\alpha_1} p_2^{\alpha_2} \cdots p_k^{\alpha_k}$$

Then the *Jacobi symbol* is defined by

$$\left(\frac{a}{n}\right) = \left(\frac{a}{p_1}\right)^{\alpha_1} \left(\frac{a}{p_2}\right)^{\alpha_2} \cdots \left(\frac{a}{p_k}\right)^{\alpha_k} \tag{1.148}$$

If n is prime, then the Jacobi symbol is just the Legendre symbol.

The Jacobi symbol has some similar properties to the Legendre symbol, but of course, there are some significant differences between the Legendre symbol and the Jacobi symbol; the interested reader is referred to e.g., [102].

Finally, we introduce two more important concepts in congruence theory: primitive roots and indices.

Definition 1.3.24 We say that a belongs to the exponent k (mod m) if k is the smallest positive integer x such that $a^x \equiv 1$ (mod m).

Remark 1.3.6 The terminology "a belongs to the exponent k" is the classical language of number theory; it is replaced by "the order of a is k" in the language of group theory (see the next section).

Example 1.3.25 Consider the powers of 5 (mod 7). We have

$$5^1 \equiv 5 \pmod 7 \qquad 5^2 \equiv 4 \pmod 7 \qquad 5^3 \equiv 6 \pmod 7$$
$$5^4 \equiv 2 \pmod 7 \qquad 5^5 \equiv 3 \pmod 7 \qquad 5^6 \equiv 1 \pmod 7$$

Thus, 5 belongs to the exponent 6 (mod 7). (Note that the powers of 5 form a reduced system of residues modulo 7).

Theorem 1.3.31 If $a^t \equiv 1$ (mod m), and a belongs to the exponent k modulo m, then $k \mid t$.

Corollary 1.3.3 If a belongs to the exponent k modulo m, then $k \mid \phi(m)$.

Example 1.3.26 Find the exponent k, modulo 47, to which 19 belongs. Since $\phi(47) = 46$, k must be one of the numbers $1, 2, 23, 46$ (these are all the divisors of 46). We then compute the powers of 19 as follows: $19^1 \equiv 19$ (mod 47), $19^2 \equiv 32$ (mod 47), $19^{23} \equiv 46$ (mod 47), $19^{46} \equiv 1$ (mod 47). So, 19 belongs to the exponent 46 modulo 47.

Definition 1.3.25 The integer a is called a *primitive root* modulo m, if a belongs to the exponent $\phi(m)$ modulo m.

Remark 1.3.7 In algebraic language (see the next section), this definition can be stated as follows: "if the order of a modulo m is $\phi(m)$, then the multiplicative group of reduced residues modulo m is a cyclic group generated by the element a". That is, $(\mathbb{Z}/m\mathbb{Z})^* = \{a, a^2, a^3, \cdots, a^{\phi(m)}\}$.

Theorem 1.3.32 If p is a prime number, then there exist $\phi(p) - 1$ primitive roots modulo p.

Corollary 1.3.4 Let g be a primitive root modulo a prime p. Then the number $g, g^2, g^3, \cdots, g^{p-1}$ form a reduced system of residues modulo p.

Note that not all moduli have primitive roots. In what follows, we shall just list some conditions for moduli to have or not to have primitive roots; interested readers are advised to consult e.g., [52] for more information.

Theorem 1.3.33 If $m = 2^\alpha$ with $\alpha \geq 3$, or $m = 2^\alpha p_1^{\alpha_1} \cdots p_k^{\alpha_k}$ with $\alpha \geq 2$ or $k \geq 2$. Then there are no primitive roots modulo m.

Theorem 1.3.34 Let m be a positive integer greater than 1. A necessary and sufficient condition for m to have a primitive root modulo m is when $m = 2, 4, p^\alpha$ or $2p^\alpha$, where p is an odd prime.

Definition 1.3.26 Let g be a primitive root modulo a prime p, and $\gcd(y, p) = 1$, then the smallest positive integer x such that $y = g^x \bmod p$ is called the *index* of y (*wrt* the primitive root g for the prime p), and is denoted by $x = \text{ind}_g y$.

Remark 1.3.8 Perhaps the most widely used name in computing science and particularly in cryptography for *index* is *discrete logarithm*, denoted by

$$x = \log_g y. \tag{1.149}$$

Discrete logarithm is an important computationally intractable problem [2]; no efficient algorithm has been found for computing discrete logarithms. Discrete logarithms have been used as a basis for certain public key cryptosystems and cryptoprotocols since 1978; the security of such cryptosystems depends precisely on the intractability of taking discrete logarithms. In Chapter 4 we shall discuss some modern algorithms for computing discrete logarithms as well as some applications of discrete logarithms in cryptosystems and cryptoprotocols.

Example 1.3.27 Construct a table of indices (discrete logarithms) for the prime 13. Let the base g be 7. Then the powers of 7 (mod 13) are as follows:

$$7^1 \equiv 7 \quad 7^2 \equiv 10 \quad 7^3 \equiv 5 \quad 7^4 \equiv 9 \quad 7^5 \equiv 11 \quad 7^6 \equiv 12$$
$$7^7 \equiv 6 \quad 7^8 \equiv 3 \quad 7^9 \equiv 8 \quad 7^{10} \equiv 4 \quad 7^{11} \equiv 2 \quad 7^{12} \equiv 1$$

Thus, the table of indices for 13 is as follows:

$y = 7^x \bmod 13$	1	2	3	4	5	6	7	8	9	10	11	12
$x = \log_7 y \bmod 13$	12	11	8	10	3	7	1	9	4	2	5	6

1.4 Groups, Rings and Fields

In this section, we introduce some basic concepts of groups, rings, and fields from modern abstract algebra.

1.4.1 Binary Operations

Definition 1.4.1 A **binary operation** $\star$ on a set S is a *rule* that assigns to each ordered pair (a, b) of elements of S some element of S.

Example 1.4.1 Ordinary addition $+$ is a binary operation on the set $\mathbb{N}$, $\mathbb{Z}$, $\mathbb{R}$, or $\mathbb{C}$. Ordinary multiplication $\cdot$ is another binary operation on the same sets.

Definition 1.4.2 Let $\star$ denote a binary operation on set S. Then S is said to be **closed** with respect to $\star$ if $a \star b \in S$ for every pair of elements $a, b \in S$.

Example 1.4.2 The ordinary addition $+$ is *not* a binary operation on the set $\mathbb{R}_{\neq 0}$ of nonzero real numbers, since e.g., $3 + (-3)$ is not in the set $\mathbb{R}_{\neq 0}$; that is, $\mathbb{R}_{\neq 0}$ is not closed under $+$.

Example 1.4.3 Let $M(\mathbb{R})$ be the set of all matrices with real entries. The ordinary matrix addition $+$ is *not* a binary operation on this set since $A + B$ is not defined for an ordered pair (A, B) of matrices having different numbers of rows and columns.

Example 1.4.4 Ordinary multiplication $\cdot$ is a binary operation on the set $M_4(\mathbb{C})$ of all 4×4 matrices with complex entries, for $\cdot$ is defined for every ordered pair (A, B) of 4×4 matrices, and the product is again a 4×4 matrix.

Definition 1.4.3 A binary operation on a set S is *commutative* if

$$a \star b = b \star a, \quad \forall a, b \in S. \tag{1.150}$$

Example 1.4.5 Ordinary addition $+$ and multiplication $\cdot$ are commutative on the set $\mathbb{R}$. That is

$$a + b = b + a \quad \text{and} \quad a \cdot b = b \cdot a, \qquad \forall a, b \in \mathbb{R}.$$

But the ordinary subtraction $-$ is *not* commutative on the set $\mathbb{R}$, since

$$a - b \neq b - a, \quad \forall a, b \in \mathbb{R}; \tag{1.151}$$

for example, if $a = 5$, $b = 1$, then $a - b = 4$ but $b - a = -4$.

Definition 1.4.4 A binary operation on a set S is *associative* if

$$(a \star b) \star c = a \star (b \star c), \quad \forall a, b, c \in S. \tag{1.152}$$

Example 1.4.6 Ordinary addition $+$ and multiplication $\cdot$ are associative on $\mathbb{R}$. But ordinary subtraction $-$ is *not* associative on $\mathbb{R}$, since

$$(a - b) - c \neq a - (b - c), \quad \forall a, b, c \in \mathbb{R}; \tag{1.153}$$

for example, if $a = 1$, $b = 2$, $c = 3$, then $(a - b) - c = -4$ but $a - (b - c) = 2$.

Definition 1.4.5 Suppose there exists an element $e \in S$ such that

$$e \star a = a \star e = a, \quad \forall a \in S, \tag{1.154}$$

then e is called an *identity* element of S with respect to the binary operation $\star$.

Example 1.4.7 The number 1 is the identity element of $\mathbb{R}$ with respect to multiplication $\cdot$, since

$$1 \cdot a = a \cdot 1, \quad \forall a \in \mathbb{R}.$$

The number 0 is the identity element of $\mathbb{R}$ with respect to addition $+$, since

$$0 + a = a + 0, \quad \forall a \in \mathbb{R}.$$

For a finite set, a binary operation on the set can be defined by means of a table (see the following example).

Example 1.4.8 The following table defines a commutative operation $\star$, for which $a \star b = b \star a = d$.

$\star$	a	b	c	d
a	b	d	a	a
b	d	c	c	b
c	a	c	d	b
d	a	b	b	c

Note that for a finite set, a binary operation on the set can also be defined by means of a finite state diagram; see Figure 1.4.

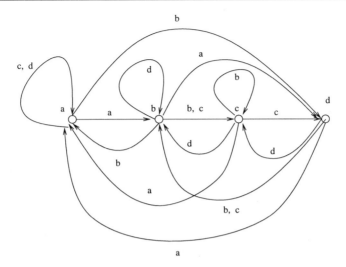

Figure 1.4: Binary Operation Defined by Automaton

1.4.2 Groups

Of all mathematical theories there is probably none which is so extensive in its applications as the theory of groups. The diverse subjects of geometry, combinatorial analysis, differential geometry, topology, and the theory of functions all deal some point with the theory of groups. In addition, the theories of numbers, fields and algebraic equations are deeply based on the theory of groups. In this section, we shall only introduce some very basic concepts and results of groups, which are related with theory of numbers.

Definition 1.4.6 A group $\langle G, \star \rangle$, is a set together with a binary operation, say $\star$, such that the following axioms are satisfied:

(i) The operation is *closed*: under a binary operation $\star$; that is,

$$a \star b \in G, \quad \forall a, b \in G. \tag{1.155}$$

(ii) The binary operation $\star$ is *associative*; that is,

$$a \star (b \star c) = (a \star b) \star c, \quad \forall a, b, c \in G. \tag{1.156}$$

(iii) The binary operation $\star$ admits an *identity* in G; that is, there is an element $e \in G$ such that

$$e \star a = a \star e = a, \quad \forall a \in G. \tag{1.157}$$

(iv) For each element $a \in G$, there exists a unique *inverse* element $a^{-1} \in G$ such that

$$a \star a^{-1} = a^{-1} \star a = e, \quad \forall a \in G. \tag{1.158}$$

Definition 1.4.7 A group G is called an *Abelian group*[19] or a *commutative group* if a further axiom is satisfied, that is,

$$a \star b = b \star a, \quad \forall a, b \in G.$$

Definition 1.4.8 A group is called a *finite group* if it has finite number of elements; otherwise it is called an *infinite group*.

Remark 1.4.1 In computer science, particularly in formal languages and automata theory, we also use the concepts of *monoid* and *semigroups*. An algebraic structure is called a *monoid* if the inverse property in Definition 1.4.6 has been dropped. An algebraic structure is called a *semigroup* if the identity property and the inverse property in Definition 1.4.6 have been dropped. Similar to groups, a monoid or a semigroup can be commutative or noncommutative, finite or infinite.

Example 1.4.9 Consider the set of positive integers $\mathbb{N}$ with operation $+$. Axioms 1 and 2 are clearly satisfied; however $\mathbb{N}$ does not contain an identity element with respect to $+$, and hence it is impossible to define the additive inverse of each element of $\mathbb{N}$. Thus, Axioms 3 and 4 are not satisfied. Therefore, $\langle \mathbb{N}, + \rangle$ is not a group.

Example 1.4.10 $\langle \mathbb{Z}_{\geq 0}, + \rangle$, the set of nonnegative integers $\mathbb{Z}_{\geq 0}$ with operation $+$ is still not a group, since although there is an identity element 0, but no additive inverse for 2.

Example 1.4.11 $\langle \mathbb{Z}, + \rangle$, the set of integers $\mathbb{Z}$ with operation $+$ is a group; since the first three axioms are clearly satisfied, for the fourth axiom, we have that $\forall a \in \mathbb{Z}, \exists -a \in \mathbb{Z}$ such that $x + (-x) = (-x) + x = 0$. Hence each element of $\mathbb{Z}$ has an additive inverse in $\mathbb{Z}$ and Axiom 4 is satisfied. Thus, $\langle \mathbb{Z}, + \rangle$ is a group.

---

[19]In honor of the Norwegian mathematician N. H. Abel (1802–1829).

Example 1.4.12 Consider $\langle \mathbb{Z}, \cdot \rangle$. Clearly, Axioms 1 and 2 are satisfied, and since $1 \in \mathbb{Z}$ is a multiplicative identity, Axiom 3 is also satisfied. However, given $a \in \mathbb{Z}$, it is not always true that its multiplicative identity $a^{-1} \in \mathbb{Z}$; consider e.g., $a = 0$ which has no multiplicative identity. Hence $\langle \mathbb{Z}, \cdot \rangle$ is not a group.

Example 1.4.13 The sets of $\mathbb{Z}$, $\mathbb{Q}$, $\mathbb{R}$ and $\mathbb{C}$ with operation $+$ are not only groups, but also (infinite) Abelian groups.

Example 1.4.14 The set $M_{m \times n}(\mathbb{R})$ of all $m \times n$ matrices with operation matrix addition is a group. The $m \times n$ matrix with all entries 0 is the identity matrix. This group is Abelian.

Example 1.4.15 The set $M_n(\mathbb{R})$ of all $n \times n$ matrices with operation matrix multiplication is *not* a group. The $n \times n$ matrix with all entries 0 has no inverse.

Example 1.4.16 If n is any positive integer, the positive integers less than and relatively prime to n together with multiplication modulo n form a group. If we denote this multiplicative group by $(\mathbb{Z}/n\mathbb{Z})^*$, then $(\mathbb{Z}/n\mathbb{Z})^* = \{x \in \mathbb{Z}/n\mathbb{Z} : \gcd(x, n) = 1\}$. For example, if $n = 9$, then 1 is the identity, 5 is the inverse of 2, 7 is the inverse of 4, and 8 is its own inverse, and $(\mathbb{Z}/9\mathbb{Z})^* = \{1, 2, 4, 5, 7, 8\}$. If $n = 12$, then 1 is the identity, 5, 7 and 11 are the other three elements in the group and they are each their own inverse.

Example 1.4.17 The multiplicative group $(\mathbb{Z}/11\mathbb{Z})^*$ can be described by Table 1.5. It is clear that $(\mathbb{Z}/11\mathbb{Z})^* = \{1, 2, 3, 4, 5, 6, 7, 8, 9, 10\}$.

Definition 1.4.9 The order of a group G, denoted by $|G|$, is the number of elements in G.

Example 1.4.18 The order of $\mathbb{Z}$ is $|\mathbb{Z}| = \infty$; and the order of $(\mathbb{Z}/n\mathbb{Z})^*$ is $|(\mathbb{Z}/n\mathbb{Z})^*| = \phi(n)$. We sometimes also denote $|(\mathbb{Z}/n\mathbb{Z})^*|$ by $\#(\mathbb{Z}/n\mathbb{Z})^*$. For example, $|(\mathbb{Z}/9\mathbb{Z})^*| = 6$, $|(\mathbb{Z}/11\mathbb{Z})^*| = 10$, and $|(\mathbb{Z}/12\mathbb{Z})^*| = 4$.

Definition 1.4.10 The order of an element x in group $G = (\mathbb{Z}/n\mathbb{Z})^*$ is the smallest positive integer r such that

$$x^r \equiv e \pmod{n}. \tag{1.159}$$

Example 1.4.19 For $x = 5$, $n = 91$. Then the order of 5 in $G = (\mathbb{Z}/91\mathbb{Z})^*$ is 12, since 12 is the smallest r satisfying

$$3^{12} \equiv 1 \pmod{91}.$$

$\odot$	1	2	3	4	5	6	7	8	9	10
1	1	2	3	4	5	6	7	8	9	10
2	2	4	6	8	10	1	3	5	7	9
3	3	6	9	1	4	7	10	2	5	8
4	4	8	1	5	9	2	6	10	3	7
5	5	10	4	9	3	8	2	7	1	6
6	6	1	7	2	8	3	9	4	10	5
7	7	3	10	6	2	9	5	1	8	4
8	8	5	2	10	7	4	1	9	6	3
9	9	7	5	3	1	10	8	6	4	2
10	10	9	8	7	6	5	4	3	2	1

Table 1.5: The Multiplication Table of $\mathbb{Z}/11\mathbb{Z}$

Exercise 1.4.1 Verify that 12 is the smallest r satisfying $5^r \equiv 1 \pmod{91}$. That is, you will need to compute $5^i \pmod{91}$ for $i = 1, 2, \cdots, 12$, in order to reach that 12 is the smallest positive integer satisfying $5^{12} \equiv 1 \pmod{91}$.

Theorem 1.4.1 (Lagrange's theorem) If x is an element of G, then the order of x divides the order of G.

Exercise 1.4.2 Let $G = (\mathbb{Z}/91\mathbb{Z})^*$ and $x = 17$. Then the order of G is $|G| = \phi(91) = 72$, and the order of $17 \in G$ modulo 91 is 6. It is clear that $6 \mid 72$, since $72 = 6 \cdot 12$.

We are now in a position to introduce some basic results of groups, which can be easily deduced from the above axioms.

Theorem 1.4.2 In any group $\langle G, \star \rangle$, the identity element e is unique.

Theorem 1.4.3 For each element a in a group $\langle G, \star \rangle$, the inverse element a^{-1} is unique.

Theorem 1.4.4 Let $a, b \in G$. Then

$$(a \star b)^{-1} = b^{-1} \star a^{-1}. \tag{1.160}$$

Theorem 1.4.5 Any group $\langle G, \star \rangle$ satisfies the following cancellation laws:

(i) Left cancellation law

$$a \star b = a \star c \Longrightarrow b = c, \quad \forall a, b, c \in G. \qquad (1.161)$$

(ii) Right cancellation law

$$b \star a = c \star a \Longrightarrow b = c, \quad \forall a, b, c \in G. \qquad (1.162)$$

Theorem 1.4.6 Let $a, b \in G$. Then the equations

$$a \star x = b, \qquad y \star a = b \qquad (1.163)$$

have unique solutions in G, given by

$$x = a^{-1} \star x, \qquad y = b \star a^{-1} \qquad (1.164)$$

respectively.

1.4.3 Rings and Fields

In this section, we introduce two further types of abstract algebraic structures: rings and fields. Let us begin with rings, the most general algebraic structure with two binary operations.

Definition 1.4.11 A *ring* $\langle R, \oplus, \odot \rangle$, is a set with *two* binary operations $\oplus$ and $\odot$, which we call addition and multiplication, defined on R such that the following axioms are satisfied:

(i) The set is closed under the operation $\oplus$, i.e., If $a \in S$, and $b \in S$, then

$$a \oplus b \in S, \qquad (1.165)$$

(ii) The associative law holds for $\oplus$, i.e.,

$$a \oplus (b \oplus c) = (a \oplus b) \oplus c, \quad \forall a, b, c \in S, \qquad (1.166)$$

(iii) The commutative law holds for $\oplus$, i.e.,

$$a \oplus b = b \oplus a, \quad \forall a, b, c \in S, \qquad (1.167)$$

(iv) The set has a unique additive identity (or the zero element) 0, i.e.,

$$a \oplus 0 = 0 \oplus a = a, \quad \forall a \in S, \tag{1.168}$$

(v) For each $a \in S$, the equation $a + 0$ has a solution in S.

(vi) The set is closed under the operation $\odot$, i.e., if $a, b \in S$, then $a \odot b \in S$,

(vii) The associative law holds for $\odot$, i.e.,

$$a \odot (b \odot c) = (a \odot b) \odot c, \quad \forall a, b, c \in S, \tag{1.169}$$

(viii) The following distributive laws (left and right) hold for $\oplus$ and $\odot$; that is,

$$a \odot (b \oplus c) = a \odot b + a \odot c, \quad \forall a, b, c \in S, \tag{1.170}$$

$$(a \oplus b) \odot c = a \odot c + b \odot c, \quad \forall a, b, c \in S. \tag{1.171}$$

Remark 1.4.2 The above definition can be briefly described, via the definition of groups, as follows: $\langle R, \oplus, \odot \rangle$ is a ring if

(i) $\langle R, \oplus \rangle$ is an Abelian group.

(ii) Multiplication is associative.

(iii) The left and right distributive laws hold.

Example 1.4.20 $\langle \mathbb{Z}, \oplus, \odot \rangle$, $\langle \mathbb{Q}, \oplus, \odot \rangle$, $\langle \mathbb{R}, \oplus, \odot \rangle$, and $\langle \mathbb{C}, \oplus, \odot \rangle$ are all rings.

Example 1.4.21 $\langle M_n(\mathbb{Z}), \oplus, \odot \rangle$, $\quad \langle M_n(\mathbb{Q}), \oplus, \odot \rangle$, $\quad \langle M_n(\mathbb{R}), \oplus, \odot \rangle$, and $\langle M_n(\mathbb{C}), \oplus, \odot \rangle$ are all rings.

Example 1.4.22 $\langle \mathbb{Z}/n\mathbb{Z}, \oplus, \odot \rangle$ is a ring.

Definition 1.4.12 A *commutative ring* is a ring R that satisfies:

$$a \odot b = b \odot a, \quad \forall a, b \in S. \tag{1.172}$$

Definition 1.4.13 A *ring with identity* is a ring R that contains an element 1 satisfying:

$$a \odot 1 = a = 1 \odot a, \quad \forall a \in S. \tag{1.173}$$

Definition 1.4.14 An *integral domain* is a commutative ring R with identity $1 \neq 0$ that satisfies:

$$a, b \in S \ \& \ ab = 0 \implies a = 0 \text{ or } b = 0. \tag{1.174}$$

Definition 1.4.15 A *division ring* is a ring R with identity $1 \neq 0$ that satisfies:

for each $a \neq 0 \in S$, the equation $ax = 1$ and $xa = 1$ have solutions in S.

Definition 1.4.16 A *field*, denoted by $\mathbb{K}$, is a division ring with commutative multiplication.

Example 1.4.23 The integer set $\mathbb{Z}$, with the usual addition and multiplication, forms a commutative ring with identity.

Figure 1.5 gives a Venn diagram view of containment for the algebraic structures having two binary operations.

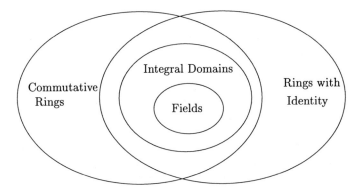

Figure 1.5: Containment of Various Rings

Example 1.4.24 Familiar examples of fields are the set of rational numbers $\mathbb{Q}$, the set of real numbers $\mathbb{R}$ and the set of complex numbers $\mathbb{C}$; since $\mathbb{Q}$, $\mathbb{R}$ and $\mathbb{C}$ are all infinite sets, they are all *infinite* fields. The set of integers $\mathbb{Z}$ is a ring but *not* a field, since 2, for example, has no multiplicative inverse, so 2 is not a unit in $\mathbb{Z}$. The only units in $\mathbb{Z}$ are 1 and -1. Another example of a ring which is not a field is the set $\mathbb{K}[x]$ of polynomials in x with coefficients belonging to a field $\mathbb{K}$.

Theorem 1.4.7 $\mathbb{Z}/m\mathbb{Z}$ is a field if and only if m is prime.

What this theorem says is that whenever m is prime, the set of congruence classes modulo m forms a field. This *prime* field is normally denoted by $\mathbb{Z}/p\mathbb{Z}$, or $\mathbb{F}_p$.

Remark 1.4.3 The ring $\mathbb{F}[x]$ of polynomials over $\mathbb{F}$ is analogous in many ways to ring $\mathbb{Z}$ of integers. Let $f(x)$ be a fixed polynomial in $\mathbb{F}[x]$. Just as $\mathbb{Z}/m\mathbb{Z}$ is a ring and $\mathbb{Z}/p\mathbb{Z}$ with p prime is a field, $\mathbb{F}[x]/f(x)$ is also a ring and $\mathbb{F}[x]/f(x)$ is a field if $f(x)$ is a irreducible polynomial in $\mathbb{F}[x]$; a polynomial $f(x)$ in $\mathbb{F}[x]$ is reducible if $f(x) = a(x)b(x)$ with $a(x), b(x) \in \mathbb{F}[x]$, and degree($a(x)$), degree($b(x)$) < degree($f(x)$), or otherwise, $f(x)$ is irreducible – it should be noted that there is a very efficient way for irreducibility testing of polynomials (although no efficient way for primality testing of integers at present), and in fact it can be shown that over any finite field, a randomly selected polynomial of degree n is irreducible with probability very close to $1/n$.

Definition 1.4.17 A *finite field* is a field which has a finite number of elements in it; we call the number the *order* of the field. The finite field of order p, say for example, is denoted by $\mathbb{F}_p$.

The following fundamental result about finite fields was proved by Évariste Galois[20]:

Theorem 1.4.8 There exists a field of order q if and only if q is a prime power (i.e., $q = p^r$ with $p \in$ Prime and $r \in \mathbb{Z}^+$. Moreover, if q is a prime power, then there is, up to relabelling, only one field of that order.

---

[20]Évariste Galois (1811–1832), French mathematician who had made major contributions to the theory of equations (e.g., he proved that the general quintic equation is not solvable by radicals) before he died at the age of 20, shot in a duel; he spent the whole night before the duel writing a letter containing notes of his discoveries.

A field of order q is often called a *Galois field*, and is denoted by $\mathrm{GF}(q)$, or just $\mathbb{F}_q$. Clearly, a Galois field is a finite field.

Example 1.4.25 The finite field $\mathbb{F}_5$ has elements $\{0, 1, 2, 3, 4\}$ and is described by the following addition and multiplication table (see Table 1.6):

$\oplus$	0	1	2	3	4
0	0	1	2	3	4
1	1	2	3	4	0
2	2	3	4	0	1
3	3	4	0	1	2
4	4	0	1	2	3

$\odot$	1	2	3	4
1	1	2	3	4
2	2	4	1	3
3	3	1	4	2
4	4	3	2	1

Table 1.6: The Addition and Multiplication for $\mathbb{F}_5$

Rings, and particularly finite field $\mathbb{F}_q$ with q a prime power, play a very important role in algebraic and number-theoretic computations, cryptography, computational complexity and analysis of algorithms.

1.4.4 Elliptic Curves over a Ring/Field

An elliptic curve is an algebraic curve given by a type of *Diophantine equation,* one given by a *cubic polynomial equation.* For example, the following three graphs (from left to right):

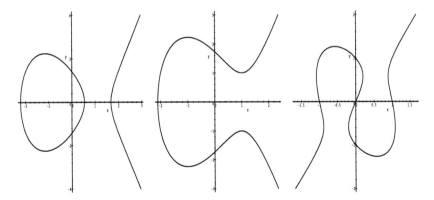

are all elliptic curves given by the following three Diophantine equations (again from left to right):

$$E_1 : \; y^2 = x^3 - 4x + 2, \quad E_2 : \; y^2 = x^3 - 3x + 3, \quad E_3 : \; y(y^2 - 1) = 2x(x^2 - 1).$$

Contrary to popular opinion, an elliptic curve is not an *ellipse*; a more accurate name for elliptic curves, in terms of *algebraic geometry*, is *Abelian varieties of dimension one*. It should be noted that *quadratic* polynomial equations are fairly well understood by mathematicians today, but cubic equations already pose enough difficulties to be topics of current research. Surprisingly enough, it is using elliptic curves that Andrew Wiles proved the famous Fermat Last Theorem (i.e., the Diophantine equation $x^n + y^n = z^n$ has no solutions in integers with $xyz \neq 0$ for all degrees $n \geq 3$). From a computational (particularly number-theoretic computational) point of view, elliptic curves are very useful tools in primality testing, integer factorization and public-key cryptography. In this subsection, we shall provide some basic concepts of elliptic curves, which will be used in Chapter 4 of this book.

Definition 1.4.18 Let $\mathbb{K}$ be a field (either the field $\mathbb{Q}$, $\mathbb{R}$, $\mathbb{C}$, or the finite field $\mathbb{F}_q$ of $q = p^\alpha$ elements), and $x^3 + ax + b$ with $a, b \in \mathbb{K}$ be a cubic polynomial. Then

(i) If $\mathbb{K}$ is a field of characteristic $\neq 2, 3$, then an *elliptic curve* over $\mathbb{K}$ is the set of points (x, y) with $x, y \in \mathbb{K}$ that satisfy the following cubic Diophantine equation:

$$E : \; y^2 = x^3 + ax + b, \tag{1.175}$$

(where the cubic on the right-hand side has no multiple roots) together with a single element, denoted by $\mathcal{O}_E$ and called the *"the point at infinity"*.

(ii) If $\mathbb{K}$ is a field of characteristic 2, then an *elliptic curve* over $\mathbb{K}$ is the set of points (x, y) with $x, y \in \mathbb{K}$ that satisfy one of the following cubic Diophantine equations:

$$E : \; y^2 + cy = x^3 + ax + b, \qquad E : \; y^2 + xy = x^3 + ax^2 + b \tag{1.176}$$

(here we do not care whether or not the cubic on the right-hand side has multiple roots) together with a *"point at infinity"* $\mathcal{O}_E$.

(iii) If $\mathbb{K}$ is a field of characteristic 3, then an *elliptic curve* over $\mathbb{K}$ is the set of points (x, y) with $x, y \in \mathbb{K}$ that satisfy the cubic Diophantine equation:

$$E : \quad y^2 = x^3 + ax^2 + bx + c, \tag{1.177}$$

(where the cubic on the right-hand side has no multiple roots) together with a *"point at infinity"* $\mathcal{O}_E$.

We are now moving on to the definition of the notion of an elliptic curve over the ring $\mathbb{Z}/N\mathbb{Z}$.

Definition 1.4.19 Let N be a positive integer with $\gcd(N, 6) = 1$. An *elliptic curve* over $\mathbb{Z}/N\mathbb{Z}$ is given by the following cubic Diophantine equation:

$$E : \quad y^2 = x^3 + ax + b, \tag{1.178}$$

where $a, b \in \mathbb{Z}$ and $\gcd(N, 4a^3 + 27b^2) = 1$. The set of points on E is the set of solutions in $(\mathbb{Z}/N\mathbb{Z})^2$ to the equation (1.178), together with a *"point at infinity"* $\mathcal{O}_E$.

We will use the well-known *tangent-and-chord* addition law on a cubic over a finite field $\mathbb{Z}/p\mathbb{Z}$ as well as over a ring $\mathbb{Z}/N\mathbb{Z}$ with N composite. If $P_1 = (x_1, y_1)$ and $P_2 = (x_2, y_2)$ are points on the elliptic curve $y^2 = x^3 + ax + b$, then the point $P_3 = P_1 + P_2 = (x_3, y_3)$ is defined as follows:

$$P_1 \neq P_2 : \quad \begin{cases} x_3 = \left(\dfrac{y_2 - y_1}{x_2 - x_1} \right)^2 - x_1 - x_2, \\[2mm] y_3 = \left(\dfrac{y_2 - y_1}{x_2 - x_1} \right) (x_1 - x_3) - y_1 \end{cases} \tag{1.179}$$

$$P_1 = P_2 : \quad \begin{cases} x_3 = \left(\dfrac{3x_1^2 + a}{2y_1} \right)^2 - 2x_1, \\[2mm] y_3 = \left(\dfrac{3x_1^2 + a}{2y_1} \right) (x_1 - x_3) - y_1 \end{cases} \tag{1.180}$$

The geometric interpretation is straightforward (see Figure 1.6): the straight line $P_1 P_2$ intersects the elliptic curve at a third point $P_3' = (x_3, -y_3)$, and $P_3 = (x_3, y_3)$ is the reflection of P_3' on the x-axis.

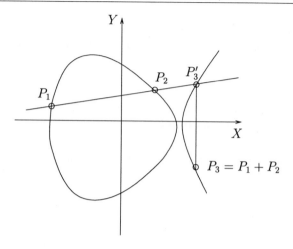

Figure 1.6: An Elliptic Curve over Field $\mathbb{K}$

Example 1.4.26 Let E be the elliptic curve $y^2 = x^3 + 17$ over $\mathbb{R}$, and let $P_1 = (x_1, y_1) = (-2, 3)$ and $P_2 = (x_2, y_2) = (1/4,\ 33/8)$ be the two points on E. To find a third point P_3 on E, we perform the following computation:

$$\begin{cases} \lambda = \frac{y_2 - y_1}{x_2 - x_1} = 1/2 \\ x_3 = \lambda^2 - x_1 - x_2 = 2 \\ y_3 = \lambda(x_1 - x_3) - y_1 = -5. \end{cases}$$

So, $P_3 = P_1 + P_2 = (x_3,\ -y_3) = (2, 5)$. As an exercise, find the points $P_1 + P_2$ and $2P_1$ on the elliptic curve $E : \ y^2 = x^3 - 36x$, where $P_1 = (-3, 9)$ and $P_2 = (-2, 8)$.

Example 1.4.27 Let E be the elliptic curve $y^2 = x^3 + 3x$ over $\mathbb{F}_5$, then

$$\mathcal{O}_E,\ (0,0),\ (1,2),\ (1,3),\ (2,2),\ (2,3),\ (3,1),\ (3,4),\ (4,1),\ (4,4)$$

are the 10 points on E. The easiest way to find these points, except the point at infinity $\mathcal{O}_E$, is to perform the following simple computation:

```
for both x and y from 0 to 4 do
    if y² mod 5 = (x³ mod 5 + 3x mod 5) mod 5
        then print (x, y)
```

But the elliptic curve $y^2 = 3x^3 + 2x$ over $\mathbb{F}_5$ has only two points: $\mathcal{O}_E$ and $(0,0)$. As an exercise, find the number of points on the following elliptic curves over $\mathbb{F}_{13}$:

$$E_1 : y^2 = x^3 + 2x + 1, \quad \text{and} \quad E_2 : y^2 = x^3 + 4x.$$

How many points are there on an elliptic curve $E : y^2 = x^3 + ax + b$ over $\mathbb{F}_p$? There are, in fact,

$$1 + p + \sum_{x \in \mathbb{F}_p} \left(\frac{x^3 + ax + b}{p} \right) = 1 + p + \epsilon \qquad (1.181)$$

number of points on E, including the point at infinity $\mathcal{O}_E$, where $\left(\dfrac{x^3 + ax + b}{p} \right)$ is the Legendre symbol. According to Hasse's theorem (1933), $|\epsilon| \le 2\sqrt{p}$. A similar question is: how many points are there on an elliptic curve $E : y^2 = x^3 + ax + b$ over $\mathbb{R}$? According to Mordell's theorem (1922), on an elliptic curve that contains a rational point, there exists a finite set of rational points such that all other rational points can be finitely generated by using the tangent-and-chord method, just as every solution to Pell's equation was obtained by starting with one basic solution and repeatedly applying a simple rule.

Further Reading

In this chapter, we have provided a survey of several branches of mathematics which are relevant to computer science. There are many good books for the same purpose on the market, say, e.g., [23], [36], [37], [38], [39], and [140]. For those who desire a more detailed exposition in elementary number theory and its applications in computer science, we recommend [10], [52], [70], [102], [103], [122], and [147].

Chapter 2

Formal Languages and Automata

The fundamental aim in linguistic analysis of language L is to separate the grammatical sequences which are sentences of L from the ungrammatical sequences which are not sentences of L and to study the structure of grammatical sequences.

– AVRAM NOAM CHOMSKY (1928–)

In this chapter, we study some important classes of formal languages and their corresponding machines. More specifically, we shall study regular languages and finite automata, context-free languages and push-down automata, and recursively enumerable languages and Turing machines.

2.1 Languages, Grammars and Automata

In this section, we shall provide an introduction to some basic concepts of languages, grammars and automata, which are fundamental to this chapter.

Definition 2.1.1 An *alphabet* Σ is a finite set of symbols. A *word* or a *string* over an alphabet Σ is a finite sequence of symbols from Σ. An empty word (or string), denoted by λ, is the sequence consisting of no symbols. The *length* of a word w, denoted by $|w|$, is the number of symbols in w.

Example 2.1.1 Let $\Sigma = \{a, b, c\}$. Then $w_1 = acb$ and $w_2 = aababc$ are two words over Σ, and $|w_1| = 3$ and $|w_2| = 6$. Let $w = \lambda$, then $|w| = 0$. Suppose $w = ab$, then $\lambda ab = ab\lambda = ab$.

Definition 2.1.2 Let Σ be an alphabet, and λ the empty word containing no symbols. Then Σ^* is defined to be the set of words obtained by concatenating zero or more symbols from Σ. If the set does not contain λ, then we denote it by Σ^+. That is,

$$\Sigma^+ = \Sigma^* - \{\lambda\}. \tag{2.1}$$

A *Language* over an alphabet Σ is a subset of Σ^*.

Example 2.1.2 Let $\Sigma = \{a, b\}$. Then

$$\Sigma^* = \{\lambda, a, b, aa, ab, ba, bb, aaa, aab, aba, baa, abb, bab, bba, bbb, \cdots\},$$

$$\Sigma^+ = \{a, b, aa, ab, ba, bb, aaa, aab, aba, baa, abb, bab, bba, bbb, \cdots\}.$$

The sets L_1 and L_2 given by

$$L_1 = \{a, b, aa, bb, aaaba\}$$

$$L_2 = \{a^n b^n : n \in \mathbb{N}\}$$

are all languages over Σ.

Definition 2.1.3 Let w_1 and w_2 be two words, and L_1, L_2 and L be sets of words.

(i) The concatenation of two words is formed by juxtaposing the symbols that form the words.

(ii) The *concatenation* of L_1 and L_2, denoted by $L_1 L_2$, is the set of all words formed by concatenating a word from L_1 and a word from L_2. That is:

$$L_1 L_2 = \{w_1 w_2 : w_1 \in L_1, w_2 \in L_2\}. \tag{2.2}$$

(iii) Powers of L are defined by the concatenation of L with itself the appropriate number of times, e.g.,

(a) $L^0 = \lambda$,

  (b) $L^1 = L$

  (c) $L^2 = LL$

  (d) $L^3 = LLL$

  (e) $L^k = \underbrace{LL \cdots L}_{k \text{ times}}.$

(iv) The complement of a language L, denoted by $\overline{L}$ is defined by

$$\overline{L} = \Sigma^* - L. \tag{2.3}$$

Example 2.1.3 The following are some examples of concatenation of two words, two sets of words, and powers of a set of words:

(i) If $w_1 = abc$ and $w_2 = aabab$, then $w_1 w_2 = abcaabab$.

(ii) If $L_1 = \{a, aba, cab, \lambda\}$ and $L_2 = \{ca, cb\}$, then

$$L_1 L_2 = \{aca, acb, abaca, abacb, cabca, cabcb, ca, cb\}.$$

(iii) If $L = \{a, b\}$, then

  (a) $L^0 = \lambda$

  (b) $L^1 = \{a, b\}$

  (c) $L^2 = LL = \{aa, ab, ba, bb\}$

  (d) $L^3 = LLL = \{aaa, aab, aba, abb, baa, bab, bba, bbb\}.$

Definition 2.1.4 Let L be a set of words. Then L^*, the *Kleene closure* of L, is defined by

$$L^* = L_0 \cup L_1 \cup L_2 \cup \cdots = \bigcup_{i=0}^{\infty} L^i, \tag{2.4}$$

and L^+, the positive closure of L is defined by

$$L^+ = L_1 \cup L_2 \cup L_2 \cup \cdots = \bigcup_{i=1}^{\infty} L^i. \tag{2.5}$$

Example 2.1.4 If $\Sigma = \{0, 1\}$ and $L = \{0, 10\}$, then L^* consists of the empty word λ and all the words that can be formed using 0 and 10 with the property that every 1 is followed by a 0.

Definition 2.1.5 A *grammar* G is defined as a quadruple

$$G = (V, T, S, P) \tag{2.6}$$

where

  (i) V is a finite set of objects called *variables*,

  (ii) T is a finite set of objects called *terminal symbols*,

  (iii) $S \in V$ is a special symbol called *start* variables,

  (iv) P is a finite set of *productions*.

Definition 2.1.6 Let $G = (V, T, S, P)$ be a grammar. Then the set

$$L(G) = \{w \in T^* : S \stackrel{*}{\Longrightarrow} w\} \tag{2.7}$$

is the language generated by G, where $S \stackrel{*}{\Longrightarrow} w$ represents an unspecified number of derivations (including zero, if not including zero, we then use $S \stackrel{+}{\Longrightarrow} w$) that can be taken from S to w.

Example 2.1.5 Find the grammar that generates the language

$$L(G) = \{a^n b^n : n \in \mathbb{N}\}.$$

Both grammar G_1 defined by

$$G_1 = (\{S\}, \{a, b\}, S, P_1)$$

with P_1 consisting of the productions

$$S \to aSb$$
$$S \to \lambda$$

and grammar G_2 defined by

$$G_2 = (\{S, A\}, \{a, b\}, S, P_2)$$

with P_2 consisting of the productions

$$S \to aAb \mid \lambda$$
$$A \to aAb \mid \lambda$$

will generate the language $L = \{a^n b^n : n \in \mathbb{N}\}$.

Exercise 2.1.1 Find the grammar that generates the language

$$L(G) = \{a^n b^{n+1} : n \in \mathbb{N}\}.$$

Find the language generated by the grammar with productions:

$$S \to Aa$$
$$A \to B$$
$$B \to Aa.$$

Automata are abstract (mathematical) machines, that can read information from input and write information to output. This input/output process is controlled by its finite state control unit (see Figure 2.1). An automaton can

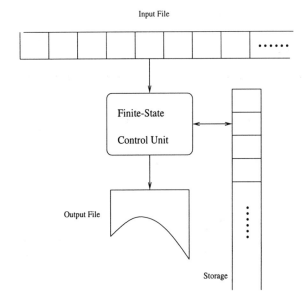

Figure 2.1: A General Automaton

be thought as a model of algorithm, or a compiler of a language, or even a general computer. In artificial intelligence, for example, they are employed to model both behavioral situations and intelligent systems, including game playing, human intelligence, machine learning, nervous system activity, and robotic motion systems.

An automaton whose output is limited to a simple answer "yes" or "no" is called a (decision) *problem solver* or a *language accepter*. On an input with a string, the accepter either accepts (recognises) the string or rejects it. A more general automaton, capable of producing strings of symbols as output, is called a *function transducer*.

There are essentially two different types of automata: deterministic automata and nondeterministic automata. In deterministic automata, each move is uniquely determined by the current internal state, the current input symbol and the information currently in the temporary storage. On the other hand, in nondeterministic automata, we cannot predict the exact future behaviour of a automaton, but only a set of possible actions. One of the very important objectives of this book is actually to study the relationship between deterministic and nondeterministic automata of various types (e.g., finite automata, push-down automata, and Turing machines).

2.2   Finite Automata and Regular Languages

Finite-State Automata, or Finite Automata (FA) for short, are the simplest automata (see Figure 2.2). In this section we shall study the basic concepts and

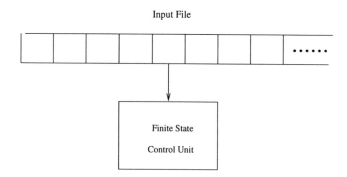

Figure 2.2: Finite Automaton

results of finite automata (both deterministic and non-deterministic), and the properties of regular languages, with an emphasis on the relationship between finite automata and regular languages.

2.2.1   Deterministic Finite Automata (DFA)

Definition 2.2.1 A *Deterministic Finite Automaton* (DFA), denoted by M, is a quintuple algebraic system (more specifically, a semigroup):

$$M = (Q, \Sigma, \delta, q_0, F), \qquad (2.8)$$

where

(i) Q is a finite set of internal *states*,

(ii) Σ is the input alphabet,

(iii) $q_0 \in Q$ is the *initial state*,

(iv) $F \subseteq Q$ is the set of *final states*, or *accepting states*,

(v) δ is the *state transition function*

$$\delta : \; Q \times \Sigma \to Q. \qquad (2.9)$$

Remark 2.2.1 The above DFA is defined without output; we can, of course, define it with output as follows: $M = (Q, \Sigma, U, \delta, \sigma, q_0, F)$, where

(vi) U is the output alphabet,

(vii) σ is the *output function*

$$\sigma : \; Q \times \Sigma \to U. \qquad (2.10)$$

Example 2.2.1 Let M be a DFA defined by

$$M = (Q, \Sigma, \delta, q_0, F)$$
$$= (\{A, B, C\}, \{0, 1\}, \delta, A, \{B\})$$

where the transition function δ is given by the following formulas:

$$
\begin{array}{ll}
\delta(A, 0) = A & \delta(A, 1) = B \\
\delta(B, 0) = C & \delta(B, 1) = B \\
\delta(C, 0) = C & \delta(C, 1) = C.
\end{array}
$$

or alternatively by the following table, called a *transition table*:

	0	1
A	A	B
B	C	B
C	C	C

Initial state: A
Final state: B

Then the DFA can be represented by a directed graph shown in Figure 2.3, where the initial state A has a starting right arrow, and the final state B has been double circled.

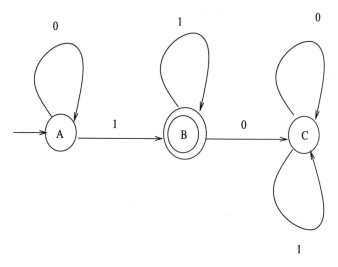

Figure 2.3: A DFA that accepts strings $0^m 1^n$ with $m \geq 0$ and $n \geq 1$

The machine defined above can read a given finite input tape containing a *word* and either accepts the word or rejects it. The word is *accepted* if after reading the tape, the machine is in any one of the accepting states.

Example 2.2.2 Consider the machine defined in Example 2.2.1. Suppose now that the machine reads the word 00011. Then the following are the actions of the machine as it reads 00011:

A
$\downarrow$

| 0 | 0 | 0 | 1 | 1 |

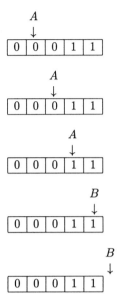

Since the machine is in the final state B after having read the input word, then the word 00011 is accepted by this machine. However, the machine cannot accept the word 000110, because

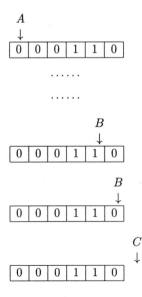

That is, the machine does not stop at the final state B after having read the word 000110. In fact, it stopped at the state C which is not a final state.

There are several other ways to describe actions of an automaton. One very useful way can described as follows (for the same automaton defined above and the same word 00011):

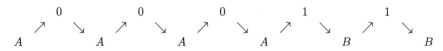

It is plain to verify that the automaton described in Figure 2.3 can accept the following words:

$$0,$$
$$1,$$
$$01,$$
$$001,$$
$$011,$$
$$0000011,$$
$$00111111111,$$
$$\cdots,$$
$$0^m1^n, \quad \text{with } m \geq 0 \text{ and } n \geq 1.$$

In set notation, the set of words L that can be accepted by the DFA is

$$L = \{0^m1^n \; : \; m \geq 0, n \geq 1\}.$$

Example 2.2.3 Figure 2.4 shows another example of a DFA, M, which has two final states D and E. The DFA, M is defined by

$$M = (Q, \Sigma, \delta, q_0, F)$$

$$= (\{A, B, C, D, E\}, \{0, 1\}, \delta, A, \{D, E\})$$

where the transition function is given by the following transition table:

	0	1
A	B	C
B	D	C
C	B	E
D	D	D
E	E	E

Initial state: A
Final states: D and E

It is clear that the following strings can be accepted by this DFA:

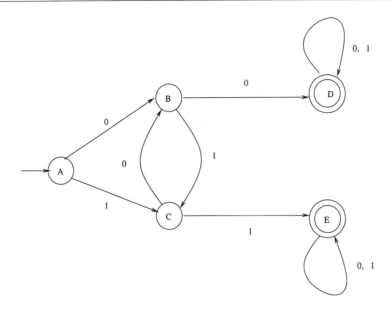

Figure 2.4: A DFA that accepts strings with two consecutive 0's or 1's

00,
11,
0011110,
01001111000,
110111101001010,
1010101010101010100
0101010101010101011

But the followings strings cannot be accepted by this DFA:

01,
10,
010101010101010101,
0101010101010101010,
1010101010101010101.

An automaton is *finite* in the sense that there are only finite states within the automaton. For example, in the automaton in Figure 2.3, there are only three states: *A*, *B* and *C*. A finite automaton is *deterministic* in the sense that for a given state and a given input, the next state of the automaton is completely determined. For example, again in the automaton in Figure 2.3, given state A and input 0, the next state can only be A.

2.2.2   Nondeterministic Finite Automata (NFA)

In contrast to deterministic automata, nondeterminism allows a machine to select arbitrarily from several possible responses to a given situation, including the possibility of selecting from several initial states. If one of the various responses to a word leaves the machine in an accepting state, then the word is said to be accepted. In this subsection, we study non-deterministic finite automata.

Definition 2.2.2 A *Non-deterministic Finite Automaton* (NFA), M, is a quintuple algebraic system:

$$M = (Q, \Sigma, \delta, S, F) \tag{2.11}$$

where

(i) Q is a finite set of *states*,

(ii) Σ is the input *alphabet*,

(iii) $S \subseteq Q$ is the set of *initial states*, usually $S = \{q_0\}$ as DFA, but it may be the case that it contains more than one state,

(iv) $F \subseteq Q$ is the set of *final states*.

(v) the *transition function* is defined by

$$\delta : Q \times (\Sigma \cup \lambda) \to 2^Q. \tag{2.12}$$

where 2^Q is the set of all subsets of Q.

Example 2.2.4 Let M be the non-deterministic finite automaton defined by

$$\begin{aligned} M &= (Q, \Sigma, \delta, S, F) \\ &= (\{A, B, C, D, E\}, \{0, 1\}, \delta, \{A, B\}, \{E\}) \end{aligned}$$

where δ is given by

$$\delta(A,0) = \{A,C\} \qquad \delta(A,1) = A$$
$$\delta(B,0) = B \qquad \delta(B,1) = \{B,D\}$$
$$\delta(C,0) = E \qquad \delta(C,1) = \lambda$$
$$\delta(D,0) = \lambda \qquad \delta(C,1) = E$$
$$\delta(E,0) = E \qquad \delta(D,1) = E$$

Then the NFA can be represented by the directed graph in Figure 2.5, or alternatively, by the following transition table:

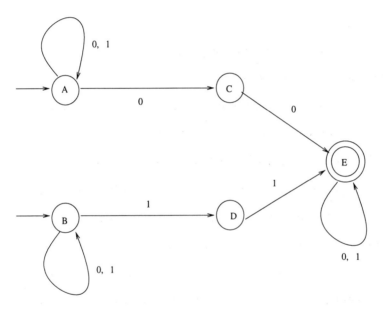

Figure 2.5: A NFA that accepts strings with two consecutive 0's or 1's

	0	1
A	$\{A,C\}$	A
B	B	$\{B,D\}$
C	E	λ
D	λ	E
E	E	E

Initial state: A
Final state: E

2.2.3 Regular Languages (L_{REG})

We have seen that finite-state automata can be used as language recognisers (or accepters). But what sets can be recognised (or accepted) by these machines? In this subsection we shall answer this question by showing that the sets which can be recognised by finite-state automata are *regular sets*.

(I) Regular Expressions

Definition 2.2.3 Let Σ be an alphabet. The *regular expressions* over Σ are defined recursively as follows:

(i) $\emptyset$ is a regular expression,

(ii) λ (empty string) is a regular expression,

(iii) x is a regular expression if $x \in \Sigma$,

(iv) $r_1 \cup r_2$ is a regular expression if r_1 and r_2 are regular expressions.

(v) $r_1 r_2$ is a regular expression if r_1 and r_2 are regular expressions.

(vi) r^* is a regular expression if r is a regular expression.

Each regular expression represents a set specifically by the following rules:

(i) $\emptyset$ represents the empty set, i.e., the set with no string,

(ii) λ represents the set $\{\emptyset\}$ containing the empty string,

(iii) x represents the set $\{x\}$ containing the string with one symbol x,

(iv) $r_1 \cup r_2$ represents the union of the sets represented by r_1 and r_2,

(v) $r_1 r_2$ represents the concatenation of the sets represented by r_1 and r_2,

(vi) r^* represents the Kleene closure of the set represented by r.

Definition 2.2.4 The language generated by a regular expression, denoted by $L(r)$, is defined recursively as follows:

(i) $L(\emptyset) = \emptyset$,

(ii) $L(\lambda) = \lambda$,

(iii) $L(r) = \{r\}$, if $r \in \Sigma$,

(iv) $L(r_1 \cup r_2) = L(r_1) \cup L(r_2)$,

(v) $L(r_1 r_2) = L(r_1) \otimes L(r_2)$,

(vi) $L(r^*) = (L(r))^*$.

So it is natural now to give a definition for regular languages:

Definition 2.2.5 The *regular languages* are defined recursively as follows:

(i) $\emptyset$ is a regular language,

(ii) $\{\lambda\}$ is a regular language,

(iii) $\{x\}$ is a regular language if $x \in \Sigma$,

(iv) $L_1 \cup L_2$ is a regular language if L_1 and L_2 are regular languages.

(v) $L_1 L_2$ is a regular language if L_1 and L_2 are regular languages.

(vi) L^* is a regular language if L is a regular language.

Thus, regular expressions are a shorthand way of describing regular languages.

Example 2.2.5 Let $\Sigma = \{a, b\}$. Then the following regular expressions represent the indicated sets of strings:

(i) a: represents the set $\{a\}$,

(ii) a^*: represents the set $\{a\}^* = \{\lambda, a, aa, aaa, \cdots\}$,

(iii) b: represents the set $\{b\}$,

(iv) ab: represents the set $\{a\}\{b\} = \{ab\}$,

(v) $a \cup b$: represents the set $\{a\} \cup \{b\} = \{a, b\}$,

(vi) $(ab)^*$: represents the set $\{ab\}^* = \{\lambda, ab, abab, ababab, \cdots\}$

(vii) $a^* \cup (ab)^*$: represents the set

$$\{a\} \cup \{ab\}^* = \{\lambda, a, aa, aaa, \cdots, ab, abab, ababab, \cdots\}$$

(viii) a^*b: represents the set $\{a\}^*\{b\} = \{b, ab, aab, aaab, \cdots\}$

(ix) $b(ab)^*$: represents the set $\{b\}\{ab\}^* = \{b, bab, babab, \cdots\}$

(x) $a^*b(ab)^*$: represents the set of all strings that begin with any number (possibly 0) of a, followed by a single b, followed by any number (possibly 0) of pair ab.

Example 2.2.6 Let $\Sigma = \{a, b\}$. By definition, $\emptyset$ and λ are regular sets. In view of the previous example, the following sets are also regular:

(i) $\{a\}$,

(ii) $\{\lambda, a, aa, aaa, \cdots\}$,

(iii) $\{b\}$,

(iv) $\{a, b\}$,

(v) $\{\lambda, ab, abab, ababab, \cdots\}$,

(vi) $\{b, ab, aab, aaab, \cdots\}$.

(II) Regular Grammar

The second way of describing regular languages is by means of a certain grammar, the *regular grammar*.

Definition 2.2.6 A *grammar* $G = (V, T, S, P)$ is said to be *right-linear* if all productions are of the form

$$A \to xB, \tag{2.13}$$
$$A \to x,$$

where $A, B \in V$, and $x \in T^*$.

A grammar $G = (V, T, S, P)$ is said to be *left-linear* if all productions are of the form

$$A \to Bx \tag{2.14}$$
$$A \to x.$$

A *regular grammar* is one that is either right-linear or left-linear.

Example 2.2.7 The grammar $G_1 = (\{S\}, \{a, b\}, S, P_1)$, with P_1 given by

$$S \to abS$$
$$S \to a$$

is right-linear, whereas the grammar $G_2 = (\{S, S_1, S_2\}, \{a, b\}, S, P_2)$, with P_2 given by

$$S \to S_1 ab,$$
$$S_1 \to S_1 ab,$$
$$S_1 \to S_2,$$
$$S_2 \to a,$$

is left-linear. Both G_1 and G_2 are regular grammars.

By G_1, we can have the following derivations:

$$S \Longrightarrow abS$$
$$\Longrightarrow aba$$

$$S \Longrightarrow abS$$
$$\Longrightarrow ababS$$
$$\Longrightarrow ababa$$
$$\Longrightarrow (ab)^2 a$$

$$S \stackrel{*}{\Longrightarrow} ababS$$
$$\Longrightarrow abababS$$
$$\Longrightarrow abababa$$
$$\Longrightarrow (ab)^3 a$$

$$\cdots\cdots$$
$$\cdots\cdots$$
$$\Longrightarrow (ab)^n a, \text{ for } n \geq 1.$$

The regular language L, denoted by $L(G_1)$, generated by the regular grammar G_1 is thus

$$L(G_1) = \{(ab)^n a : \text{ for } n \geq 1\}.$$

Similarly, by G_2, we have

$$S \Longrightarrow S_1 ab$$
$$\Longrightarrow S_2 ab$$
$$\Longrightarrow aab$$

$$S \Longrightarrow S_1 ab$$

$$\implies S_1 abab$$
$$\implies S_2 abab$$
$$\implies aabab$$
$$\implies a(ab)^2$$

$$S \stackrel{*}{\implies} S_1 abab$$
$$\implies S_1 ababab$$
$$\implies S_2 ababab$$
$$\implies aababab$$
$$\implies a(ab)^3$$

$$\cdots\cdots$$
$$\cdots\cdots$$

$$\implies a(ab)^n, \text{ for } n \geq 1.$$

The regular language L, denoted by $L(G_2)$, generated by the regular grammar G_2 is thus

$$L(G_2) = \{a(ab)^n : \text{ for } n \geq 1\}.$$

Theorem 2.2.1 Let $G = (V, T, S, P)$ be a regular grammar (either right-linear or left-linear). Then $L(G)$ is a regular language.

Theorem 2.2.2 A language L is regular if and only if there exists a regular grammar (either left-linear or right-linear) G, such that $L = L(G)$.

Thus, regular languages and regular grammars are, in fact, equivalent concepts. From a regular language, we can get it's regular grammar. From a regular grammar, we can also generate it's regular languages.

(III) The Kleene's and Moore's Theorems

The third way to describe regular languages is by finite automata (FA). In 1956, Stephen Kleene[1] proved that regular sets are the sets that are accepted by a finite automaton. Consequently, this result is called the Kleene's Theorem.

---

[1]Stephen C. Kleene (1909–1994) was born in Hartford, Connecticut, and earned his PhD with Alonzo Church from Princeton University in the 1930. Kleene made significant contributions to the theory of recursive functions while investigating questions of computability and decidability, and proved one of the central results in automata theory, now known as the *Kleene's Theorem.*

Theorem 2.2.3 (The Kleene's Theorem) A language L over an alphabet Σ is regular if and only if it is acceptable (recognisable) by a finite automaton FA, $M = (Q, \Sigma, \delta, q_0, F)$.

The proof of the *only if* part of the theorem involves showing that

(i) $\emptyset$ is accepted by a finite automata;

(ii) $\{\lambda\}$ is accepted by a finite automata;

(iii) For each $x \in \Sigma$, x is accepted by a finite automata;

(iv) AB is accepted by a finite automata if both A and B are;

(v) $A \cup B$ is accepted by a finite automata if both A and B are;

(vi) A^* is accepted by a finite automata if A is.

The proof of the *if* part of the theorem can be done by induction on the number of states in a finite automaton FA that accepts L.

The Kleene's theorem is one of the central results in automata theory. It outlines the limitations as well as the capabilities of finite automata, because there are certainly many languages that are not regular, and hence not accepted by finite automata.

Finally, we introduce another important result about regular sets, the equivalence theorem, discovered by E. F. Moore in 1956:

Theorem 2.2.4 (The Moore's Theorem) There exists an algorithm to determine whether or not two given regular sets over Σ are equivalent.

The Moore's theorem is one of the results of decidability for regular languages. We shall study some more decidability results for regular languages in Chapter 3.

(IV) Pumping Theorems and Closure Properties for L_{REG}

As we have seen, a language L is regular if and only if there exists a finite automata (FA) to accept it; if no FA can accept it, it is then not a regular language. Our next result will provide another technique showing languages nonregular.

Theorem 2.2.5 (Pumping Theorem for Regular Languages) Let L be a regular language. There exists a positive integer N (depending on L) such that for any $x \in L$ and $|x| \geq n$, there exist strings u, v and w, satisfying the following conditions:

$$x = uvw, \tag{2.15}$$
$$|v| > 0, \tag{2.16}$$
$$|uv| \leq N, \tag{2.17}$$
$$uv^i w \in L, \quad \forall i \geq 0. \tag{2.18}$$

The number N is called the *pumping number* for the regular language L.

This theorem describes a property that is common to all regular languages. So we can use it to show a language nonregular if we can show that the property fails to hold for the languages.

Example 2.2.8 Use the pumping theorem to show that

$$L = \{a^n b^n : n \in \mathbb{Z}^+\}$$

is not a regular language.

PROOF. Suppose

$$L = \{a^n b^n : n \in \mathbb{Z}^+\} = \underbrace{aa \cdots a}_{n \text{ times}} \underbrace{bb \cdots b}_{n \text{ times}}$$

is regular and let N be the pumping number for L. We must show that no matter what N is, we may find x with $|x| \geq N$, that produces a contradiction. Let $x = a^N b^N$. According to Theorem 2.2.5, there are strings u, v and w, such that (2.15), (2.16), (2.17) and (2.18) in the theorem hold. From (2.15) and (2.16) we can see that $uv = a^k$ for some k, so from (2.17) we have $v = a^j$ form some $j > 0$. Then (2.18) says that $uv^m w \in L, \forall m \geq a$. But

$$\begin{aligned}
uv^m w &= (uv)v^{m-1}w \\
&= a^k(a^j)m - 1a^{N-k}b^N \\
&= a^{N+j(m-1)}b^N \quad (w = 0^{N-k}b^N, \text{ since } uv = a^k) \\
&= a^{N+t}b^N \quad (\text{let } t = j(m-1) \text{ when } m > 1)
\end{aligned}$$

Clearly, there are t more consecutive 0's than there are consecutive 1's in x. Since this string is not in the form $a^n b^n$, then it is not regular. ∎

Exercise 2.2.1 Use the pumping theorem to show that

$$L = \{a^i b^j : i \neq j, \ i,j \in \mathbb{Z}^+\}$$

and

$$L = \{x \in \{0,1\}^* : x \text{ contains equal numbers of } 0's \text{ and } 1's\}$$

are not regular languages.

Finally we present some closure properties for regular languages.

Theorem 2.2.6 The family of regular languages is closed under the operations union, intersection, difference, concatenation, right-quotient, complementation, and star-closure. That is,

$$L_1 \text{ and } L_2 \text{ are regular} \implies$$
$$L_1 \cup L_2, \ L_1 \cap L_2, \ L_1 - L_2, \ L_1 L_2, \ \overline{L_1}, \ L_1^* \text{ are regular.} \qquad (2.19)$$

Exercise 2.2.2 Let L_1 and L_2 are regular languages. Show that L_1/L_2 and $L_1 \ominus L_2$ are regular, where L_1/L_2 and $L_1 \ominus L_2$ are defined as follows:

$$L_1/L_2 = \{x : xy \in L_1 \text{ for some } y \in L_2\}.$$
$$L_1 \ominus L_2 = \{x : x \in L_1 \text{ or } x \in L_2, \text{ but } x \text{ is not in both } L_1 \text{ and } L_2\}.$$

2.2.4 Applications of Finite Automata

One of the applications of automata theory in computer science is compiler construction. For instance, a compiler must be able to recognize which strings of symbols in the source program should be considered as representations of single objects, such as variables, names, numerical constants, and reserved words. This pattern-recognition task is handled by the lexical analyzer within the compiler. The design of lexical analysers depends more or less on automata theory. In fact, a lexical analyzer is a finite automaton.

Example 2.2.9 Suppose we wish to design a lexical analyzer for identifiers; an identifier is defined to be a letter followed by any number of letters or digits, i.e.,

$$\text{identifier} = \{\{\text{letter}\}\{\text{letter}, \text{digit}\}^*\}.$$

It is easy to see that the DFA in Figure 2.6 will accept the above defined identifier. The corresponding transition table for the DFA is given as follows:

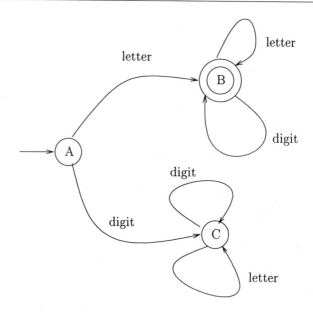

Figure 2.6: DFA that Accepts Identifier

state / symbol	letter	digit
A	B	C
B	B	B
C	C	C

Initial state: A

Final state: B

For example, all the elements in set S_1 are acceptable identifiers by the DFA, whereas all the elements in set S_2 are unacceptable identifiers:

$$S_1 = \{C,\ A21,\ x2w101,\ s13579\}, \quad S_2 = \{87,\ 2add,\ 7w101\}.$$

Example 2.2.10 Suppose we now want to design a lexical analyzer for real numbers; a real number can be either in decimal form (e.g., $45, 79$) or in exponential form (e.g., $34.0E-9$). The DFA described in Figure 2.7 will accept the real numbers just defined. The corresponding transition table for the DFA is given as follows:

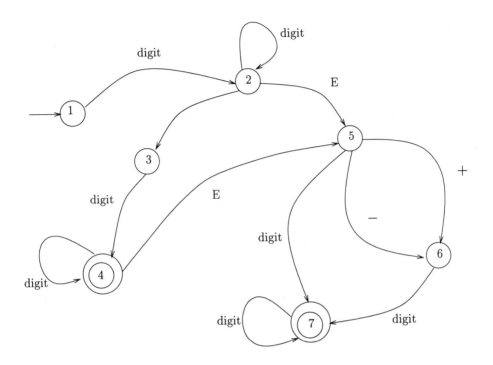

Figure 2.7: DFA that Accepts Real Numbers

state / symbol	digit	·	E	+	−
1	2				
2	2	3	5		
3	4				
4	4		5		
5	7			6	6
6	7		5		
7	7				

Initial state: 1
Final state: 4 and 7

For example, all elements in the set

$$S = \{54.3, \ 54.3E7, \ 54.3E+7, \ 54.3E-7, \ 54E7, 54E+7, \ 54E-7\}$$

are acceptable real numbers by the DFA defined above.

Exercise 2.2.3 Pattern Matching: On UNIX operating system, we can e.g., use the following command

$$\text{rm *sy*}$$

to remove all the files with letters sy in the middle of their names; for example, files with names ysy011, 100sypaper, and 01syreport will be removed. Design an automaton for an operating system, that can accept all the patterns of the form:

$$\{\{\text{letter, digit}\}^*\{\text{letter, digit}\}^*\}.$$

For example, the patterns 123abd, doc311, d1f22 are all accepted by the automaton.

2.2.5   Variants of Finite Automata

In this subsection, we shall provide a brief survey of some variants of finite-state automata, including stochastic automata, fuzzy automata, Petri nets, connectionist machines, and cellular automata. These automata are the natural extensions of the classical finite automata (particularly nondeterministic finite automata) and are very useful in certain areas of computer science.

(I) Stochastic Automata

Intelligent behavior is very often characterised by a lack of deterministic predictability. Given the same input, an intelligent being (e.g., a robot's brain) might appear to act in varying ways. The apparent uncertainty in behavior requires models that reflect that uncertainty. One way of achieving such a model is through the use of probability. Stochastic automata are types of probabilistic automata, which are, in fact, very similar to nondeterministic automata (NFA) discussed in the previous sections.

Definition 2.2.7 A *stochastic automaton*, M, is a six-tuple:

$$M = (Q, \Sigma, V, \delta, q_0, F) \tag{2.20}$$

where

(i) Q is a finite set of *states*,

(ii) $q_0 \in Q$ is the *initial state*,

(vi) $F \subseteq Q$ is the set of *final state* or *accepting state*, denoted by a double circle.

(iii) Σ is a finite set of inputs or instructions,

(iv) V is the valuation space $[0, 1]$,

(v) δ is transition function:

$$\delta: Q \times \Sigma \times Q \to V, \tag{2.21}$$

It is required that for any fixed non-final state q and any fixed instruction a

$$\sum_{q' \in Q} \delta(q, a, q') = 1 \tag{2.22}$$

This requirement allows us to interpret

$$\delta(q, a, q') = x \tag{2.23}$$

as meaning that x is the probability of the machine going from state q to state q' utilising the instruction a and the sum of the probability must be 1.

Example 2.2.11 Let $M = (Q, \Sigma, V, \delta, q_0, F)$ be a stochastic automaton with

$\Sigma = \{a, b\}$
$Q = \{A, B, C\}$ $\qquad\qquad q_0 = A$ $\qquad\qquad F = C$
$\delta(A, a, A) = 0.7$ $\qquad \delta(B, a, A) = 1$ $\qquad \delta(C, a, C) = 1$
$\delta(A, a, C) = 0.1$ $\qquad \delta(B, b, B) = 0.6$ $\qquad \delta(C, b, C) = 1$
$\delta(A, a, B) = 0.2$ $\qquad \delta(B, b, C) = 0.4$
$\delta(A, b, B) = 0.9$
$\delta(A, b, C) = 0.1$

where $\sum_{q' \in Q} \delta(q, a, q') = 1$. This stochastic automaton can be diagrammatically shown in Figure 2.8. Suppose we now wish to calculate the probability that the

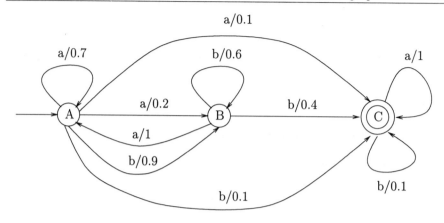

Figure 2.8: A Stochastic Automaton

automaton will go to state C from A given instructions a and b:

$$
\begin{aligned}
\delta'(A, ab, C\} &= \sum_{q \in Q} \delta(A, a, q) \cdot \delta(q, b, C) \\
&= \delta(A, a, A) \cdot \delta(A, b, C) + \delta(A, a, B) \cdot \delta(B, b, C) \\
&\quad + \delta(A, a, C) \cdot \delta(C, b, C) \\
&= 0.7 \times 0.1 + 0.2 \times 0.4 + 0.1 \times 1 \\
&= 0.25
\end{aligned}
$$

Stochastic automata have an important application in modelling the robot behaviour; readers are referred to [37] for more information.

(II) Fuzzy Automata

In stochastic automata, the uncertainty was modelled by probability. We now introduce another similar automata in which the uncertainty was modelled by fuzziness, rather than by probability. A fuzzy automaton is again similar to a nondeterministic automaton in that several destination states may be entered simultaneously; however, it is also similar to a stochastic automaton in that there is a measure of the degree to which the automaton transitions between states, that measure being between 0 and 1.

Definition 2.2.8 A *fuzzy automaton*, M, is a six-tuple

$$M = (Q, \Sigma, V, \delta, q_0, F) \tag{2.24}$$

where

(i) Q is a finite set of *states*,

(ii) $q_0 \in Q$ is the *initial state*,

(iii) Σ is a finite set of *inputs* or *instructions*,

(vi) $F \subseteq Q$ is the set of *final states* or *accepting states*, denoted by a double circle.

(iv) V is the *valuation space* $[0, 1]$,

(v) δ is transition function:

$$\delta : \ Q \times \Sigma \times Q \to V. \tag{2.25}$$

Example 2.2.12 Let $M = (Q, \Sigma, V, \delta, q_0, F)$ be a fuzzy automaton with

$\Sigma = \{a, b\}$
$Q = \{A, B, C\}$ $q_0 = A$ $F = C$
$\delta(A, a, A) = 0.8$ $\delta(B, a, C) = 0.9$ $\delta(C, b, B) = 0.4$
$\delta(A, a, B) = 0.7$
$\delta(A, a, C) = 0.5$
$\delta(A, b, C) = 0.4$

Then M can be graphically described in Figure 2.9. Note that a fuzzy automata is not necessarily stochastic, say, e.g., $\sum_{q' \in Q} \delta(C, b, q') = 0.4 \neq 1$. Suppose now we also wish to calculate the certainty that the automaton will go to state C from A given instructions a and b:

$$
\begin{aligned}
\delta'(A, ab, C\} &= \bigvee_{q \in Q} [\delta(A, a, q) \wedge \delta(q, b, C)] \\
&= [\delta(A, a, A) \wedge \delta(A, b, C)] \vee [\delta(A, a, B) \wedge \delta(B, b, C)] \\
&\quad \vee [\delta(A, a, C) \wedge \delta(C, b, C)] \\
&= (0.8 \wedge 0.4) \vee (0.7 \wedge 0.4) \vee (0.5 \wedge 0.7) \\
&= 0.4 \vee 0.4 \vee 0.5 \\
&= 0.5
\end{aligned}
$$

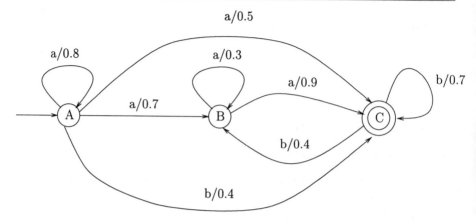

Figure 2.9: A Fuzzy Automaton

Note that "∧" (resp. "∨") means that the minimum (resp. maximum) is being taken over all the possible states.

Fuzzy automata are an important tool for modelling uncertainty in artificial intelligence, particularly in fuzzy logic based expert systems.

(III) Petri Nets

Automata of the kind discussed so far are useful models for *uniprocessing* systems but are not well-suited for modelling multiprocessing systems where the state of the system depends on the states of independently and concurrently executing processes. In what follows, we shall introduce an abstract model, called *Petri net*, which is suitable for representing parallel systems and processes. Recently, it has been also used to represent productions and knowledge in expert systems.

Definition 2.2.9 A *Petri net, M*, is a seven-tuple

$$M = (Q, E, V, f, g, m_0) \tag{2.26}$$

where

(i) Q is a finite set of states,

(ii) E is a finite set of events that is disjoint from Q,

(iii) V is in the valuation space $\{0, 1\}$,

(iv) f is the binary function used in determining the connections from Q to E; thus:

$$f : Q \times E \to V \qquad (2.27)$$

and if $f(q, e) = 1$, then state p connects to event e; otherwise, it does not,

(v) g is the binary function used in determining the connections from E to Q; thus:

$$g : E \times Q \to V \qquad (2.28)$$

and if $g(e, q) = 1$, then event e connects to state q; otherwise, it does not,

(vi) $N = \{0, 1, 2, ...\}$, is the set of markings,

(vii) m_0 is the initial marking function:

$$m_0 : q \to N. \qquad (2.29)$$

Example 2.2.13 Figure 2.10 shows a simple Petri net, M,

$$M = (Q, E, V, f, g, m_0)$$

where

$$Q = \{q_1, q_2, q_3, q_4\}$$
$$E = \{e_1, e_2, e_3, e_4\}$$
$$f(q_1, e_1) = f(q_1, e_3) = f(q_1, e_4) = f(q_2, e_1) = f(q_2, e_2)$$
$$= f(q_3, e_2) = f(q_3, e_3) = f(q_4, e_2) = f(q_4, e_3) = 1$$
$$g(e_1, q_1) = g(e_1, q_2) = g(e_1, q_3) = g(e_2, q_4) = g(e_3, q_4) = 1$$
$$m(q_1) = 1, \quad m(q_2) = m(q_4) = 2, \quad m(q_3) = 0$$

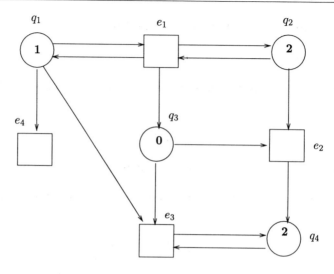

Figure 2.10: A Simple Petri Net

At the beginning, the net is labeled with a initial marking given by m_0, and different markings occur due to the firing of a specific single event. Once a firing has occurred, a new Petri net, with new markings, results. Moreover, event e fires for a given marking m provided each input state for e possesses at least one token: e fires if, for any q such that $f(q,e) = 1$, $m(q) \geq 1$. Thus, in the above simple net, only one event, i.e., e_1 can fire, since $f(q_1, e_1) = 1$, $f(q_2, e_1) = 1$ and $m(q_1) \geq 1$; whereas e_2 cannot fire since $f(q_3, e_2) = 1$ but $m(q_3) = 0$, e_3 also cannot fire since $f(q_3, e_3) = 1$ but $m(q_3) = 0$.

We remark, in conclusion, that Petri nets also serve as useful models for data flow computer architectures, where instructions are not executed in prescribed sequence, but rather executed upon the occurrence of specific events.

(IV) Connectionist Machines

Connectionist machines (also known as massively parallel machines or neural networks were first studied by McCulloch and Pitts [90] in 1943. They are possibly the earliest models of finite state automata.

Definition 2.2.10 A *connectionist machine* or a neural network is a 4-tuple algebraic system defined by

$$M = (n, \ N, \ B, \ A) \tag{2.30}$$

where

(i) $n \in N$ defines the size of the model,

(ii) $N = \{C_1, C_2, \cdots, C_n\}$ is a set of n formal neurons,

(iii) $B = (b_1, b_2, \cdots, b_n)$ is a real vector of dimension n, which serves as the *threshold* vector of the model.

(iv) $A = (a_{ij})$, $i, j, = 1, 2, \cdots, n$, is a matrix of real numbers whose entry a_{ij} represents the connection weight, with which C_i affects C_j.

The machine defined above is allowed to act synchronously, through successive state transitions. At any time $t = 0, 1, 2, 3, \cdots$, each neuron C_i has a Boolean state value $v(i, t)$, which is either 0 or 1. If the value is 0, we say the cell is *off*, otherwise, it is *on*. Given the state values $v(i, t)$, $i = 1, 2, 3, \cdots, n$, for some t, then the value $v(j, \ t+1)$ of the next state is determined by

$$v(j, \ t+1) = \begin{cases} 1, & \text{if } \sum_{i=1}^{n} a_{ij} v(i, t) > b_j, \\ 0, & \text{otherwise.} \end{cases} \tag{2.31}$$

In other words, if the total sum of the connection weights to C_j from all other cells that are currently *on* is greater than the threshold b_j of the cell C_j at the time $t + 1$, then the cell will be *on*; otherwise, it will be *off*.

Connectionist machines are very useful in machine learning and parallel computation. More importantly, connectionist machines can also be regarded as general computation models which have the same computation power as any other computation models, such as Turing machines. We shall discuss this point further in Chapter 5.

(V) Cellular Automata

Cellular automata, also known as *tessellation structures*, and *iterative circuit computers*, is a model of parallel computation [22]. The basic ideas for cellular automata are due to John von Neumann[2], who introduced them, probably not because of interest in them, but rather as vehicles to study the feasibility of designing automata to reproduce themselves. This is why we call cellular automata *self-reproducing automata*. We shall only give a very brief introduction to this powerful automata.

Consider a two-dimensional lattice of cells extending indefinitely in all directions (see Figure 2.11). Suppose that each cell (e.g., A) is a finite automaton

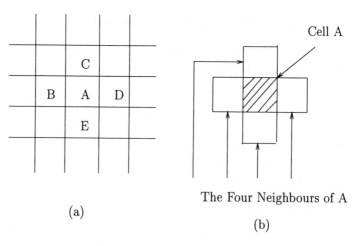

Cell A

The Four Neighbours of A

(a)

(b)

Figure 2.11: Two-Dimensional Lattice of Cells

that can assume any one of the n states, $q_1, q_2, \cdots, q_n$. All cells are identical in their structure, but at any given time they may be in different states. At each time step, every machine assumes a state which is determined by the states of its four neighboring cells B, C, D, E and of itself at the preceding time step. One of the possible states is a "quiescent" one and if a given cell and its four neighbours are in this quiescent state at time t then the given cell remains in the same state at time $t+1$. If all cells in the array arrive at the quiescent state

---

[2]John von Neumann (1903–1957) was born in Budapest, Hungary, but lived in the USA from 1930. He was one of the legendary figures of 20th century mathematics. He made important contributions to logic, quantum physics, optimisation theory and game theory. His lifelong interest in mechanical devices led to his being involved crucially in the initial development of the modern electronic computer and the important concept of the stored program. von Neumann was also involved in the development of the first atomic bomb.

at the same instant, no further changes of state are possible. In the following, we shall present a formal definition for a simple linear cellular automaton:

Definition 2.2.11 A *bi-infinite linear cellular automaton*, M, is a 5-tuple algebraic system defined by

$$M = (\mathbb{Z}, \ \Sigma, \ q, \ N, \ \delta) \tag{2.32}$$

where

(i) $\mathbb{Z} = \{\cdots, -i, \cdots, 0, \cdots, i, \cdots\}$ is the set of *cells*; $i \in \mathbb{Z}$ is the *location* of the cell i.

(ii) $\Sigma = \{0, 1, 2, \cdots, k-1\}$ is a finite non-empty set of *(cell-)states*; At each step of the computation, each cell is in a particular state.

(iii) $N = (a_1, \cdots, a_r)$ is the *neighborhood*; it is a strictly increasing sequence of signed integers for some $r \geq 1$, giving the addresses of the neighbours related to each cell. This means that the *neighbours* of cell i are indexed by $i + a_1, \cdots, i + a_r$. We call $r = |N|$ the *size* of the neighborhood. Cells are simultaneously changing their states at each time step according to the states of their neighbours.

(iv) δ is the *local transition function* defined by

$$\delta: \ \Sigma^{|N|} \to \Sigma. \tag{2.33}$$

If at a given step the neighbours of a cell are respectively in states $p_1, \cdots, p_r$, then at the next step the state of the cell will be $\delta(p_1, \cdots, p_r)$.

(v) $q \in \Sigma$ is the distinguished *quiescent* state, which satisfies the condition $\delta(q, \cdots, q) = q$.

We remark, in conclusion, that a cellular automaton is, in a way, a neural network except that the atomic cells have the power of an arbitrary finite-state automaton but are, on the other hand, restricted to be identical to each other. In fact, cellular automata can be simulated by a particular neural network defined as follows: A *bi-infinite neural network* is a 5-tuple algebraic system:

$$M = (\mathbb{Z}, \ \Sigma, \ W, \ B, \ \delta_i) \tag{2.34}$$

where

(i) $\mathbb{Z} = \{\cdots, -i, \cdots, 0, \cdots, i, \cdots\}$ is the set of *neurons.* of the cell i.

(ii) $\Sigma = \{0, 1, 2, \cdots, k-1\}$ is a finite non-empty set of *(neuron-)states.*

(iii) $W = (w_{ij})_{i,j} \in \mathbb{Z}$, $w_{ij} \in \mathbb{R}$ is the bi-infinite *connection matrix*, satisfying

$$\forall x = (x_i)_{i \in \mathbb{Z}} \in \Sigma^{\mathbb{Z}} \text{ and } \forall i \in \mathbb{Z} \Longrightarrow \sum_j w_{ij} x_j \text{ is convergent.} \qquad (2.35)$$

(iv) $B \in \mathbb{R}^{\mathbb{Z}}$ is the threshold vector.

(iv) δ_i is the *activation function* of neuron i, defined by

$$\delta_i : \mathbb{R} \to \Sigma. \qquad (2.36)$$

We refer interested readers to [24] and [45] for more information about cellular automata and neural-like cellular automata.

2.3   Push-Down Automata and Context-Free Languages

Push-Down Automata (PDA) form the most important class of automata between finite automata and Turing machines. As we have seen from the previous section, DFAs cannot accept even very simple languages such as

$$\{x^n y^n : n \in \mathbb{N}\},$$

but fortunately, there exists a more powerful machine, push-down automata, which can accept it. Just as DFA and NFA, there are also two types of push-down automata: Deterministic Push-Down Automata (DPDA) and Non-Deterministic Push-Down Automata (NPDA). The languages which can be accepted by PDA are called Context-Free Languages (CFL), denoted by L_{CF}. Diagrammatically, a PDA is a finite state automaton (see Figure 2.12), with memories (push-down stacks). In this section, we shall study push-down automata (PDA) and their associated languages, context-free languages L_{CF}.

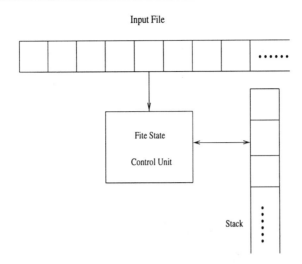

Figure 2.12: Push-Down Automata

2.3.1 Push-Down Automata (PDA)

We first give a formal definition of non-deterministic push-down automata (NPDA).

Definition 2.3.1 A *Non-deterministic Push-Down Automata* (NPDA) is defined by

$$M = (Q, \Sigma, \Gamma, \delta, q_0, z, F) \tag{2.37}$$

where

(i) Q is a finite set of *internal states*,

(ii) Σ is a finite set called the *input alphabet*,

(iii) Γ is a finite set of symbols called the *stack alphabet*,

(iv) δ is the transition function, which is defined as:

$$\delta : \ Q \times (\Sigma \cup \{\lambda\}) \times \Gamma \to \text{ finite subsets of } Q \times \Gamma^*, \tag{2.38}$$

(v) $z \in \Gamma$ is the *stack initial symbol*,

(vi) $q_0 \in Q$ is the *initial state*,

(vii) $F \subseteq Q$ is the set of *final states*.

Example 2.3.1 Let an NPDA, M, be $M = (Q, \Sigma, \Gamma, \delta, q_0, z, F)$ with

$$Q = \{q_0, q_1, q_2, q_3\},$$
$$\Sigma = \{a, b\},$$
$$\Gamma = \{0, 1\},$$
$$z = 0,$$
$$F = q_3,$$

and

$$\delta(q_0, a, 0) = \{(q_1, 10), (q_3, \lambda)\},$$
$$\delta(q_0, \lambda, 0) = \{(q_3, \lambda)\},$$
$$\delta(q_1, a, 1) = \{(q_1, 11)\},$$
$$\delta(q_1, b, 1) = \{(q_2, \lambda)\},$$
$$\delta(q_2, b, 1) = \{(q_2, \lambda)\},$$
$$\delta(q_2, \lambda, 0) = \{(q_3, \lambda)\}.$$

Then the automata accepts the language

$$L = L(M) = \{a^n b^n : n \in \mathbb{N}\} \cup \{a\}.$$

A deterministic push-down automata (DPDA) is an automaton that never has a choice in its move:

Definition 2.3.2 A push-down automata (PDA) is said to be *deterministic* if it is an automaton, $M = (Q, \Sigma, \Gamma, \delta, q_0, z, F)$, as in Definition 2.3.1, subject to the following two restrictions: for every $q \in Q$, $a \in \Sigma \cup \{\lambda\}$ and $b \in \Gamma$,

(i) $\delta(q, a, b)$ contains at most one element,

(ii) if $\delta(q, \lambda, b)$ is not empty, then $\delta(q, c, b)$ must be empty for every $c \subset \Sigma$.

2.3.2   Context-Free Languages (L_{CF})

This subsection establishes the relationship between push-down automata and context-free languages. We first discuss context-free grammars and context-free languages.

Definition 2.3.3 A grammar $G = (V, T, S, P)$ is said to be *context-free* if all productions have the form

$$A \to x, \tag{2.39}$$

where $A \in V$, and $x \in (V \cup T)^*$.

A language L is said to be *context-free* if there is a context-free grammar G such that $L = L(G)$.

Example 2.3.2 The grammar $G(V, T, S, P)$, with productions

$$S \to abB,$$
$$A \to aaBb,$$
$$B \to bbAa,$$
$$A \to \lambda,$$

is context-free. Some typical derivations in this grammar are:

$$S \Longrightarrow abB$$
$$\Longrightarrow abbbAa$$
$$\Longrightarrow abbba$$

$$S \Longrightarrow abB$$
$$\Longrightarrow abbbAa$$
$$\Longrightarrow abbbaaBba$$
$$\Longrightarrow abbbaabbAaba$$
$$\Longrightarrow abbbaabbaba$$

$$S \stackrel{*}{\Longrightarrow} abbbaabbAaba$$
$$\Longrightarrow abbbaabbaaBbaba$$
$$\Longrightarrow abbbaabbaabbAababa$$
$$\Longrightarrow abbbaabbaabbababa$$
$$\Longrightarrow ab(bbaa)^2 bba(ba)^2$$

$$S \stackrel{*}{\Longrightarrow} abbbaabbaabbAababa$$
$$\Longrightarrow abbbaabbaabbaaBbababa$$
$$\Longrightarrow abbbaabbaabbaabbAabababa$$
$$\Longrightarrow ab(bbaa)^3 bba(ba)^3$$

$$\cdots\cdots\cdots$$
$$\cdots\cdots\cdots$$

$$S \Longrightarrow ab(bbaa)^n bba(ba)^n, \text{ for } n \geq 0$$

Thus, the language generated by this grammar is

$$L(G) = \{ab(bbaa)^n bba(ba)^n : n \geq 0\}.$$

Remark 2.3.1 Every regular grammar is a context-free grammar, so a regular language is a context-free language. For example, we know that $L = \{a^n b^n : h \geq 0\}$ is not a regular language, but this language can be generated by the grammar $G = (\{S\}, \{a, b\}, S, P)$ with P given by $S \to aSb$ and $S \to \lambda$, which is apparently a context-free grammar. So, the family of context-free languages is the superset of the family of regular languages, whereas the family of regular languages is the proper subset of the family of context-free languages.

We call a string $x \in (V \cup T)^*$ a *sentential form* of G if there is a derivation $S \overset{*}{\Rightarrow} x$ in G. But notice that there may be several variables in a sentential form, in such a case, we have a choice of order to replace the variables. A derivation is said to be *leftmost* if in each step the leftmost variable in the sentential form is replace. If in each step the rightmost variable is replace, then we called the derivation *rightmost*.

Example 2.3.3 Let $G = (\{S, A\}, \{a, b\}, S, P)$ with P given by

(i) $S \to AA$,
(ii) $A \to AAA$,
(iii) $A \to bA$,
(iv) $A \to Ab$,
(v) $A \to a$.

Then we have the following four distinct derivations for string $L(G) = ababaa$:

$S \overset{i}{\Rightarrow} AA$	$S \overset{i}{\Rightarrow} AA$	$S \overset{i}{\Rightarrow} AA$
$\overset{v}{\Rightarrow} aA$	$\overset{v}{\Rightarrow} Aa$	$\overset{v}{\Rightarrow} aA$
$\overset{ii}{\Rightarrow} aAAA$	$\overset{ii}{\Rightarrow} AAAa$	$\overset{ii}{\Rightarrow} aAAA$
$\overset{iii}{\Rightarrow} abAAA$	$\overset{iii}{\Rightarrow} AAbAa$	$\overset{v}{\Rightarrow} aAAa$
$\overset{v}{\Rightarrow} abaAA$	$\overset{v}{\Rightarrow} AAbaa$	$\overset{iii}{\Rightarrow} abAAa$
$\overset{iii}{\Rightarrow} ababAA$	$\overset{iii}{\Rightarrow} AbAbaa$	$\overset{iii}{\Rightarrow} abAbAa$
$\overset{v}{\Rightarrow} ababaA$	$\overset{v}{\Rightarrow} Ababaa$	$\overset{v}{\Rightarrow} ababAa$
$\overset{v}{\Rightarrow} ababaa$	$\overset{v}{\Rightarrow} ababaa$	$\overset{v}{\Rightarrow} ababaa$
Derivation (1)	Derivation (2)	Derivation (3)

It is clear that derivation (1) is left-most, (2) right-most, whereas (3) is neither.

Theorem 2.3.1 (Pumping Theorem for Context-Free Languages) Let L be a context-free language. There exists a positive integer $N \in \mathbb{Z}^+$ (depending on L) such that for any $z \in L$ and $|z| \geq N$, there exist strings u, v, w, x and y satisfying the following conditions:

$$z = uvwxy, \tag{2.40}$$
$$|v| + |x| > N, \tag{2.41}$$
$$uv^iwx^iy \in L, \quad \forall i \geq 0. \tag{2.42}$$

The number N is called *pumping number* for the context-free language L.

Like its counter-part for regular languages, the pumping theorem for context-free languages provides a tool for demonstrating that languages are not context-free.

Exercise 2.3.1 Use the above pumping theorem to show that

$$L = \{a^nb^nc^n : n \geq 0\}$$

is not a context-free language.

Now we investigate the relationship between push-down automata and context-free languages.

Theorem 2.3.2 A Language L is context-free if and only if it is acceptable (recognisable) by some PDA. A Language L is deterministic context-free if and only if it is acceptable (recognisable) by some DPDA.

Remark 2.3.2 It is interesting to note that nondeterminism does not add more computing power to deterministic finite automata (DFAs). That is, DFAs and NFAs accept exactly the same languages. In contrast, this is not the case for push-down automata (PDA). There are languages that can be accepted by NPDA but that cannot be accepted by DPDA. So the class of deterministic context-free languages forms a proper subclass of the class of context-free languages. Since the languages of logic, mathematics and programming (with some exceptions) are readily described by context-free grammars, push-down automata provide an appropriate mechanism for parsing sentences in programming languages.

Finally we present some closure/nonclosure properties for context-free languages.

Theorem 2.3.3 The family of context-free languages is closed under the operations union, concatenation, and star-closure. That is,

$$L_1 \text{ and } L_2 \text{ are context-free} \implies L_1 \cup L_2, \ L_1 L_2, \ L_1^* \text{ are context-free.} \quad (2.43)$$

Theorem 2.3.4 The family of context-free languages is *not* closed under intersection and complementation. That is,

$$L_1 \text{ and } L_2 \text{ are context-free} \not\implies L_1 \cap L_2, \ \overline{L_1} \text{ are context-free.} \quad (2.44)$$

Theorem 2.3.5 Let L_1 be a context-free language and L_2 be a regular language. Then $L_1 \cap L_2$ is context-free, but not necessarily regular. That is, the family of context-free languages is closed under *regular intersection*.

Exercise 2.3.2 Show that the family of context-free languages is not closed under the operation difference in general, but it is closed under *regular difference*, that is,

$$L_1 \text{ is are regular are } L_2 \text{ context-free} \implies L_1 - L_2 \text{ is context-free.} \quad (2.45)$$

2.3.3   Applications of Context-Free Languages

Context-free grammars and languages have important applications in programming language definition and compiler construction. The most popular language definition method, *Backus-Naur Form* (BNF), after John Backus[3] who invented the method and Peter Naur[4] who refined it for the programming language ALGOL, directly corresponds to context-free grammar. In fact, many parts of a ALGOL-like or Pascal-like programming languages are susceptible to definition by restricted forms of context-free grammars.

---

[3]John Backus (1924–) received his BSc and MSc in mathematics from Columbia University. He joined IBM as a programmer in 1950 and became an IBM fellow in 1963. Backus is the 1977 Turing Award recipient for leading the development of FORTRAN and the creation of the syntax description language Backus-Naur Form.

[4]Peter Naur (1928–) studied astronomy at the University of Copenhagen. He changed from astronomy to computing in 1959, and his first full-time job in computing was involved in the development of the programming language ALGOL. From 1960 to 1967 he worked on the development of compliers for ALGOL and COBOL. Since 1969, he has been a professor of computer science at the University of Copenhagen.

Example 2.3.4 The following grammar (context-free grammar, but using BNF notation) defines a language of even, non-negative integers.

⟨ even-integer ⟩   ::= ⟨ even-digit ⟩ | ⟨ integer ⟩ ⟨ even-digit ⟩
⟨ integer ⟩   ::= ⟨ digit ⟩ | ⟨ digit ⟩ ⟨ integer ⟩
⟨ digit ⟩   ::= ⟨ even-digit ⟩ | ⟨ odd-digit ⟩
⟨ even-digit ⟩   ::= 0 | 2 | 4 | 6 | 8
⟨ odd-digit ⟩   ::= 1 | 3 | 5 | 7 | 9

With this grammar, we can easily generate the even integers, and show their parse trees, which are useful in syntax analysis and code generation in compiler construction.

2.4   Turing Machines and Recursively Enumerable Languages

As we have seen, finite automata (FA) can recognise regular languages L_{REG}, but not non-regular languages, such as $L = \{a^n b^n : n \in \mathbb{N}\}$, which is known to be context-free language. Push-down automata (PDA), however, can recognise all the context-free languages L_{CF} generated by context-free grammars G_{CF}. There are languages, however, say for example, context-sensitive languages L_{CS}, such as $L = \{a^n b^n c^n : n \in \mathbb{N}\}$, that cannot be generated by context-free grammars. Fortunately, there are other machines, called linear bounded automata (LBA), more powerful than push-down automata, that can recognise all the languages generated by context-sensitive grammars G_{CS}. However, LBA cannot recognise all languages generated by phrase-structure grammars G_{PS}. To avoid the limitations of the above mentioned three special types of automata, a Turing machine (TM), named after the British mathematician Alan Turing[5] is used. Turing machines can recognise all the languages generated by phrase-structure grammars,

---

[5] Alan M. Turing (1912–1954) was born in London, England. He was educated in Sherborn, an English boarding school and King's College, Cambridge. In 1935, Turing became fascinated with the decision problem, a problem posed by the great German mathematician David Hilbert, which asked whether there is a general method that can be applied to any assertion to determine whether the assertion is true. The paper which made him famous "On Computable Numbers, with an Application to the Entscheidungsproblem (problem of decidability)" was published in the *Proceedings of the London Mathematics Society*, Vol 42, November 1936 [141]. It was in this paper that he proposed the very general computation model, now widely known as the *Turing machine*, which can perform any computable function. The paper

called the recursively enumerable languages L_{RE}, that includes, of course, all the regular languages, context-free languages and context-sensitive languages. In addition, Turing machines can also model all the computations that can be performed on any computing machine. In this section, we shall study Turing machines and their associated languages L_{RE}. In the next chapter, we shall study the computability/noncomputability and decidability/undecidability of Turing machines.

2.4.1   Standard Turing Machines

A standard Turing machine (see Figure 2.13) has the following features:

(i) The Turing machine has a tape that is unbounded in both directions.

(ii) The Turing machine is deterministic.

(iii) There are no special input and output files.

In this section, we shall present the precise definition of standard Turing machines, and provide some examples of Turing machines, particularly in contrast to finite state machines and push-down machines.

Definition 2.4.1 A *Turing Machine (TM)* is defined by

$$M = (Q, \Sigma, \Gamma, \delta, q_0, \Box, F) \tag{2.46}$$

where

(i) Q is a finite set of *internal states*,

(ii) Σ is a finite set of symbols called the *input alphabet*, we assume that $\Sigma \subseteq \Gamma - \{\Box\}$.

---

attracted immediate attention and led to an invitation to Princeton, where he worked with Alonzo Church. He took his PhD there in 1938; the subject of his thesis is "Systems of Logic based on Ordinals". During the Second World War Turing also led the successful effort in Bletchley Park (Government Code and Cipher School) to crack the German "Enigma" code, an effort central to the defeat of the Nazi Germany. To commemorate Turing's original contribution, the *Association for Computing Machinery* in the USA created the Turing Award in 1966. The award is presented annually to an individual selected for contributions of a technical nature to the computing community that are judged to be of lasting and major importance to the field of computer science.

(iii) Γ is a finite set of symbols called the *tape alphabet*,

(iv) δ is the transition function, which is defined as

$$\delta : \ Q \times \Gamma \rightarrow Q \times \Gamma \times \{L, R\}, \tag{2.47}$$

(v) $\square \in \Gamma$ is a special symbol called the *blank*,

(vi) $q_0 \in Q$ is the *initial state*,

(vii) $F \subseteq Q$ is the set of *final states*.

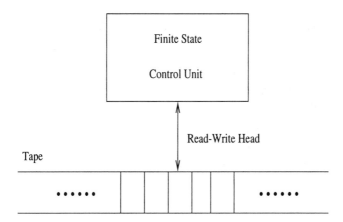

Figure 2.13: Standard Turing Machine

2.4.2 Turing Machines as Language Accepters

A Turing machine can be viewed as an accepter in the following sense. A string w is written on the tape, with blanks filling out the unused portions. The machine is started in the initial state q_0 with the read-write head positioned on the leftmost symbol of w. If, after a sequence of moves, the Turing machine enters a final state and halts, then w is considered to be accepted by the Turing machine. We shall provide a precise definition for the above descriptions and present some examples of how Turing machines accept strings that can not be accepted by a DFA or PDA.

Definition 2.4.2 Let $M = (Q, \Sigma, \Gamma, \delta, q_0, \Box, F)$ be a Turing machine. Then the languages that can be accepted by M are defined by

$$L(M) = \{w \in \Sigma^* : q_0 w \overset{*}{\vdash} w_1 q_f w_2, \text{ for } q_f \in F, \text{ and } w_1, w_2 \in \Gamma^*\}. \quad (2.48)$$

Example 2.4.1 Let $\Sigma = \{a, b\}$. Design a Turing machine that accepts the language

$$L = \{a^n b^n : n \geq 1\}.$$

As we have seen from the preceding section that this language is a context-free language and can be accepted by a push-down automata. In this example, we shall see that this language can be accepted by a Turing machine as well. Let q_0 be the initial state, and suppose that we use the x's to replace a' and y's to replace b'. Then we can design the transitions as follows:

$$\delta(q_0, a) = (q_1, x, R)$$
$$\delta(q_1, a) = (q_1, a, R)$$
$$\delta(q_1, y) = (q_1, y, R)$$
$$\delta(q_1, b) = (q_2, y, L)$$

$$\delta(q_2, y) = (q_2, y, L)$$
$$\delta(q_2, a) = (q_2, a, L)$$
$$\delta(q_2, x) = (q_0, x, R)$$

$$\delta(q_0, y) = (q_3, y, R)$$

$$\delta(q_3, y) = (q_3, y, R)$$
$$\delta(q_3, \Box) = (q_4, \Box, R)$$

So finally, the designed Turing machine is as follows:

$$M = (Q, \Sigma, \Gamma, \delta, q_0, \Box, F)$$
$$= (\{q_0, q_1, q_2, q_3, q_4\}, \{a, b\}, \{a, b, x, y, \Box\}, \delta, q_0, \Box, \{q_4\}).$$

For a particular input $aaabbb$, we have the following successive *instantaneous descriptions* (IDs) of the designed Turing machine:

$$q_0 aaabbb \vdash x q_1 aabbb$$
$$\vdash x a q_1 abbb$$
$$\vdash x a a q_1 bbb$$
$$\vdash x a q_2 aybb$$

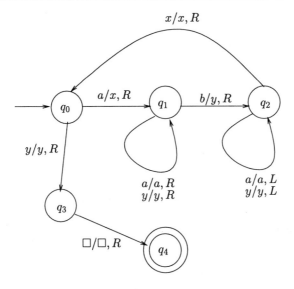

Figure 2.14: A Turing Machine That Accepts $\{a^n b^n : n \geq 1\}$

$$\vdash xq_2 aaybb$$
$$\vdash q_2 xaaybb$$
$$\vdash xq_0 aaybb$$
$$\vdash xxq_1 aybb$$
$$\vdash xxaq_1 ybb$$
$$\vdash xxayq_1 bb$$
$$\vdash xxaq_2 yyb$$
$$\vdash xxq_2 ayyb$$
$$\vdash xq_2 xayyb$$
$$\vdash xxq_0 ayyb$$
$$\vdash xxxq_1 yyb$$
$$\vdash xxxyq_1 yb$$
$$\vdash xxxyyq_1 b$$

$\vdash xxxyq_2yy$

$\vdash xxxq_2yyy$

$\vdash xxq_2xyyy$

$\vdash xxxq_0yyy$

$\vdash xxxyq_3yy$

$\vdash xxxyyq_3y$

$\vdash xxxyyyq_3\square$

$\vdash xxxyyy\square q_4\square$

At this point the Turing machine halts in a final state, so the string *aaabbb* is accepted by the Turing machine. The above successive instantaneous descriptions can also be showed diagrammatically as follows:

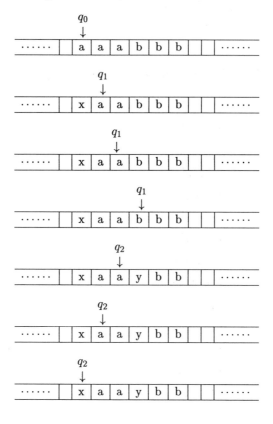

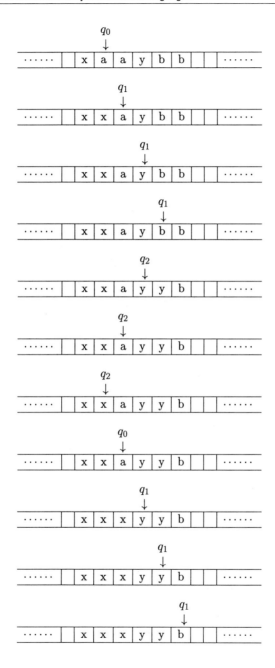

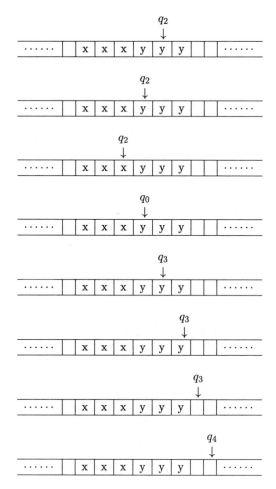

Remark 2.4.1 The above example shows that Turing machines can accept languages that can be accepted by push-down automata. It is, of course the case that Turing machines can accept languages that can be accepted by finite automata. For example, the following regular language

$$L_{\text{REG}} = \{w \in \{a, b\}^* : w \text{ contains the substring } aba\}.$$

can be accepted by both Turing machines and finite automata; Figure 2.15 gives a Turing machine and a finite automaton that accept the above language.

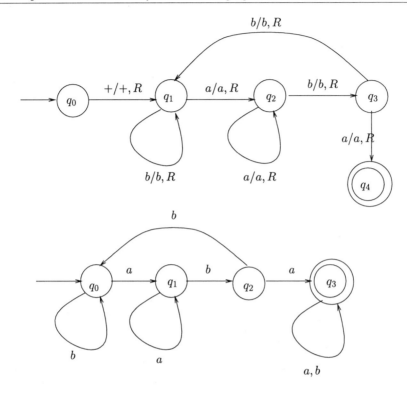

Figure 2.15: A TM and a DFA that accept the language $\{a,b\}^*\{aba\}\{a,b\}^*$

Exercise 2.4.1 Show that the Turing machine constructed in Example 2.4.1 cannot accept the language $L = \{a^n b^m : m \geq 1, n > m\}$.

Exercise 2.4.2 Construct Turing machines that accept the languages $L_1 = \{a^n b^{2n} : n \geq 1\}$ and $L_2 = \{a^{2^n} : n \geq 1\}$ over $\Sigma = \{a, b\}$.

Example 2.4.2 Design a Turing machine that accepts the language

$$L = \{a^n b^n c^n : n \geq 1\}.$$

As we already know that this language is not a context-free language, thus it cannot be accepted by a push-down automata. In this example, we shall show that it is possible to design a Turing machine that accepts this language.

$$\delta(q_0, a) = (q_1, x, R)$$
$$\delta(q_1, a) = (q_1, a, R)$$
$$\delta(q_1, y) = (q_1, y, R)$$
$$\delta(q_1, b) = (q_2, y, L)$$

$$\delta(q_2, y) = (q_2, y, L)$$
$$\delta(q_2, a) = (q_2, a, L)$$
$$\delta(q_2, x) = (q_0, x, R)$$

$$\delta(q_0, y) = (q_3, y, R)$$

$$\delta(q_3, y) = (q_3, y, R)$$
$$\delta(q_3, \square) = (q_4, \square, R)$$

We design the Turing as follows:

$$M = (Q, \Sigma, \Gamma, \delta, q_0, \square, F)$$

where

$$Q = \{q_0, q_1, q_2, q_3, q_4, q_5\}$$
$$\Sigma = \{a, b, c\}$$
$$\Gamma = \{a, b, cx, y, z, \square\}$$
$$F = \{q_4\}$$
$$\delta : Q \times \Gamma \to Q \times \Gamma \times \{L, R\} \text{ is defined by}$$
$$\delta(q_0, a) = (q_1, x, R)$$
$$\delta(q_1, a) = (q_1, a, R)$$
$$\delta(q_1, y) = (q_1, y, R)$$
$$\delta(q_1, b) = (q_2, y, R)$$
$$\delta(q_2, z) = (q_2, z, R)$$
$$\delta(q_2, b) = (q_2, b, R)$$
$$\delta(q_2, c) = (q_3, z, L)$$
$$\delta(q_3, a) = (q_3, a, L)$$
$$\delta(q_3, b) = (q_3, b, L)$$
$$\delta(q_3, y) = (q_3, y, L)$$
$$\delta(q_3, z) = (q_3, z, L)$$
$$\delta(q_3, x) = (q_0, xR)$$
$$\delta(q_0, y) = (q_4, y, R)$$
$$\delta(q_4, \square) = (q_5, \square, R)$$

For the particular input $aabbcc$, we have the following successive instantaneous descriptions of the designed Turing machine:

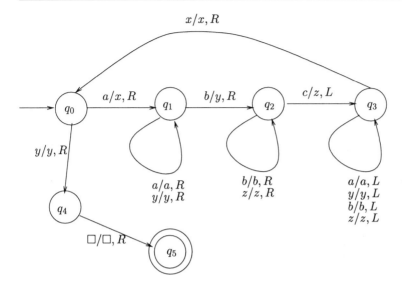

Figure 2.16: A Turing Machine That Accepts $\{a^n b^n c^n : n \geq 1\}$

$$
\begin{aligned}
q_0 aabbcc \;&\vdash\; x q_1 abbcc \\
&\vdash\; xa q_1 bbcc \\
&\vdash\; xay q_2 bcc \\
&\vdash\; xayb q_2 cc \\
&\vdash\; xay q_3 bzc \\
&\vdash\; xa q_3 ybzc \\
&\vdash\; x q_3 aybzc \\
&\vdash\; q_3 xaybzc \\
&\vdash\; x q_0 aybzc \\
&\vdash\; xx q_1 ybzc \\
&\vdash\; xxy q_1 bzc \\
&\vdash\; xxyy q_2 zc \\
&\vdash\; xxyyz q_2 c \\
&\vdash\; xxyy q_3 zz \\
&\vdash\; xxy q_3 yzz \\
&\vdash\; xx q_3 yyzz
\end{aligned}
$$

$$\vdash xq_3xyyzz$$
$$\vdash xxq_0yyzz$$
$$\vdash xxyq_4yzz$$
$$\vdash xxyyq_4zz$$
$$\vdash xxyyzq_4z$$
$$\vdash xxyyzzq_4\Box$$
$$\vdash xxyyzz\Box q_4\Box$$

Example 2.4.2 actually tells us that

Theorem 2.4.1 The class of Turing-acceptable languages properly includes the classes of regular languages and context-free languages.

Exercise 2.4.3 Construct a Turing machine that accepts the language

$$L = \{a^k b^m c^n : k, m, n > 0, k = m \text{ or } k = n \text{ or } m = n\}$$

over $\Sigma = \{a, b, c\}$.

2.4.3   Variants of Turing Machines

The Turing machines we have just studied in the previous section are the standard (one-tape, deterministic) Turing machines. There are, however, various versions of Turing machines. In this section, we shall briefly review some of the variants of Turing machines.

(I) Multitape Turing Machines

The standard Turing machine has only equipped with one tape. It is natural to extend one tape Turing machines to multitape Turing Machines.

Definition 2.4.3 A *Multitape Turing Machine* (see Figure 2.17) is defined by

$$M = (Q, \Sigma, \Gamma, \delta, q_0, \Box, F) \tag{2.49}$$

where

(i) Q is a finite set of *internal states*,

(ii) Σ is a finite set of symbols called the *input alphabet*, we assume that $\Sigma \subseteq \Gamma - \{\Box\}$.

(iii) Γ is a finite set of symbols called the *tape alphabet*,

(iv) δ is the transition function, which is defined as

$$f: \ Q \times \Gamma^k \to Q \times \Gamma^k \times \{L, R\}^n, \tag{2.50}$$

(v) $\square \in \Gamma$ a special symbol called the *blank*,

(vi) $q_0 \in Q$ is the *initial state*,

(vii) $F \subseteq Q$ is the set of *final states*.

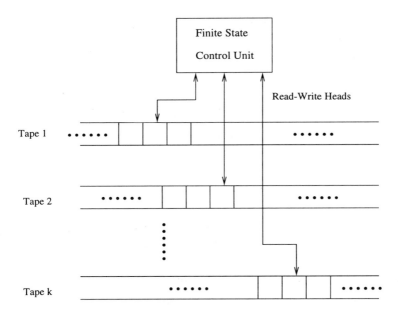

Figure 2.17: Multitape Turing Machine

Theorem 2.4.2 For any multitape Turing machine, there exists a standard (i.e., single-tape) Turing machine to simulate it.

(II) Multidimensional Turing Machines

A multidimensional Turing machine is one in which the tape can be viewed as extending infinitely in more than one dimension.

Definition 2.4.4 A *Multidimensional Turing Machine* (see Figure 2.18) is defined by

$$M = (Q, \Sigma, \Gamma, \delta, q_0, \Box, F) \tag{2.51}$$

where

(i) Q is a finite set of *internal states*,

(ii) Σ is a finite set of symbols called the input alphabet, we assume that $\Sigma \subseteq \Gamma - \{\Box\}$.

(iii) Γ is a finite set of symbols called the *tape alphabet*,

(iv) δ is the transition function, which is defined as

$$f : \ Q \times \Gamma \to Q \times \Gamma \times \{L, R, U, D\} \tag{2.52}$$

where U and D specify movement of the read-write head up and down.

(v) $\Box \in \Gamma$ is a special symbol called the *blank*,

(vi) $q_0 \in Q$ is the *initial state*,

(vii) $F \subseteq Q$ is the set of *final states*.

Theorem 2.4.3 For any multidimensional Turing machine, there exists a standard (i.e., single-tape) Turing machine to simulate it.

(III) Nondeterministic Turing Machines

If we want to argue that nondeterminism adds nothing to the power of a standard Turing machine, we must study nondeterminism of Turing machines.

Definition 2.4.5 A *Nondeterministic Turing Machine* is defined by

$$M = (Q, \Sigma, \Gamma, \delta, q_0, \Box, F), \tag{2.53}$$

where

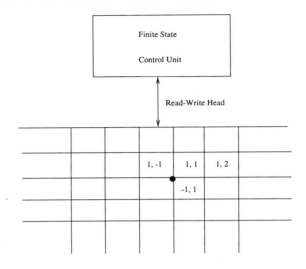

Figure 2.18: Multidimensional Turing Machine

(i) Q is a finite set of *internal states*,

(ii) Σ is a finite set of symbols called the *input alphabet*, we assume that $\Sigma \subseteq \Gamma - \{\square\}$.

(iii) Γ is a finite set of symbols called the *tape alphabet*,

(iv) δ is the transition function, which is defined as

$$f : Q \times \Gamma \to 2^{Q \times \Gamma \times \{L,R\}} \tag{2.54}$$

(v) $\square \in \Gamma$ is a special symbol called the *blank*,

(vi) $q_0 \in Q$ is the *initial state*,

(vii) $F \subseteq Q$ is the set of *final states*.

Obviously, A DTM is a special type of an NTM, therefore, any language that can be accepted by a DTM can be accepted by an NTM. It is interesting to note that the converse is also true:

Theorem 2.4.4 For any nondeterministic Turing machine, there exists a standard (i.e., deterministic) Turing machine to simulate it.

(IV) Probabilistic Turing Machine

A *probabilistic Turing machine* (PTM short) is a Turing machine with the ability to make random decisions, which may be informally defined as follows:

Definition 2.4.6 A *probabilistic Turing machine*, M,

$$M = (Q, \Sigma, \Gamma, \delta, q_0, \Box, F)$$

is a Turing machine with distinguished states called *coin-tossing states*. For each coin-tossing state, the finite control unit specifies two possible legal next states. The computation of a probabilistic Turing machine is deterministic except that in coin-tossing states the machine tosses an unbiased coin to decide between the two *possible legal* next states.

Intuitively, a probabilistic Turing machine is a Turing machine somewhere between the deterministic Turing machine and the nondeterministic Turing machine. Just as standard Turing machines, a probabilistic Turing machine M accepts a language, denoted by $L(M)$, it must accept all strings $w \in L(M)$, and rejects all strings $w \notin L(M)$, except that the machine may make a small probability of error in accepting or rejecting w.

Definition 2.4.7 Let the probability of branch b of M's computation on input w be

$$\text{Prob}(b) = 2^{-k} \tag{2.55}$$

where k is the number of coin-tossing steps that occur on branch b (one coin-tossing step can be regarded as one move step in the probabilistic computation). Let also the probability that M accepts w be

$$\text{Prob}(M \text{ accepts } w) = \sum_{\substack{b \text{ is an} \\ \text{accepting branch}}} \text{Prob}(b) \tag{2.56}$$

and the probability that M rejects w be

$$\text{Prob}(M \text{ rejects } w) = 1 - \text{Prob}(M \text{ accepts } w). \tag{2.57}$$

When a probabilistic Turing machine accepts a language, it must accept all strings in the language, and reject all strings not in the language, except that a small probability of error is allowed. More precisely, a probabilistic Turing machine M is said to accept language $L(M)$, with error probability ϵ, $0 \leq \epsilon \leq \frac{1}{2}$, if

$$w \in L_{\text{PTM}} \Longleftrightarrow \text{Prob}(M \text{ accepts } w) \geq 1 - \epsilon \qquad (2.58)$$

$$w \notin L_{\text{PTM}} \Longleftrightarrow \text{Prob}(M \text{ rejects } w) \geq 1 - \epsilon. \qquad (2.59)$$

M is said to be a bounded-error probabilistic Turing machine if it has bounded-error probability.

Probabilistic Turing machines are a very important and useful model of randomised computations. But in computer science, we are more interested in polynomial time probabilistic Turing machines and algorithms. We shall discuss this topic further in chapters 3, 4 and 5.

(V) Equivalence of Various Turing Machines

We could, of course, list many more different types of Turing machines. However, all the different types of Turing machines have the same power. This establishes the following important result about the equivalence of the various Turing machines:

Theorem 2.4.5 A Language L is accepted by a multitape, or multidimensional, or nondeterministic, or probabilistic Turing Machine, if and only if it is accepted by a standard Turing machine.

We now establish another important result for Turing machines.

Let $\Sigma = \{a, b, c\}$. We said the set $S = \Sigma^+$ is countable if we can find an enumeration procedure that produces its elements in some order, e.g., dictionary order.

Theorem 2.4.6 The set of all Turing Machines, although infinite, is countable.

Theorem 2.4.7 Let S be an infinite countable set. Then its power set 2^S is not countable.

Since there are only countably many Turing machines, but uncountably many sets (languages), then there must be some sets (languages) that are not acceptable by Turing machines. We shall discuss this issue in the next subsection of this chapter, and particularly in Chapter 3.

2.4.4 Recursively Enumerable Languages L_{RE}

In this subsection, we shall study languages associated with Turing machines.

Definition 2.4.8 A language L over an input alphabet Σ is said to be *recursively enumerable*, denoted by L_{RE}, if there exists a Turing machine that accepts it. Recursively enumerable languages are also called *Turing acceptable languages*, or *Turing recognisable languages*.

Definition 2.4.9 A language L over an input alphabet Σ is said to be *recursive*, denoted by L_{REC}, if there exists a Turing machine that accepts L, and that *halts* on every input $w \in \Sigma^+$. Recursive languages are also called *Turing decidable languages*, or recursively decidable languages; we shall discuss the concept *"decidable"* in Chapter 3.

The term *"recursive"* comes from the theory of recursive functions. It is clear that a recursive language is also a recursively enumerable, but on the other hand, a recursively enumerable language is not necessarily recursive. That is,

Theorem 2.4.8 There exists a recursively enumerable language that is not recursive. That is,

$$L_{REC} \subset L_{RE}. \qquad (2.60)$$

From a Turing machine point of view, both recursively enumerable languages and recursive languages are Turing acceptable; the only difference between the two types of languages is that recursive languages will halt on every input $w \in \Sigma^+$; whereas recursively enumerable languages may not halt on every input $w \in \Sigma^+$, that is, they may fall into an infinite loop on some input $w \in \Sigma^+$.

We list in the following some important properties about recursive and recursively enumerable languages:

Theorem 2.4.9 A language L is recursive if and only if both L and its complement $\overline{L}$ are recursively enumerable.

Theorem 2.4.10 There is a recursively enumerable language L whose complement $\overline{L}$ is not recursively enumerable.

Interestingly, recursively enumerable languages are not the highest languages; there exist languages that cannot be accepted by any Turing machine:

Theorem 2.4.11 For any nonempty alphabet Σ, there exist languages over Σ that are not recursively enumerable.

There is yet another approach to studying Turing acceptable languages, namely, the grammatical approach:

Definition 2.4.10 A grammar $G = (V, T, S, P)$ is called a *phrase-structure grammar* or a *unrestricted grammar* if all productions have the form

$$x \to y, \tag{2.61}$$

where $x \in (V \cup T)^+$, and $y \in (V \cup T)^*$.

Definition 2.4.11 Any language generated by an unrestricted grammar is *recursively enumerable*.

Theorem 2.4.12 For every recursively enumerable language L, there is an unrestricted grammar G, such that $L = L(G)$.

2.4.5 Context-Sensitive Languages L_{CS}

The context-sensitive grammar represents an intermediate step between context-free grammars and unrestricted grammars. No restrictions are placed on the left-hand side of a production, but the length of the right-hand side is required to be at least as long as the left.

Definition 2.4.12 A phrase-structure grammar $G = (V, T, S, P)$ is called a *context-sensitive grammar*, denoted by G_{rmCS}, if all productions have the form

$$x \to y, \tag{2.62}$$

where $x, y \in (V \cup T)^+$, and length(x) $\leq$ length(y) (or briefly as $|x| \leq |y|$).

A language L is called a *context-sensitive language*, denoted by L_{CSL}, if there exists a context-sensitive grammar G, such that $L = L(G)$ or $L = L(G) \cup \lambda$.

Example 2.4.3 Design a context-sensitive grammar to generate the context-sensitive language

$$L = \{a^n b^n c^n : n > 0\}.$$

We can construct the grammar $G(V, T, S, P)$ with the following productions:

(i) $S \to abc$,
(ii) $S \to aAbc$,
(iii) $A \to abC$,
(iv) $A \to aAbC$,
(v) $Cb \to bC$,
(vi) $Cc \to cc$.

By Definition 2.4.12, it is context-sensitive. Some typical derivations in this grammar are:

$$S \xrightarrow{(i)} abc$$

Derivation (1)

$$S \xrightarrow{(ii)} aAbc$$
$$\xrightarrow{(iii)} aabCbc$$
$$\xrightarrow{(v)} aabbCc$$
$$\xrightarrow{(vi)} aabbcc$$

Derivation (2)

$$S \xrightarrow{(ii)} aAbc$$
$$\xrightarrow{(iv)} aaAbCbc$$
$$\xrightarrow{(iii)} aaabCbCbc$$
$$\xrightarrow{(v)} aaabbCCbc$$
$$\xrightarrow{(v)} aaabbCbCc$$
$$\xrightarrow{(v)} aaabbbCCc$$
$$\xrightarrow{(vi)} aaabbbCcc$$
$$\xrightarrow{(vi)} aaabbbccc$$

Derivation (3)

Exercise 2.4.4 Find a context-sensitive grammar for the languages $L_1 = \{a^n b^n a^{2n} : n > 0\}$ and $L_2 = \{a^n b^m c^n d^m : n, m > 0\}$.

We have examined several variants of the standard Turing machines that do not alter the set of languages accepted by the machines. Restricting the amount of available tape for computation decreases the capabilities of a Turing machine computation. A linear bounded automata is a *restricted* Turing machine in which the amount of available tape is determined by the length of the input string. The input alphabet contains two symbols "⟨" and "⟩", that designate the left and right boundaries of the tape.

Definition 2.4.13 A *linear bounded automata* (LBA) is an algebraic structure

$$M = (Q, \Sigma, \Gamma, \delta, q_0, \langle, \rangle, F) \tag{2.63}$$

where $Q, \Sigma, \Gamma, \delta, q_0$ and F are the same as for a nondeterministic Turing machine. The symbols $\langle$ and $\rangle$ are distinguished elements of Γ.

Theorem 2.4.13 For every context-sensitive language L, denoted by L_{CS}, not including λ, there is some LBA, M, that accepts L_{CS}. That is, $L_{CS} = L(M)$.

Theorem 2.4.14 If a language L is accepted by some LBA, M, then there exists a context-sensitive grammar G that accepts L. That is $L = L(G)$.

Theorem 2.4.15 Every context-sensitive language L is recursive. That is, $\forall L_{CS} \in L_{REC}$.

Theorem 2.4.16 There exists a recursive language that is not context-sensitive. That is, $L_{CS} \subset L_{REC}$.

2.4.6 Hierarchy of Machines and Grammars/Languages

In this subsection, we shall provide a brief review of the Chomsky hierarchy of formal languages and their generating grammars and their corresponding machines.

(I) Hierarchy of Machines

All the classes (families) of machines we have studied so far are finite (state) machines, but some of the machines have exactly the same power (here by the same power, we mean they accept exactly the same language), whilst some of the machines have more power than others. For example, deterministic finite automata (DFA) have the same power as nondeterministic finite automata (NFA); nondeterministic push-down automata (NPDA) have more power than deterministic push-down automata (DPDA); push-down automata (PDA) with two push-down stores have more power than the push-down automata (PDA) with only one push-down store; but push-down automata (PDA) with more than two push-down stores have the same power as push-down automata with two push-down stores. Interestingly enough, push-down automata with two or more push-down stores have the same power as Turing machines; All different types of Turing machines (such as deterministic, nondeterministic, probabilistic, multitape and multidimensional, etc.) have the same power. But, however, restricting the amount of available tape for computation decreases the capabilities of a Turing machine; linear bounded automata is such a type of *restricted*

Turing machines in which the amount of available tape is determined by the length of the input string. The relation between the various classes of finite machines over the same alphabet Σ can be summarized as follows:

Deterministic Finite Automata (DFA)
$\Updownarrow$
Nondeterministic Finite Automata (NFA)
$\cap$
Deterministic Push-Down Automata (DPDA)
$\cap$
Nondeterministic Push-Down Automata (NPDA)
$\cap$
Linear-Bounded Automata (LBA)
$\cap$
Deterministic Push-Down Automata (DPDA)
with two push-down stores
$\Updownarrow$
Nondeterministic Push-Down Automata (NPDA)
with two push-down stores
$\Updownarrow$
Deterministic Turing Machines (DTM)
$\Updownarrow$
Nondeterministic Turing Machines (NTM)
$\Updownarrow$
Probabilistic Turing Machines (PTM)
$\Updownarrow$
Multitape Turing Machines
$\Updownarrow$
Multidimensional Turing Machines

So, there are essentially four main classes of machines: finite automata (FA), push-down automata (PDA), linear-bounded automata (LBA) and Turing machines (TM). The hierarchy of these classes of machines can be described as follows:

Finite Automata (FA)

$\cap$

Push-Down Automata (PDA)

$\cap$

Linear-Bounded Automata (LBA)

$\cap$

Turing Machines (TM).

(II) Hierarchy of Grammars and Languages

In this subsection, we shall study the Chomsky[6] hierarchy of formal grammars and their generating languages. First let us recall that two grammars are called *equivalent* if they generate the same language.

Definition 2.4.14 A generative grammar $G = (V, T, S, P)$ is said to be of type i if it satisfies the corresponding restrictions in the following list:

$i = 0$: No restrictions. That is, every production in P is just in the general form $x \to y$, where $x \in (V \cup T)^+$ and $y \in (V \cup T)^*$. Type 0 grammars are often called *unrestricted* grammars, or *phrase-structure* grammars, denoted by G_{PS}. The languages generated by Type 0 grammars are called Type 0 languages, or recursively enumerable languages, denoted by L_{RE}.

$i = 1$: Every production in P has the form $x \to y$, where $x, y \in (V \cup T)^+$, and $|x| \leq |y|$. Type 1 grammars are also called *context-sensitive* grammars, denoted by G_{CS}. The languages generated by Type 1 grammars are called Type 1 languages, or context-sensitive languages, denoted by L_{CS}.

$i = 2$: Every production in P has the form $A \to x$, where $A \in V$, and $x \in (V \cup T)^*$. Type 2 grammars are also called *context-free* grammars, denoted by G_{CF}. The languages generated by Type 2 grammars are called Type 2 languages, or context-free languages, denoted by L_{CF}.

---

[6] Avram N. Chomsky (1928–), somewhat the father of formal languages, received his BA, MA and PhD in linguistics, all from the University of Pennsylvania. He is responsible for the hierarchy of grammars that bears his name. Chomsky's work on the syntax of languages coincided neatly with the development of programming languages and thus his work found ready application to the formal style of artificial languages rather than those of his original interest – natural languages. He joined the faculty at MIT in 1955 and latter became the Ferrari P. Ward Professor of foreign languages and linguistics there. Prof. Chomsky is currently still at MIT, giving talks every now and then.

$i = 3$: Every production in P has the form either $A \to Bx$ and $A \to x$, and $A \to xB$ and $A \to x$, where $A, B \in V$, and $x \in T^*$. Type 3 grammars are called *regular* grammars, denoted by G_{REG}. The languages generated by Type 3 grammars are called Type 3 languages, or regular languages, denoted by L_{REG}.

We have, in fact, already studied all the above listed grammars and their generating languages in the previous sections of this chapter. What we are interested in here is the hierarchy of the grammars and their corresponding languages, which are described as follows:

$$
\begin{array}{ccc}
\text{Type 0 Grammars } (G_{\text{PS}}) & \Longleftrightarrow & \text{Type 0 Languages } (L_{\text{RE}}) \\
\cap & & \cap \\
\text{Type 1 Grammars } (G_{\text{CS}}) & \Longleftrightarrow & \text{Type 1 Languages } (L_{\text{CS}}) \\
\cap & & \cap \\
\text{Type 2 Grammars } (G_{\text{CF}}) & \Longleftrightarrow & \text{Type 2 Languages } (L_{\text{CF}}) \\
\cap & & \cap \\
\text{Type 3 Grammars } (G_{\text{REG}}) & \Longleftrightarrow & \text{Type 3 Languages } (L_{\text{REG}})
\end{array}
$$

(III) Relations between Machines and Languages/Grammars

We have already seen that languages and grammars are actually equivalent concepts; on one hand, given a language, we can find the grammar which generates the language; on the other hand, given a grammar, we can find the set of languages generated by the given grammar. Remarkably enough, languages also have a one-to-one correspondence with the various machines. Figure 2.19 shows the hierarchical relationships between various formal languages and their accepting machines. That is, regular languages (L_{REG}) generated by regular grammars (G_{REG}) are acceptable by finite-state automata (FA), context-free languages (L_{CF}) generated by context-free grammars (G_{CF}) are acceptable by push-down automata (PDA), context-sensitive languages (L_{CS}) generated by context-sensitive grammars (G_{CS}) are acceptable by linear bounded automata (LBA), and recursively enumerable languages (L_{RE}) generated by phrase-structure grammars (G_{PS}) are acceptable by Turing machines. Thus, we finally arrive at the hierarchical relations between various languages, grammars and machines as follows:

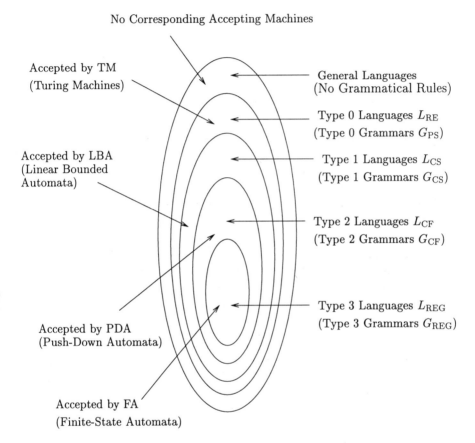

Figure 2.19: Hierarchical Relations between Various Languages/Grammars and their Accepting Machines

$$G_{\mathrm{REG}} \quad \subset \quad G_{\mathrm{CF}} \quad \subset \quad G_{\mathrm{CS}} \quad \subset \quad G_{\mathrm{RE}}$$

$$\Updownarrow \qquad\qquad \Updownarrow \qquad\qquad \Updownarrow \qquad\qquad \Updownarrow$$

$$L_{\mathrm{REG}} \quad \subset \quad L_{\mathrm{CF}} \quad \subset \quad L_{\mathrm{CS}} \quad \subset \quad L_{\mathrm{RE}}$$

$$\Updownarrow \qquad\qquad \Updownarrow \qquad\qquad \Updownarrow \qquad\qquad \Updownarrow$$

$$\mathrm{FA} \quad \subset \quad \mathrm{PDA} \quad \subset \quad \mathrm{LBA} \quad \subset \quad \mathrm{TM}$$

Literally, the relationships between the various grammars, languages and machines can also be summarized as follows:

Grammars	Languages	Accepting Machines
Type 0 grammars (or phrase-structure grammars, unrestricted grammars)	Recursively enumerable languages	Turing Machines
Type 1 grammars (or context-sensitive grammars, monotonic grammars)	Context-sensitive languages	Linear-bounded automata
Type 2 grammars (or context-free grammars)	Context-free languages	Push-down automata
Type 3 grammars (or regular grammars, linear grammars)	Regular languages	Finite automata

If we wish to include some other (small) language families such as deterministic context-free languages and recursive languages, we will then arrive at an extended hierarchy as shown in Figure 2.20.

Further Reading

Formal languages and automata theory is an important as well as a popular subject of computer science. This chapter has only provided a brief introduction to the subject. For those who desire a more detailed exposition, we recommend [21], [22], [26], [31], [32], [42], [54], [61], [68], [82], [86], [87], [96], [97], [114], [124], [136], [138], and [141]. The following two books, [36] and [37] also provide very readable introductory materials in formal languages and automata.

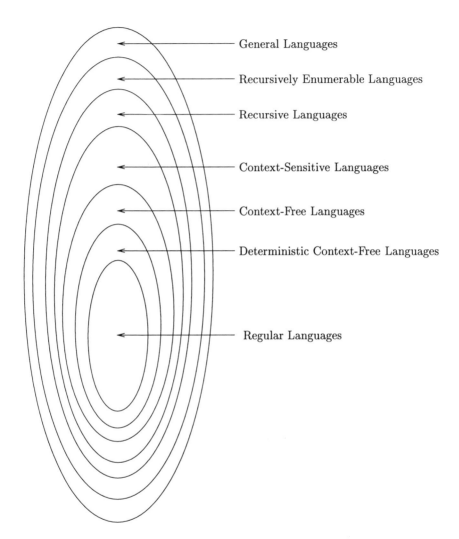

Figure 2.20: Extended Hierarchy of Formal Languages

Chapter 3

Turing Computability and Complexity

To build a theory, one needs to know a lot about the basic phenomena of the subject matter. We simply do not know enough about these, in the theory of computation, to teach the subject very abstractly.

– MARVIN LEE MINSKY (1927–)

Computability and complexity are the most important topics in computer science. In this chapter, we shall study the theory of computability and complexity within the framework of Turing machines.

3.1 Computability and Noncomputability

Turing machines can be used in at least three different types of applications in theoretical computer science:

(i) **Language accepter**: Turing machine is used as an *accepter* to accept a recursive or a recursive enumerable language.

(ii) **Function transducer**: Turing machine is used as a *transducer* to compute a total recursive or a partial recursive function.

(iii) **Decision problem solver**: Turing machine is used as an *algorithm* to solve or partially solve a class of decision problems.

We have discussed the first type of applications of Turing machines in Section 2.4 of Chapter 2. In this section, we shall mainly be concerned with the second type of applications of Turing machines.

3.1.1   Computation with Turing Machines

In the preceding chapter, we have seen that a Turing machine is an abstract computational model, which can be used as a language accepter. In this section, we shall see that a Turing machine can also be used as a transducer, which can simulate computation that can be performed on an ordinary digital computer.

Recall that a Turing machine M is defined as

$$M = (Q, \Sigma, \Gamma, \delta, q_0, \Box, F) \tag{3.1}$$

where

(i) Q is a finite set of *internal states,*

(ii) Σ is a finite set of symbols called the *input alphabet,* we assume that $\Sigma \subseteq \Gamma - \{\Box\}$.

(iii) Γ is a finite set of symbols called the *tape alphabet,*

(iv) δ is the transition function, which is defined by

    (a) if M is deterministic, then

$$\delta : Q \times \Gamma \to Q \times \Gamma \times \{L, R\}, \tag{3.2}$$

    (b) if M is non-deterministic, then

$$\delta : Q \times \Gamma \to 2^{Q \times \Gamma \times \{L,R\}}, \tag{3.3}$$

(v) $\Box \in \Gamma$ is a special symbol called the *blank,*

(vi) $q_0 \in Q$ is the *initial state,*

(vii) $F \subseteq Q$ is the set of *final states.*

The computation of a Turing machine is formalized by using the notion of an *instantaneous description:*

Definition 3.1.1 Let $M = (Q, \Sigma, \Gamma, \delta, q_0, \square, F)$ be a Turing machine. Then any string $a_1...a_{k-1}q_1a_ka_{k+1}...a_n$, with $a_i \in \Gamma$ and $q_1 \in Q$, is an *instantaneous description* (ID) of M. A move

$$a_1...a_{k-1}q_1a_ka_{k+1}...a_n \vdash a_1...a_{k-1}bq_2a_{k+1}...a_n \qquad (3.4)$$

is possible if

$$\delta(q_1, a_k) = (q_2, b, R). \qquad (3.5)$$

A move

$$a_1...a_{k-1}q_1a_ka_{k+1}...a_n \vdash a_1...q_2a_{k-1}ba_{k+1}...a_n \qquad (3.6)$$

is possible if

$$\delta(q_1, a_k) = (q_2, b, L). \qquad (3.7)$$

M is said to halt starting from some initial configuration $x_1q_ix_2$ if

$$x_1q_ix_2 \overset{*}{\vdash} y_1q_jay_2 \qquad (3.8)$$

for any q_j and a, for which $\delta(q_j, a)$ is undefined. The sequence of configurations leading to a halt state is called a computation.

If M will never halt, then we represent it by

$$x_1q_ix_2 \overset{*}{\vdash} \infty, \qquad (3.9)$$

indicating that, starting from the initial configuration $x_1q_ix_2$, the machine never halts.

So Turing machines provide us with the simplest possible abstract model for digital computers in general.

Example 3.1.1 Given two positive integers x and y, design a Turing machine that computes $x + y$.

We first have to choose some convention for representing natural numbers. There are three common representations for natural numbers:

N	Unary Form	1st Binary Form	2nd Binary Form
0	ϵ	0	ϵ
1	1	1	0
2	11	10	1
3	111	11	00
4	1111	100	01
5	11111	101	10
6	111111	110	11
7	1111111	111	000
$\vdots$	$\vdots$	$\vdots$	$\vdots$

For simplicity, we will use the *unary representation* for natural numbers, in which any positive integer x is represented by $w(x) \in \{1\}^*$, such that

$$|w(x)| = x.$$

We must also decide how x and y are placed on the tape initially and how their sum is to appear at the end of the computation. We will assume that $w(x)$ and $w(y)$ are on the tape in unary notation, separated by a single 0, with the read-write head on the leftmost symbol of $w(x)$. After the computation, $w(x + y)$ will be on the tape followed by a single 0, and the read-write head will be positioned at the left end of the result. We therefore want to design a Turing machine for performing the computation

$$q_0 w(x) 0 w(y) \overset{*}{\vdash} q_f w(x + y) 0$$

where q_f is a final state. Constructing a program for this is relatively simple. All we need to do is to move the separating 0 to the right end of $w(y)$, so that the addition amounts to nothing more than the coalescing of the two strings. To achieve this, we construct

$$M = (Q, \Sigma, \Gamma, \delta, q_0, \square, F),$$

with

$$
\begin{aligned}
Q &= \{q_0, q_1, q_2, q_3, q_4\}, \\
F &= \{q_4\}, \\
\delta(q_0, 1) &= (q_0, 1, R), \\
\delta(q_0, 0) &= (q_1, 1, R),
\end{aligned}
$$

$$\delta(q_1, 1) = (q_1, 1, R),$$
$$\delta(q_1, \square) = (q_2, \square, L),$$
$$\delta(q_2, 1) = (q_3, 0, L),$$
$$\delta(q_3, 1) = (q_3, 1, L),$$
$$\delta(q_3, \square) = (q_4, \square, R).$$

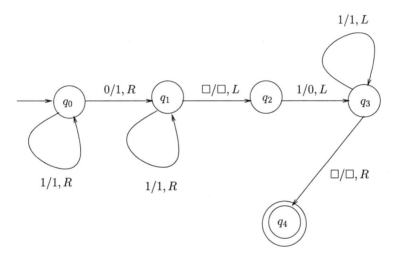

Note that in moving the 0 to the right we temporarily create an extra 1, a fact that is remembered by putting the machine into state q_1. The transition $\delta(q_2, 1) = (q_3, 0, R)$ is needed to remove this at the end of the computation. This can be seen from the sequence of instantaneous descriptions for adding 111 to 11:

$$q_0 1110011 \vdash 1q_0 11011$$
$$\vdash 11q_0 1011$$
$$\vdash 111q_0 011$$
$$\vdash 1111q_1 11$$
$$\vdash 11111q_1 1$$
$$\vdash 111111q_1$$
$$\vdash 11111q_2 1$$

$\vdash 1111q_310$

$\vdash 111q_3110$

$\vdash 11q_31110$

$\vdash 1q_311110$

$\vdash q_3111110$

$\vdash q_4111110.$

The above successive instantaneous descriptions can also be showed diagrammatically as follows:

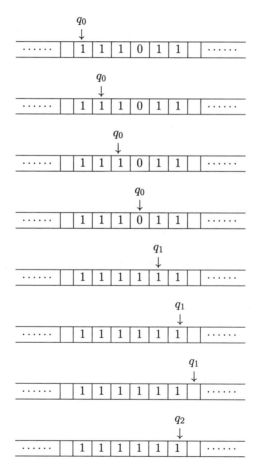

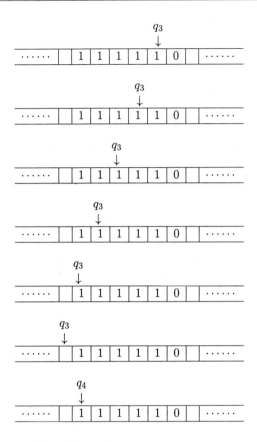

3.1.2 Computable Functions

In this subsection, we study what functions can be computed by Turing machines.

Definition 3.1.2 A function $f : \Sigma^* \to \Sigma^*$ is said to be *Turing-computable* or just *computable* if there exists some Turing machine $M = (Q, \Sigma, \Gamma, \delta, q_0, \Box, F)$, on every input w, which halts with just $f(w)$ on its tape.

Normally, we can just consider the computation of the number-theoretic functions for Turing machines:

$$f : \mathbb{N}^k \to \mathbb{N}, \quad k \geq 1.$$

This is actually not a unduly restriction, since, e.g., more general function from string to string, can be discussed by finding some suitable encoding method for both the arguments and the function values as numbers. A number-theoretic function is said to be *partial* from $\mathbb{N}^k$ to $\mathbb{N}$ if f is undefined at some points not in D, where D is the domain of f, some subset of $\mathbb{N}^k$. If f is defined at each point in $\mathbb{N}^k$, then it is a *total*. By function, we mean it *partial function*. If f is a total function, we will explicitly say that it is total.

Example 3.1.2 The Usual arithmetic operations on integers are computable functions; on input (m, n), we can design a Turing machine to compute $m \pm n$, $m \cdot n$ and $m \div n$.

It is claimed in the following so-called *Church-Turing Thesis*, or just *Church Thesis*, formulated by Church[1], that Turing machines can compute any effectively computable functions.

> **Church-Turing Thesis for number-theoretic functions**: A number-theoretic function is computable by an algorithm if and only if it is Turing computable. That is, a number-theoretic function is effectively computable if and only if it can be computed by a Turing machine.

Remark 3.1.1 It may be very difficult to actually construct a Turing machine to compute a particular number-theoretic function, nevertheless, such a Turing machine can always be found.

Remark 3.1.2 Church-Turing Thesis is a *thesis* not a *theorem*, because the solvability by an effective procedure is informal and imprecise, so it cannot be formally proved. Nevertheless, a tremendous amount of evidence has shown that Church-Turing Thesis is true. More importantly, Church-Turing Thesis is an essential tool in proofs of many results concerning undecidability and noncomputability.

---

[1]Alonzo Church (1903–1995) was born in Washington, D. C. He received his BA and PhD from Princeton University in 1924 and 1927, respectively. He spent three years at Harvard, Göttingen and Amsterdam as a National Research Fellow after his PhD, and then returned to Princeton in 1929 as a faculty member until 1967 when he moved to the University of California at Los Angeles (UCLA). He retired from teaching at UCLA in 1990 and died on 11 August 1995. His most recent paper was published in 1995. Among Church's 31 doctoral students were S. Kleene, A. Turing, M. Davis, M. Rabin, and D. Scott. Church made many substantial contributions to the theory of computability including his solution to the decision problem, his invention of lambda-calculus, and his statement known as the Church-Turing thesis.

Next, we turn to an important class of number-theoretic functions of the form called *recursive functions*, introduced by Kleene in 1936, which can be evaluated algorithmically, and which are equivalent to Turing computable functions. We will specifically show that Turing computable functions are partial recursive functions. We begin with the definition of the very basic primitive recursive functions:

Definition 3.1.3 The set of *basic primitive recursive functions* consists of

(i) The *zero function z*:

$$z(x) = 0, \quad \forall x \in \mathbb{Z}^+. \tag{3.10}$$

(ii) The *successor function s*:

$$s(x) = x + 1. \tag{3.11}$$

(iii) The *projection functions* $p_i^{(n)}$:

$$p_i^{(n)}(x_1, x_2, \cdots, x_n) = x_i, \quad 1 \le i \le n. \tag{3.12}$$

A *primitive recursive function* can be constructed from the basic primitive recursive functions by application of two operations that preserve computability:

(i) *Composition*:

$$
\begin{aligned}
f(x_1, x_2, \cdots, x_n) \;=\; & h(g_1(x_1, x_2, \cdots, x_n),\, g_2(x_1, x_2, \cdots, x_n), \\
& \cdots,\, g_k(x_1, x_2, \cdots, x_n)).
\end{aligned} \tag{3.13}
$$

(ii) *Primitive recursion*:

$$f(x_1, x_2, \cdots, x_n, 0) = g(x_1, x_2, \cdots, x_n), \tag{3.14}$$
$$f(x_1, x_2, \cdots, x_n, y + 1) = h(x_1, x_2, \cdots, x_n, y, f(x_1, x_2, \cdots, x_n, y)). \tag{3.15}$$

Definition 3.1.4 A function is *primitive recursive* if it can be obtained from the basic primitive recursive functions z, s and p_i by a finite number of applications of composition and primitive recursion.

Example 3.1.3 Addition of integers can be implemented with the function $add(x, y)$ defined by

$$\text{add}(x,0) = x$$
$$\text{add}(x,y+1) = \text{add}(x,y) + 1.$$

Example 3.1.4 Multiplication of integers can be implemented with the function $\text{mult}(x,y)$ defined by

$$\text{mult}(x,0) = 0$$
$$\text{mult}(x,y+1) = \text{add}(x,\text{mult}(x,y)).$$

Theorem 3.1.1 Every primitive recursive function is Turing computable.

It was once conjectured that the class of primitive recursive functions contains all computable total number-theoretic functions. However it is not true, since for example, the Ackermann's function, proposed by W. Ackermann in 1928, $A : \mathbb{N}^2 \to \mathbb{N}$ defined by

$$A(0,y) = y + 1$$
$$A(x+1,0) = A(x+1)$$
$$A(x+1,y+1) = A(x,A(x+1,y))$$

is computable and total, but not primitive recursive.

Theorem 3.1.2 There is a computable total number-theoretic function from $\mathbb{N}^2 \to \mathbb{N}$ that is not primitive recursive.

Corollary 3.1.1 The set of primitive recursive functions is a proper set of the set of effectively computable total number-theoretic functions.

We have just studied some classes of computable functions, namely basic primitive recursive functions, primitive recursive functions, computable total functions, all of them are *total*. We now extend our study to include the computable partial functions, i.e., partial μ-recursive functions, or just partial recursive functions. We begin with a introduction to a third rule for constructing recursive functions:

Definition 3.1.5 (Minimalization) Let $g : \mathbb{N}^{n+1} \to \mathbb{N}$ be a total function, not necessarily primitive recursive. Define the function $g : \mathbb{N}^n \to \mathbb{N}$ so that

$$f(x_1, x_2, \cdots, x_n) = \mu y\big[g(x_1, x_2, \cdots, x_n, y) = 0\big] \qquad (3.16)$$

is the least integer $y \geq 0$ for which $g(x_1, x_2, \cdots, x_n, y) = 0$. Then f is obtained from g by *Minimalization*.

The class of *partial μ-recursive functions* or just *partial recursive functions* is the class of partial functions that can be constructed from the basic primitive recursive functions by applying a finite number of compositions, primitive recursions, and Minimalization. A partial function that is computed by some Turing machine is said to be *Turing-computable*. It is clear that not every partial function is Turing-computable. But the partial recursive functions which are the proper subset of partial functions are Turing-computable.

Definition 3.1.6 The class of μ-recursive functions is defined as follows:

(i) The zero, successor, and projection functions are μ-recursive.

(ii) If h is an n-variable μ-recursive function and $g_1, g_2, \cdots, g_n$ are k-variable μ-recursive functions, then $f = h \circ (g_1, g_2, \cdots, g_n)$ is μ-recursive.

(iii) If g and h are n and $n+2$-variable μ-recursive functions, then the functions defined from g and h by primitive recursion is μ-recursive.

(iv) If $p(x_1, x_2, \cdots, x_n, y)$ is a total μ-recursive predicate, then $f = \mu(p(x_1, x_2, \cdots, x_n, y))$ is μ-recursive.

(v) A function is μ-recursive only if it can be obtained from (i) by a finite number of applications of the rules in (ii), (iii), and (iv).

Theorem 3.1.3 Every μ-recursive function is Turing-computable.

Theorem 3.1.4 Every Turing-computable function $f : \mathbb{N}^n \to \mathbb{N}$ is μ-recursive.

Corollary 3.1.2 A function is Turing computable if and only if it is μ-recursive.

Theorem 3.1.5 Every partial recursive function is Turing-computable.

Theorem 3.1.6 Any computational process performed by a Turing machine is actually the process of computing a partial recursive function.

Church-Turing's Thesis for μ-recursive functions: A partial function is computable if and only if it is μ-recursive.

Finally, we present in Figure 3.1 a hierarchy of various functions in terms of the recursion function theory. Functions from basic primitive recursive up to partial recursive are Turing computable; there are however functions that are not Turing computable.

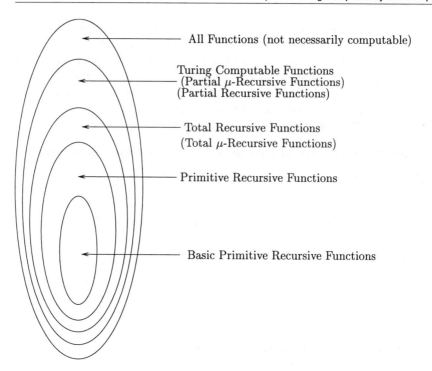

All Functions (not necessarily computable)

Turing Computable Functions
(Partial μ-Recursive Functions)
(Partial Recursive Functions)

Total Recursive Functions
(Total μ-Recursive Functions)

Primitive Recursive Functions

Basic Primitive Recursive Functions

Figure 3.1: Hierarchy of Functions

Remark 3.1.3 It is interesting to note that from a *learnability* (the analogue of computability) point of view [48], basic primitive recursive functions and primitive recursive functions are (Turing) learnable, whereas the total recursive functions are not in general (Turing) learnable. It is also interesting to note that learnability can also be defined in terms of *Kolmogorov complexity*[2], interested

---

[2]Let M be a Turing machine, we say that p is the *description* of x, if on input p, M outputs x. Write $M(p, y) = x$ to mean that the Turing machine M with input p and y terminates with output x, where y is some extra information to help generate x. The *Kolmogorov complexity*, denoted by $K_M(x \mid y)$, is the length of the minimal description of x. That is,

$$K_M(x \mid y) = \min\{|p| : p \in \{0, 1\}^*, M(p, y) = x\} \qquad (3.17)$$
$$= \infty, \text{ if no such a } p \text{ exists.}$$

It was named after the Russian mathematician Andrei Nikolaevich Kolmogorov (1903–1987), who was the first to study the theory of descriptive complexity in the 1960's. Kolmogorov

readers are referred to [80].

3.1.3 Noncomputable Functions

We have just shown that a number-theoretic function is Turing computable only if there is a Turing machine that computes it. However, there exist some number-theoretic functions that are noncomputable by a Turing machine. The existence of number-theoretic functions that are *not* Turing computable can be shown by a simple counting argument.

Definition 3.1.7 A set S is *countably infinite* if there is a bijection from $\mathbb{N}$ to S, and *countable* if it is either finite or countably infinite. A set S is *uncountably infinite* or simply *uncountable* if it is not countable.

Theorem 3.1.7 If the set S is a finite set, then the set S^* is countable.

Theorem 3.1.8 Let $S_0, S_1, \cdots, S_n$ be countable sets. Then

(i) The union of these countable sets are countable, that is,

$$S_0 \cup S_1 \cup \cdots \cup S_n = \bigcup_{i=0}^{n} S_i \qquad (3.18)$$

is countable.

(ii) The Cartesian product of these countable sets are countable, that is,

$$S_0 \times S_1 \times \cdots \times S_n = \prod_{i=0}^{n} S_i \qquad (3.19)$$

is countable.

(iii) The set of finite subsets of a countable set is countable.

(iv) The set of finite-length sequences consisting of elements of a nonempty countable set is countably infinite.

---

also made significant contributions to probability theory by giving the subject a rigorous foundation using the language and notation of set theory. In computer science, Kolmogorov complexity is also known as algorithmic information theory, Kolmogorov-Chaitin randomness, Solomonoff-Kolmogorov-Chaitin complexity, descriptive complexity, algorithmic complexity, and program-size complexity, etc.

Theorem 3.1.9 The set of computable number-theoretic functions is countably infinite.

But there exist uncountably many total unary number-theoretic functions, so,

Theorem 3.1.10 There exists a total unary number-theoretic function that is not computable.

Thus, computable functions comprise a countable set, whereas noncomputable functions an uncountable set. The above result actually shows that there exists noncomputable functions that cannot be computed by any Turing machine. Now we are in a position to present a *real* noncomputable function, the so-called busy-beaver function, which was first studied by Tibor Rado in 1962.

Definition 3.1.8 Let $\mathbb{B}$ be the class of Turing machines with $\Sigma = \{0, 1\}$. For $n \geq 0$, consider $\mathbb{B}_n$ be the class of n-states Turing machines in $\mathbb{B}$ with the following restrictions:

(i) the tape is infinite in both directions,

(ii) the tape alphabet is $\Sigma = \{0, 1\} = \{\text{blank}, \text{mark}\}$,

(iii) no cell in the tape is empty, initially the tape is blank, i.e., containing all 0's,

(iv) an act consisting of printing one symbol and making one move to an adjacent cell before changing to the next state,

(v) the halt state is not included in the n states of the machine, i.e., we need to add an external "zero" state to denote the act of halting.

The *busy-beaver function* $\Sigma(n)$

$$\Sigma : \mathbb{N} \to \mathbb{N} \tag{3.20}$$

is defined to be the maximum numbers of (ones) on the final tape of any member of $\mathbb{B}_n$ that halts when started on blank tape. That is,

$$\Sigma(n) = \max \{\text{MARKS(TM)} : \text{ TM} \in \mathbb{B}_n\}. \tag{3.21}$$

The machine with n-states, started on a blank tape and halted with $\Sigma(n)$ marks on its tape is called the n-state *busy-beaver*. The problem *what is the largest numbers of 1's it may leave on the tape when it halts* is called the *busy-beaver problem*. A related function is the *shift function* $s(n)$:

$$S : \mathbb{N} \to \mathbb{N} \qquad (3.22)$$

it is the maximum number of *shifts* an n-state Turing machine, started on a blank tape, may make before halting, That is,

$$S(n) = \max \{ \text{SHIFTS(TM)} : \text{TM} \in \mathbb{B}_n \}. \qquad (3.23)$$

(The move to the halt state is also counted).

The function $\Sigma(n)$ is called the busy-beaver function since, for a given n, it indicates how many symbols an n-state Turing machine can possibly print, that is, how "busy" it can be, before halting for blank tape. It is clear that $\Sigma(n_1) \geq \Sigma(n_2)$ if $n_1 \geq n_2 \geq 0$. That is, the busy-beaver function is monotonic increasing. In fact, the busy-beaver function is a very rapidly increasing function, with some of the known values as follows:

n	$\Sigma(n)$
0	0
1	1
2	4
3	6
4	13
5	> 4098
7	> 20000
8	> 10^{44}

It can be seen from the above list that with 7 states it is sufficient to perform exponentiation. With more than 7 states, $\Sigma(n)$ is not computable because its value grows faster than any computable function. So we have the following important noncomputable result:

Theorem 3.1.11 The busy-beaver function $\Sigma : \mathbb{N} \to \mathbb{N}$ is noncomputable.

PROOF. To prove the theorem, it is sufficient to prove that for some computable function $f(n)$, the inequality

$$\Sigma(n) > f(n) \tag{3.24}$$

is satisfied for all sufficiently large n. Hence, the function $\Sigma(n)$ is not recursive (i.e., $\Sigma(n)$ is noncomputable).

Given a computable function $f(n)$, we consider function $g(n)$ defined by

$$g(n) = \max\{f(2n+2),\ f(2n+3)\}. \tag{3.25}$$

Clearly, $g(n)$ is computable by a Turing machine TM with say m states.

For $n = 0, 1, 2, \cdots$, we construct a Turing machine TM_n such that

(i) TM_n first prints $n+1$ marks when started on a blank tape,

(ii) then TM_n simulates TM, initially scanning the leftmost 4.

Clearly, $n = m+x+1$ is the adequate number of states for TM_n. When started on a blank tape, TM_n halts with at least $g(n)+1$ marks on the tape. So by the definition of Σ and g, we have the following inequalities:

$$\begin{cases} \Sigma(m+n+2) > f(2x+2) \\ \Sigma(m+n+2) > f(2x+3) \end{cases} \tag{3.26}$$

Since Σ is monotonic, then assume $k \geq m$, we have

$$\begin{cases} \Sigma(2k+2) > f(2k+2) \\ \Sigma(2k+3) > f(2x+3) \end{cases} \tag{3.27}$$

This implies that $\Sigma(n) > f(n)$ holds whenever $n \geq 2m+2$. Thus, $\Sigma(n)$ is *noncomputable*. ∎

Exercise 3.1.1 Use proof by contradiction to prove that it is impossible to construct an n-state Turing machine that computes $\Sigma(n)$.

Exercise 3.1.2 Show that the following two statements are equivalent:

(i) $\Sigma(n)$ is noncomputable.

(ii) $S(n)$ is noncomputable.

3.1.4 Horn Clause Computability

In subsection 1.1.3 of Chapter 1, we have introduced some basic concepts of logic clauses and some basic ideas of programming in Horn clauses. In this subsection we shall study the computability of Horn clauses. Recall that a Horn clause is of one of the following three forms:

$A \vee \neg B_1 \vee \neg B_2 \vee \cdots \vee \neg B_n$ (one positive and some negative literals)

A (one positive and no negative literals)

$\neg B_1 \vee \neg B_2 \vee \cdots \vee \neg B_n$ (no positive and all negative literals).

Definition 3.1.9 A *binary Horn clause* is a Horn clause with at most one negative literal. That is, a binary Horn clause is of one of the following three forms:

$$A \vee \neg B, \qquad\qquad A, \qquad\qquad \neg B. \qquad\qquad (3.28)$$

Remark 3.1.4 Just as Horn clauses are a special type of general logic clauses, a binary Horn clause is a special kind of Horn clause.

We first present some axiomatic definitions for Turing machines:

Definition 3.1.10 Let M be a Turing machine TM, and the finite set of quintuples $\langle q, s, q', s', d \rangle$ be a description of M, where q and s are the present state and accepted symbol of M, q' and s' are the next state and symbol, and $d = \langle H, L, R \rangle$ the direction (i.e., halt, left, right) of the head of M, respectively. Then the *Turing machine relation* may be defined as follows:

$\text{machine}(x) \iff x = \emptyset \vee$
$\qquad [x = m(q, s, q', s', d, z) \wedge \text{state}(q) \wedge \text{symbol}(s) \wedge$
$\qquad\qquad \text{next-state}(q') \wedge \text{next-symbol}(s') \wedge \text{direction}(d) \wedge \text{machine}(z)].$

Definition 3.1.11 Let U be a universal Turing machine UTM that simulates the behaviour of a Turing machine M; the memory of a UTM is an infinite tape which has a head that points to a *cell* for reading and writing a symbol. Then the *tape relation* of U may be defined as follows:

$\text{tape}(t(u, v, w)) \iff$
$\qquad [u = l(x', x) \wedge \text{tape}(t(x', x, r(s, w))) \vee w = r(z, z') \wedge \text{tape}(t(l(u, s), z, z'))]$
$\qquad\qquad \wedge \text{head-cell}(v) \wedge \text{left-tape}(l(x', x)) \wedge \text{right-tape}(r(z, z')) \wedge \text{symbol}(s)$
$\text{left-tape}(l(x', x)) \iff \text{symbol}(s) \wedge \text{left-tape}(x')$
$\text{right-tape}(r(z, z')) \iff \text{symbol}(z) \wedge \text{right-tape}(z').$

Definition 3.1.12 Suppose a universal Turing machine operates in two phases: *locate* and *copy*. Then *universal Turing machine relation* is defined as follows:

$$\text{locate}(t, q, s, m(q'', s'', q', s', d, z), m) \iff \big[q'' = q \ \wedge \ s'' = s \ \wedge \ q' = H\big] \ \vee$$
$$\big[q'' = q \ \wedge \ s'' = s \ \wedge \ q' \neq H \ \wedge \ \text{copy}(t, q', s', d, m)\big] \ \vee$$
$$\big[q'' \neq q \ \wedge \ s'' \neq s \ \wedge \ \text{locate}(t, q, s, z, m)\big]$$
$$\text{copy}(t(u, v, w), q, s, d, m) \iff$$
$$\big[d = R \wedge w = r(z, z') \ \wedge \ \text{locate}(t(l(u, s), z, z'), q, z, m, m)\big] \ \vee$$
$$\big[d = L \wedge u = r(x', x) \ \wedge \ \text{locate}(t(x', x, r(s, w)), q, x, m, m)\big].$$

Note that the UTM tape is divided into five parts: a tape, a state, a symbol, and two Turing machine descriptions (one of them is an auxiliary description). There are two cases for *locate*: (i) the present q and s are found in the present quintuple of M, the next state is either H which makes M halt or q' with new symbol s' both of them are in the *copy relation* $\text{copy}(t, q', s', d, m)$ where t is the tape, d the direction of the head and m the Turing machine, (ii) the present q and s are not found in the present quintuple, they are in the recursive *locate* relation $\text{locate}(t, , q, s, z, m)$ where t is the tape, z is the tail of a TM description and m an auxiliary description of the TM. There are also two cases for *copy*: when $d = R$ the head moves one cell to the right, whereas $d = L$ to the left. Let the tape to the right of the head be $w = r(z, z')$, then the UTM writes s at the head cell and reads z, both symbols are in relation to $\text{locate}(t(l(u, s), z, z'), q, z, m, m)$ where the term $l(u, s)$ is the tape to the left of the next head cell z, z' is the tape to the right of the next head cell, q is the present state, z the new accepted symbol and m the Turing machine, the second m is an auxiliary description.

Now we are in a position to present the computability result of Horn clauses.

Theorem 3.1.12 (Tärnlund, 1977) A function f is computable in binary Horn clauses if it is computable by a Turing machine M.

PROOF. As pointed out in [139], it will suffice to just use the "if part" of the definition of the UTM relation, the reflexivity $x = x$ for equality and the definition of $\vee$ to show that there exists a UTM in binary Horn clauses. We have:

locate$(t, q, s, m(q, s, H, s', d, z), m)$
locate$(t, q, s, m(q, s, q', s', d, z), m) \lor \neg\text{copy}(t, q', s', d, m)$
locate$(t, q, s, m(q'', s'', q', s', d, z), m) \lor \neg\text{locate}(t, q, s, z, m)$
copy$(t(u, v, r(z, z')), q, s, R, m) \lor \neg\text{locate}(t(l(u, s), z, z'), q, z, m, m)$
copy$(t(l(x', x), v, w), q, s, L, m) \lor \neg\text{locate}(t(x', x, r(s, w))), q, x, m, m).$ ∎

Note first that Turing employed the standard logic formulas (i.e., non-clausal formulas) to specify Turing machines [141], whereas here we just use binary Horn clauses. Note also that it is possible to reduce the auxiliary functions in the binary Horn clauses without losing the computation power. That is:

Theorem 3.1.13 A Turing computable function f is computable in binary Horn clauses using at most one function symbol.

The proof of this theorem is left as an exercise. There are some other ways to establish this result; the following is one of these ways, due to J. Sebelik and P. Stepanek, by showing that every partial recursive function can be computed by a Horn clause program [85]:

Definition 3.1.13 Let P be a logic program and G a goal, all based on Horn clauses. A *computed answer* θ for $P \cup \{G\}$ is the substitution obtained by restricting the composition $\theta_1\theta_2\cdots\theta_n$ to the variables of G, where $\theta_1\theta_2\cdots\theta_n$ is the sequence of mgu's used in an SLD-refutation of $P \cup \{G\}$.

Theorem 3.1.14 (Computational Adequacy of Horn Clauses) Let f be an n-ary partial recursive function. Then there exists a Horn clause program P_f such that all computed answers for

$$P_f \cup \left\{ \leftarrow p_f\left(s^{k_1}(0), s^{k_2}(0), \cdots, s^{k_n}(0), x \right) \right\} \qquad (3.29)$$

have the form $\{x/s^k(0)\}$ and for nonnegative integers $k_1, k_2, \cdots, k_n$ and k, we have $f(k_1, k_2, \cdots, k_n) = k$ if and only if $s^k(0)$ is a computed answer for (3.29).

The proof of the above theorem is done by induction on the number k of applications of *composition, primitive recursion* and *Minimalization* needed to define f. For example, if $k = 0$, then f must be either the *zero* function, the *successor* function, or a *projection* function:

(i) *zero* function: suppose that f is the zero function defined by $f(x) = 0$; define P_f to be the set of the program clauses of the form $p_f(x, 0) \leftarrow$.

(ii) *successor* function: suppose that f is the successor function defined by $f(x) = x + 1$; define P_f to be the set of the program clauses of the form $p_f(x, s(x)) \leftarrow$.

(iii) *projection* function: suppose that f is projection function defined by $f(x_1, x_2, \cdots, x_n) = x_j$, where $1 \leq j \leq n$; define P_f to be the set of the program clauses of the form $p_f(x_1, x_2, \cdots, x_n, x_j) \leftarrow$.

It is clear that for each of the basic functions, the program P_f defined has the desired properties. Now assume that the partial recursive function f is defined by $k = m$ applications of composition, primitive recursion and Minimalization, we will then need to show that when $k = m + 1$, for each of the functions (i.e., composition, primitive recursion and Minimalization), P_f has the desired properties; the detailed proof is left as an exercise (interested readers could also consult [85] for a full proof of the theorem).

3.2   Decidability and Undecidability

3.2.1   Decision Problems

In the previous section, we studied Turing computability for number-theoretic functions, where Turing machines are used as a function transducer. In this section, we shall study Turing decidability for decision problems, where Turing machines are used as a problem-solver or an algorithm.

Generally speaking, a *decision problem* is a set of questions whose answers are either "yes" or "no". A solution to a decision problem is an effective procedure that determines the answers for each question in the set. More formally, we have:

Definition 3.2.1 An abstract *problem* P is a binary relation on a set I of problem *instances* and a set S of problem *solutions*:

$$\text{P} : I \to S. \tag{3.30}$$

Example 3.2.1 Consider the problem SHORTEST-PATH of finding a shortest path of two vertices in a undirected graph $G = (V, E)$. An instance of SHORTEST-PATH is a triple (G, u, v), where $u, v \in V$. A solution is a sequence (with possible length zero) of vertices in the graph. The problem itself is a relation because it relates each instance of a graph and two vertices with a shortest path in the graph that connects the two vertices. Since shortest paths are not necessarily unique, a given problem instance may have more than one solution.

Definition 3.2.2 A *decision problem* is a function in which the only possible answers are "yes" and "no". That is, a decision problem DP is a function that maps the instance set I to solution set $\{0, 1\}$:

$$DP : I \to \{0, 1\}. \tag{3.31}$$

Example 3.2.2 Consider the decision problem PATH related to the shortest-path problem:

$$\begin{cases} \text{Input} & i = (G, u, v, k), \text{ where } G = (V, E), u, v \in V, k \in \mathbb{N} \\ \text{Output} & \begin{cases} \text{PATH}(i) = 1, & \text{if a shortest path from u to v has} \\ & \quad\quad\quad\quad \text{length at most k} \\ \text{PATH}(i) = 0, & \text{otherwise.} \end{cases} \end{cases}$$

Definition 3.2.3 Let the alphabet $\Sigma = \{0, 1\}$. A *language* over the alphabet Σ is a set of strings whose symbols are chosen from Σ. That is, a language is any subset of $\{0, 1\}^*$.

Example 3.2.3 Let the alphabet $\Sigma = \{0, 1\}$. Then

$$L_{\text{Primes}} = \{10, 11, 101, 111, 1011, 1101, 10001, 10011, 10111, 11101,$$
$$11111, 100101, 101001, 101011, 101111, 110101, 111011, \cdots\}$$

is the language of binary representations of prime numbers, and

$$L_{\text{Composite}} = \{100, 110, 1000, 1001, 1010, 1100, 1110, 1111,$$
$$10000, 10010, 10100, 10101, 10110, 11000, 11001, \cdots\}$$

is the language of binary representations of composite numbers.

Definition 3.2.4 Let L be a language. Then

(i) the decision problem D_L corresponding to L is

$$D_L = \{(x, \text{yes}) : x \in L\} \cup \{(x, \text{no}) : x \notin L\}. \tag{3.32}$$

(ii) the complementary language of L is defined by

$$\text{co-}L = \{0,1\}^* - L. \tag{3.33}$$

Example 3.2.4 The decision problem $D_{L_{\text{Primes}}}$ corresponding to L_{Primes} defined above is

$$D_{L_{\text{Primes}}} = \{(x, \text{yes}) : \ x \in L_{\text{Primes}}\} \cup \{(x, \text{no}) : \ x \notin L_{\text{Primes}}\}.$$

Thus,

$$D_{L_{\text{Primes}}} = (100101, \text{yes})$$

$$D_{L_{\text{Primes}}} = (10101, \text{no}).$$

Example 3.2.5 By the definition of the complementary language of L, we have

$$\text{co-}L_{\text{Primes}} = \{0,1\}^* - \{0,1\} - L_{\text{Composite}}$$

$$\text{co-}L_{\text{Composite}} = \{0,1\}^* - \{0,1\} - L_{\text{Primes}}.$$

The reason for not including 0 and 1 in both cases is that the numbers 0 and 1 are neither prime, nor composite.

A decision problem is said to be *decidable*[3] if there is an algorithm (i.e., a Turing) that solves the problem, or otherwise, it is said to be *undecidable*. By Church-Turing's thesis, a Turing machine can be considered to be the most general possible computing device. That is, a Turing machine can be used to solve any decision problem that is solvable by any effective procedure.

---

[3]Instead of saying "decidable", one can also speak of "solvable". Moreover, the additional attributes "effectively", "recursively", or "algorithmically" do not alter the meaning. That is, "effectively decidable", "recursively decidable", "algorithmically decidable" are all synonymous. Similarly, one can freely use any one of the following synonymous terms for "undecidable": unsolvable, recursively undecidable, recursively unsolvable, effectively undecidable, effectively unsolvable, algorithmically undecidable, algorithmically unsolvable. But of course, the "decidable" is Turing-decidable, and the "undecidable" is Turing-undecidable, because the definition of all the terms are based on Turing machines.

3.2.2   Decidable Problems

A decision problem may or may not be decidable. In this subsection, we study some decidable problems, and in the next two subsections, we shall study some undecidable problems.

Let us consider a simple decision problem P_{SQF} of determining whether $p_n = 2^n - 1, \forall n \in \mathbb{N}$ is square-free (an integer n is square-free, denoted by $n \in SQF$, if it is not divisible by a perfect square):

$$P_{SQF} = \{(p_n, \text{ yes}) : p_n \in SQF) \cup (p_n, \text{ no}) : p_n \notin SQF)\}$$

$$(3.34)$$

The problem has of course infinitely many instances as follows:

Input	$(p_n = 2^n - 1, \quad n \in \mathbb{Z}^+)$	Output
$p_1 = 1$		yes
$p_2 = 3$		yes
$p_3 = 7$		yes
$p_4 = 15 = 3 \cdot 5$		yes
$p_5 = 31$		yes
$p_6 = 63 = \mathbf{3^2} \cdot 7$		no
$p_7 = 127$		yes
$p_8 = 255 = 3 \cdot 5 \cdot 17$		yes
$p_9 = 511 = 7 \cdot 73$		yes
$p_{10} = 1023 = 3 \cdot 11 \cdot 31$		yes
$p_{11} = 2047 = 23 \cdot 89$		yes
$p_{12} = 4095 = \mathbf{3^2} \cdot 5 \cdot 7 \cdot 13$		no
$p_{13} = 8191 = 8191$		yes
$p_{14} = 16383 = 3 \cdot 43 \cdot 127$		yes
$p_{15} = 32767 = 7 \cdot 31 \cdot 151$		yes
$p_{16} = 65535 = 3 \cdot 5 \cdot 17 \cdot 257$		yes
$p_{17} = 131071 = 131071$		yes
$p_{18} = 262143 = \mathbf{3^3} \cdot 7 \cdot 19 \cdot 73$		no
$p_{19} = 524287$		yes
$p_{20} = 1048575 = 3 \cdot \mathbf{5^2} \cdot 11 \cdot 31 \cdot 41$		no
$\ldots\ldots$		$\ldots$
$\ldots\ldots$		$\ldots$

A solution to a decision problem P is an algorithm that determines the appropriate answer to every question $p_n \in P$. It is clear that The problem P_{SQF} is decidable.

If we change the input $p_n = 2^n - 1$ to $p_{\text{ithprime(i)}} = 2^{\text{ithprime(i)}} - 1$ where $i \in \mathbb{N}$, e.g., the first prime is 2, the second is 3, the third is 5 and so on (numbers in this form are called Mersenne numbers [147]), then we have

Input	$p_{\text{ithprime(i)}} = 2^{\text{ithprime(i)}} - 1, \quad i \in \mathbb{N}$	Output
$p_1 = 3$		yes
$p_2 = 7$		yes
$p_3 = 31$		yes
$p_4 = 127$		yes
$p_5 = 2047 = 23 \cdot 89$		yes
$p_6 = 8191$		yes
$p_7 = 131071$		yes
$p_8 = 524287$		yes
$p_9 = 8388607 = 47 \cdot 178481$		yes
$p_{10} = 536870911 = 233 \cdot 1103 \cdot 2089$		yes
$p_{11} = 2147483647$		yes
$p_{12} = 137438953471 = 223 \cdot 616318177$		yes
$p_{13} = 2199023255551 = 164511353 \cdot 13367$		yes
$p_{14} = 8796093022207 = 431 \cdot 2099863 \cdot 9719$		yes
$p_{15} = 140737488355327 = 13264529 \cdot 2351 \cdot 4513$		yes
$p_{16} = 9007199254740991 = 69431 \cdot 20394401 \cdot 6361$		yes
$p_{17} = 576460752303423487 = 3203431780337 \cdot 179951$		yes
$p_{18} = 2305843009213693951$		yes
$p_{19} = 147573952589676412927 = 761838257287 \cdot 193707721$		yes
$p_{20} = 2361183241434822606847 = 212885833 \cdot 48544121 \cdot 228479$		yes
$\cdots\cdots$		$\cdots$
$\cdots\cdots$		$\cdots$

This problem again is decidable, although at present we do not know whether all numbers in this form are square-free (this is an important open problem in number theory).

The Church-Turing Thesis for number-theoretic functions can be extended to decision problems:

> **Church-Turing Thesis for decision problems**: There is an effective procedure to solve a decision problem if and only if there is a Turing machine accepting a recursive language that solves the problem.

Thus, as we can see, computability and decidability are actually the *same* concept with respect to Turing machines.

3.2.3 Decidable Problems in Formal Languages

In Chapter 2, we have studied regular languages L_{REG}, context-free languages L_{CF}, context-sensitive languages L_{CS}, recursive languages L_{REC}, recursively enumerable languages L_{RE} and their accepting machines. Now we present some decidability results for these classes of languages.

Definition 3.2.5 The set of languages accepted by a machine M with n states, denoted by $L(M)$, is

(i) *nonempty*, if $L(M) \neq \emptyset$.

(ii) *infinite*, if $|L(M)| = l$, where $n \leq l < 2n$.

Two machines M_1 and M_2 are *equivalent* if they accept the same language, that is, $L(M_1) = L(M_2)$.

Note that the machine M in the above definition can be a FA, a PDA, a LBA or a TM, depending on the accepting languages.

(I) Decidability in L_{REG}

Theorem 3.2.1 Let $L(M)$ be the language accepted by a DFA, $M = (Q, \Sigma, \delta, q_0, F)$, i.e.,

$$L(M) = \{(M, w) : M \text{ is a DFA that accepts input string } w \in \Sigma\}.$$

Then it is decidable whether or not

(i) $L(M) = \emptyset$.

(ii) $L(M)$ is finite.

(iii) $L(M)$ is infinite.

Theorem 3.2.2 Let $L(M_1)$ and $L(M_2)$ be the languages accepted by a DFA, $M_1 = (Q_1, \Sigma, \delta_1, q_{0_1}, F_1)$, and a DFA $M_2 = (Q_2, \Sigma, \delta_2, q_{0_2}, F_2)$. Then it is decidable whether or not $L(M_1) = L(M_2)$.

Theorem 3.2.3 It is decidable whether or not two given regular grammars $G_1 = (V_1, T, S_1, P_1)$ and $G = (V_2, T, S_2, P_2)$ are equivalent. That is, it is decidable whether or not $L(G_1) = L(G_2)$.

(II) Decidability in L_{CF}

Theorem 3.2.4 Given any push-down automaton PDA, M, it is decidable whether or not

(i) $L(M) = \emptyset$.

(ii) $L(M)$ is finite.

(iii) $L(M)$ is infinite.

Theorem 3.2.5 Given any context-free grammar G_{CF}, it is decidable by a TM whether or not

(i) $L(G) = \emptyset$.

(ii) $L(G)$ is finite.

(iii) $L(G)$ is infinite.

Theorem 3.2.6 Given a context-free grammar G_{CF} and a string w in the same alphabet, it is decidable whether or not $w \in L(G_{\text{CF}})$.

(III) Decidability in L_{REC}

Theorem 3.2.7 Given a context-sensitive grammar G_{CS} and a string w in the same alphabet, it is decidable by a TM whether or not $w \in L(G_{\text{CS}})$.

In fact, every recursive language L_{REC} (including L_{REG}, L_{CF}, and L_{CS}, by definition) is decidable (see Figure 3.2). But unfortunately, this cannot be carried out any further to recursively enumerable languages L_{RE}, because L_{RE} is undecidable. Remarkably enough, there are also some very simple undecidable problems related to context-free grammars. We shall study some undecidable problems in the next subsection.

3.2.4 Undecidable Problems

Just as there are noncomputable functions, many decision problems are undecidable. Certainly, the best known of all undecidable problems is concerned with the Turing machine itself:

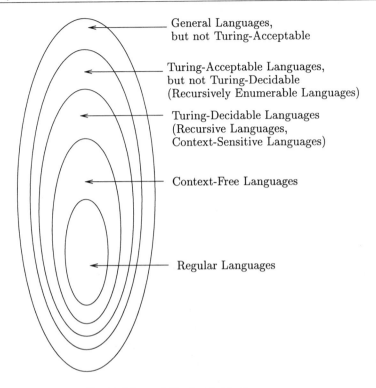

General Languages,
but not Turing-Acceptable

Turing-Acceptable Languages,
but not Turing-Decidable
(Recursively Enumerable Languages)

Turing-Decidable Languages
(Recursive Languages,
Context-Sensitive Languages)

Context-Free Languages

Regular Languages

Figure 3.2: Decidable and Undecidable Languages

Definition 3.2.6 The halting problem for Turing machines: Given a Turing machine M and a string w, does M halt on input w?

Theorem 3.2.8 The Halting Problem (HP) for Turing machines is undecidable.

PROOF. To prove the theorem, we first note that we can express halting by saying that a Turing Machine $TM_i(w)$ (or $f_i(w)$ if we talk about functions) *converges*, as opposed to that it *diverges*. So, the halting problem (HP) is decidable if and only of the set

$$\{\phi(i, w) : \ TM_i(w) \text{ converges}\}$$

is recursive, or alternatively, HP is decidable if and only if the set

$$g(\phi(i, w)) = \begin{cases} 1, & \text{if } TM_i(w) \text{ converges} \\ 0, & \text{if } TM_i(w) \text{ diverges} \end{cases}$$

is not recursive. The proof is by contradiction. Assume g is recursive. Then the function $g_1(i, w) = g(\phi(i, w))$ is recursive. Moreover, the function g_2 defined by

$$g_2(x) = \begin{cases} 0, & \text{if } g_1(x, x) = 0, \\ \text{undefined}, & \text{otherwise} \end{cases}$$

is partial recursive. Now let j be the index for g_2. We ask whether $TM_j(j)$ converges or diverges. Assume firstly that $TM_j(j)$ converges. Then

$$g(\phi(j, j)) = g_1(j, j) = 1$$

and, hence, $g_2(j)$ is undefined. Hence, $TM_j(j)$ diverges. Assume secondly that $TM_j(j)$ diverges. Then

$$g(\phi(j, j)) = g_1(j, j) = 0$$

and, hence $TM_j(j)$ converges. Thus, a contradiction occurs in both cases. This means that g is not recursive, and hence, HP is not decidable. This completes the proof of the above theorem. ∎

The following exercise will show us that noncomputable functions and the undecidable problems are actually equivalent.

Exercise 3.2.1 Show that the following statements are equivalent:

(i) The busy-beaver function $\Sigma(n)$ is noncomputable.

(ii) The halting problem on an initially blank tape is undecidable.

The halting problem has important implications in both theoretical and practical computer science, e.g.,

(i) if the halting problem were decidable, then every recursively enumerable language would be recursive.

(ii) if the halting problem were decidable, then it would be possible to prove the correctness of any computer program.

Unfortunately, the halting problem is undecidable, so there are recursively enumerable languages, as we already know, that are not recursive, and also to prove the correctness of programs in their general forms (the dream of software engineers for many years) is impossible! It should be noted that it is possible to solve the halting problem for some specific cases, for example, it is possible to prove the correctness for some specific programs (this is exactly the subject matter of mechanical theorem proving, that many computer scientists are currently working on).

The second most famous undecidable problem is the Post Correspondence Problem (PCP), proposed by Emil Post in 1946:

Definition 3.2.7 *The Post Correspondence Problem (PCP):* Given two sequences of n strings over an alphabet Σ, say

$$A = \langle\langle x_1, x_2, \cdots, x_n \rangle\rangle$$
$$B = \langle\langle y_1, y_2, \cdots, y_n \rangle\rangle. \tag{3.35}$$

does there exist a nonempty sequence $\langle\langle i_1, i_2, \cdots, i_k \rangle\rangle$ of numbers in $\{1, 2, \cdots, n\}$ such that

$$\langle\langle x_{i_1}, x_{i_2}, \cdots, x_{i_k} \rangle\rangle = \langle\langle y_{i_1}, y_{i_2}, \cdots, y_{i_k} \rangle\rangle? \tag{3.36}$$

The sequence

$$\langle\langle (x_{i_1}, y_{i_1}), (x_{i_2}, y_{i_2}), \cdots, (x_{i_k}, y_{i_k}) \rangle\rangle \tag{3.37}$$

is called a solution to the PCP instance. The PCP is to devise an algorithm that will tell us, for any pair (A, B), whether or not there exists a PCP solution.

Example 3.2.6 Let $\Sigma = \{0, 1\}$. Take A and B as follows:

$$A = \langle\langle 11, 100, \cdots, 111 \rangle\rangle$$
$$B = \langle\langle 111, 001, \cdots, 11 \rangle\rangle.$$

Then there exists a PCP-solution to this instance as follows:

$$\frac{A}{B} = \frac{11}{111}\frac{100}{001} \cdots \frac{111}{11}$$

$$= \frac{11}{111}\left(\frac{100}{001}\right)^{*}\frac{111}{11}$$

$$= \frac{11}{111}\frac{100}{001}\frac{111}{11}$$

$$= \frac{11}{111}\frac{100}{001}\frac{100}{001}\frac{111}{11}$$

$$= \cdots\cdots$$

$$= \cdots\cdots$$

in which the top and the bottom represent e.g., the same sequence "11100111". But if we take

$$A = \langle\langle 00, 001, \cdots, 1000\rangle\rangle$$
$$B = \langle\langle 0, 11, \cdots, 011\rangle\rangle.$$

Then there is not any PCP-solution to this instance of PCP.

Example 3.2.7 Consider another instance of PCP as follows:

$$A = \langle\langle ba, na, \cdots, s\rangle\rangle$$
$$B = \langle\langle b, an, \cdots, as\rangle\rangle.$$

Then every sequence of the form

$$\frac{A}{B} = \frac{ba}{b}\left(\frac{na}{an}\right)^{*}\frac{s}{as}$$

is a solution to to this PCP instance, e.g.,

$$\frac{A}{B} = \frac{ba}{b}\frac{na}{an}\frac{na}{an}\frac{s}{as}$$

in which both the top and the bottom are the word "bananas".

But in general, there is no algorithm that determines whether or not an arbitrary Post Correspondence Problem has a solution. That is:

Theorem 3.2.9 The Post Correspondence Problem is undecidable.

Note that the undecidability of the Post Correspondence Problem is very often used to prove the undecidability of other decision problems; for example, as we shall introduce later, it is undecidable whether or not $L(G_1) \cap L(G_2) = \emptyset$, where G_1 and G_2 are two context-free grammars; because $L(G_1) \cap L(G_2) \neq \emptyset$ if and only if PCP has a solution, but since PCP is undecidable, then whether or not $L(G_1) \cap L(G_2) \neq \emptyset$ is therefore undecidable.

3.2.5   Semi-decidable and Hilbert's Tenth Problem

There are many problems which are not computable or not decidable, and some have been presented in the previous sections. However, some noncomputable/undecidable are less computable/decidable than others; for example, the halting problem for Turing machines is *unsolvable* but not *absolutely* unsolvable. It is, in fact, *semidecidable* or *partially decidable*; because on input L, the Turing machine M will answer *"yes"* if $M(L)$ halts, although it loops for ever if $M(L)$ does not halt. The difference between computable/decidable and partially computable/decidable is illustrated in Figure 3.3. More generally, we can say that recursive languages L_{REC} are decidable (as we have already seen), whereas recursively enumerable languages L_{RE} are semidecidable. Now we actually arrive at the extended Church-Turing Thesis for semidecidable/semicomputable problems:

> **Church-Turing Thesis for semidecidable/semicomputable problems**: A problem P is semidecidable/semicomputable if and only if there is a Turing machine that accepts precisely the elements of P whose answer is *"yes"*.

In this subsection, we shall study another (actually the earliest) *undecidable* problem, Hilbert's Tenth Problem, which is semidecidable.

There is rather a long history about the problem; at the International Congress of Mathematicians in Paris in 1900, the 38-year old German mathematician, David Hilbert proposed 23 unsolved problems. The tenth of the 23 problems is so "short" that we can cite it here in full:

> 10. DETERMINATION OF THE SOLVABILITY OF A
> DIOPHANTINE EQUATION
> Given a Diophantine equation with any number of unknown
> quantities and with rational integral coefficients: *To devise a process*

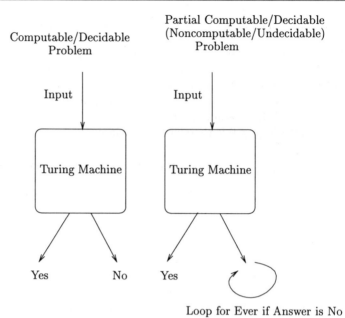

Figure 3.3: Decidable/Undecidable Problem

according to which it can be determined by a number of operations whether the equation is solvable in rational integers.

As we have seen in Section 1.3 of Chapter 1, for certain types of Diophantine equations, we do have an algorithm to determine whether or not a given Diophantine equation has integer solutions. For example, consider the following quadratic Diophantine equation:

$$x^2 = 991y^2 + 1.$$

Since

$$\sqrt{991} = [31, \overline{2, 12, 10, 2, 2, 2, 1, 1, 2, 6, 1, 1, 1, 1, 3, 1, 8, 4, 1, 2, 1, 2, 3, 1, 4, 1, 20, 6, 4, 31,}$$
$$\overline{4, 6, 20, 1, 4, 1, 3, 2, 1, 2, 1, 4, 8, 1, 3, 1, 1, 1, 1, 6, 2, 1, 1, 2, 2, 2, 10, 12, 2, 62}]$$

then the period m of the expansion of the continued fraction for $\sqrt{991}$ is $m = 60$ and of course is even. So the equation is soluble with

$$\begin{cases} x = P_{m-1} = P_{59} = 379516400906811930638014896080 \\ y = Q_{m-1} = Q_{59} = 12055735790331359447442538767 \end{cases}$$

as the smallest positive solution. But in general, there is no algorithm that can tell us whether or not an arbitrary Diophantine equation is soluble. That is:

Theorem 3.2.10 Hilbert's Tenth Problem is undecidable.

We shall not give a complete proof of this theorem here, rather, we just present a sketch of the proof of the theorem. Recall that an arbitrary Diophantine equation is an equation of the form

$$D(x_1, \cdots, x_m) = 0 \qquad (3.38)$$

where D is a polynomial with integer coefficients. What one needs to prove about this theorem is that no algorithm exists for testing a polynomial with integer coefficients to determine whether or not it has positive integer solutions. From a decidability point of view, it is irrelevant whether or not the solutions are integers or positive integers, since by the Four Squares Theorem (i.e., every positive integer is the sum of four squares) in Number Theory, Equation (3.38) has a solution in positive integers if and only if

$$D(p_1^2 + q_1^2 + r_1^2 + s_1^2, \cdots, p_n^2 + q_n^2 + r_n^2 + s_n^2) = 0 \qquad (3.39)$$

has an integral solution. On the other hand, Equation (3.38) has a solution in integers if and only if

$$D(p_1 - q_1, \cdots, p_n - q_n) = 0 \qquad (3.40)$$

has a solution in positive integers. Thus, we shall only consider the positive integer solutions to Equation (3.38). To prove the theorem, we will need some more definitions and lemmas:

Definition 3.2.8 A *family of Diophantine equations* is a *relation* of the form

$$D(a_1, \cdots, a_n, x_1, \cdots, x_m) = 0 \qquad (3.41)$$

where D is a polynomial with integer coefficients with respect to all the variables $a_1, \cdots, a_n, x_1, \cdots, x_m$ with $a_1, \cdots, a_n$ parameters and $x_1, \cdots, x_m$ unknowns. Fixing values of the parameters results in the particular Diophantine equations that comprise the family. Families of Diophantine equations are also called *parametric equations*, or just *relations*. The parametric equation (3.41) defines a set $\mathcal{M}$, consisting of the n-tuples of values of the parameters $a_1, \cdots, a_n$ for which there are values of the unknowns $x_1, \cdots, x_m$ satisfying (3.41):

$$\langle a_1, \cdots, a_n \rangle \iff \exists x_1 \cdots \exists x_m [D(a_1, \cdots, a_n, x_1, \cdots, x_m) = 0]. \quad (3.42)$$

The number n is called the *dimension* of the set $\mathcal{M}$, and equivalence (3.42) is called a *Diophantine representation* of $\mathcal{M}$; by taking some liberties, (3.41) and (3.42) can both be regarded as a Diophantine representation. Sets having Diophantine representations are also called *Diophantine*.

The central question here is which sets are Diophantine? It is easy to see that the union and/or the intersection of two Diophantine sets with the same dimension is Diophantine. It is also not so difficult to see that a set $\mathcal{M}$ of positive integers is Diophantine if and only if there is a polynomial D such that $\mathcal{M}$ is precisely the set of positive integers in the range of D.

Lemma 3.2.1 Every Diophantine set is Turing semidecidable.

This lemma can be demonstrated by giving any parametric Diophantine equation

$$D(a_1, \cdots, a_n, x_1, \cdots, x_m) = 0 \quad (3.43)$$

one can construct a Turing machine M that will eventually halt, beginning with a representation of the tuple. If Equation (3.43) has no solution, then the machine M will continue forever.

Lemma 3.2.2 Every Turing semidecidable set is Diophantine.

To prove this lemma, simply let M be a Turing machine that semidecides a set $\mathcal{M}$ consisting of n-tuples of natural numbers and let

$$\{\alpha_1, \cdots, \alpha_w\} \quad (3.44)$$

be the alphabet of M, and then try to obtain the Diophantine representation of the set $\mathcal{M}$.

What the above two lemmas have shown is that

$$\text{Diophantine sets} \iff \text{Turing semi-decidable sets} \qquad (3.45)$$

which actually implies that

$$\text{Diophantine sets} \iff \text{recursively enumerable sets} \qquad (3.46)$$

which, in turn, implies the unsolvability of Hilbert's Tenth Problem (note that Turing semidecidable is undecidable).

Remark 3.2.1 The first serious attempt to obtain a result about Hilbert's Tenth Problem was made by Martin Davis (now Professor Emeritus in the Department of Computer Science at Columbia University in New York) in 1950. His strategy was as follows: prove that for every listable set $\mathcal{M}$ there is a corresponding polynomial $P(x_1, \cdots, x_n)$, with integer coefficients, such that a positive integer k belongs to $\mathcal{M}$ if and only if the Diophantine equation $P(x_1, \cdots, x_n) = 0$ has a solution. Unfortunately, though the strategy works in principle, Davis was unable to prove that such a polynomial always exists. The key to the problem turned out to lie in some work begun by Julia Robinson, a noted American mathematician. She collaborated with Martin Davis and Hilary Putnam to show in 1960 that if just one Diophantine equation could be found whose solutions behaved exponentially in an appropriate sense, then it would be possible to describe every listable set by a Diophantine equation in the manner sought by Davis, and hence to solve Hilbert's tenth problem. But again unfortunately they were unable to to find such an equation. Almost exactly 10 years later in 1970, the 22-year-old Russian mathematician Yuri Matijasevich succeeded where the three Americans had failed. He did so by making use of the famous Fibonacci numbers: to solve Hilbert's tenth problem, it was sufficient to find a Diophantine equation whose solutions were appropriately related to the Fibonacci numbers. It is surprising to note that nothing was known about undecidability in 1900 and, in view of Hilbert's later attempts to find general decision methods in logic, it remains questionable whether in 1900 Hilbert really had in mind the possibility of a negative solution for the tenth problem.

3.2.6 Undecidability in Formal Languages

Now we present some more undecidable problems for formal languages, more specifically, for L_{CF}, L_{CS}, and L_{RE}.

Definition 3.2.9 A context-free grammar G_{CF} is *ambiguous* if there is a string $w \in L(G)$ which has two leftmost derivations in G_{CF}. If G_{CF} is not ambiguous, it is said to be *unambiguous*.

Theorem 3.2.11 Given a context-free grammar G_{CF}, it is undecidable whether or not G_{CF} is ambiguous.

Theorem 3.2.12 *Given two context-free grammars G_1 and G_2 it is undecidable whether or not $L(G_1) \cap L(G_2) = \emptyset$.*

Theorem 3.2.13 Given context-sensitive grammar G_{CS}, it is undecidable whether or not $L(G_{CS}) = \emptyset$.

For recursively enumerable languages L_{RE}, we have the following undecidable results:

Theorem 3.2.14 The following properties of L_{RE} that are acceptable by a Turing Machine M with n states (i.e., $L_{RE} = L(M)$) are all undecidable:

(i) emptiness, i.e., whether or not $L(M) = \emptyset$;

(ii) finiteness, i.e., whether or not $n \leq |L(M)| \leq 2n$;

(iii) regularity. i.e., whether or not $L(M)$ is a regular language;

(iv) context-freedom; i.e., whether or not $L(M)$ is a context-free language;

(v) equivalence; i.e., whether or not $L(M_1) = L(M_2)$, where $L(M_1)$ and $L(M_2)$ are two recursively enumerable languages accepted by Turing machines M_1 and M_2, respectively.

3.3   Computational Complexity

The computability and decidability we have just discussed in the previous sections are actually only the "theoretical" computability and decidability of Turing machines or digital computers. However theoretical computability or decidability does not imply "practically" computability or decidability. By "practically" computability we mean Turing "polynomial" computability. There are some problems, called intractable problems, which are "theoretical" computable or decidable in principle, but not "practical" computable or decidable in general, since they would e.g., take too long to complete the involved computations. The study of such tractable problems is the subject matter of computational complexity. One of the most important tasks for complexity theorists is to identify and to advise (practical) computer scientists which classes of problems are tractable and which are intractable. In this section, we shall introduce the basic concepts and results of computational complexity, within the theoretical framework of Turing polynomial-bounded computability.

3.3.1 Time and Space Complexity Classes

We begin this section with the definition of time complexity for Turing machines:

Definition 3.3.1 If M is a Turing machine, which halts for all $x \in \Sigma^*$, then the *time complexity* of M is the function $t_M : \mathbb{N} \to \mathbb{N}$ given by

$$t_M(n) = \max \Big\{ \, t : \ \exists x \in \Sigma^* \text{ with } |x| = n$$
$$\text{such that the time taken by } M \text{ on input } x \text{ is } t \, \Big\}.$$

A function f is *computable in polynomial time* or has *polynomial time complexity* if there exists some Turing machine M that computes f, and some polynomial function in p such that

$$t_M(n) \leq p(n), \quad \forall n \in \mathbb{N}. \tag{3.47}$$

A function f is *computable in exponential time* or has *exponential time complexity* if there exists some Turing machine M that computes f, and some exponential function exp (or e for short) such that

$$t_M(n) \leq e(n), \quad \forall n \in \mathbb{N}. \tag{3.48}$$

Similarly, we can define the space complexity for Turing machines as follows:

Definition 3.3.2 If M is a Turing machine, which halts for all $x \in \Sigma^*$, then the *space complexity* of M is the function $s_M : \mathbb{N} \to \mathbb{N}$ given by

$$s_M(n) = \max \Big\{ \, t : \ \exists x \in \Sigma^* \text{ with } |x| = n \text{ such that}$$
$$\text{the tape space taken by } M \text{ on input } x \text{ is } s \, \Big\}.$$

A function f is *computable in polynomial space* or has *polynomial space complexity* if there exists some Turing machine M that computes f, and some polynomial function in p such that

$$s_M(n) \leq p(n), \quad \forall n \in \mathbb{N}. \tag{3.49}$$

A function f is *computable in exponential space* or has *exponential space complexity* if there exists some Turing machine M that computes f, and some exponential function exp (or e for short) such that

$$s_M(n) \leq \exp(n), \quad \forall n \in \mathbb{N}. \tag{3.50}$$

The theory of computational complexity is mainly concerned with the classes of problems that can be solved by a Turing machine in polynomial time or polynomial space.

Definition 3.3.3 (DTIME(n), NTIME(n), DSPACE(n) and NSPACE(n))

(i) DTIME(n) is the class of languages decidable[4] by a deterministic Turing machine (DTM) of time-complexity $\mathcal{O}(t(n))$. That is,

$$\begin{aligned}
\text{DTIME}(n) \quad = \quad &\{L : \; L \text{ is a language decided by} \\
&\text{a DTM of } \mathcal{O}(t(n)) \text{ time}\}. \tag{3.51}
\end{aligned}$$

(ii) NTIME(n) is the class of languages decidable by a non-deterministic Turing machine (NDTM) of time-complexity $\mathcal{O}(t(n))$. That is,

$$\begin{aligned}
\text{NTIME}(n) \quad = \quad &\{L : \; L \text{ is a language decided by} \\
&\text{a NDTM of } \mathcal{O}(t(n)) \text{ time}\}. \tag{3.52}
\end{aligned}$$

(iii) DSPACE(n) is the class of languages decidable by a deterministic Turing machine (DTM) of space-complexity $\mathcal{O}(s(n))$. That is,

$$\begin{aligned}
\text{DSPACE}(n) \quad = \quad &\{L : \; L \text{ is a language decided by} \\
&\text{a DTM of } \mathcal{O}(t(n)) \text{ space}\}. \tag{3.53}
\end{aligned}$$

---

[4]As explained previously, it is customary to use many terms synonymously with the terms "decidable", "computable" and "solvable". So, instead of saying "the class of languages decidable by a TM", we can also speak of "the class of problems solvable by a TM" or "the class of problems (functions) computable by a TM".

(iv) NSPACE(n) is the class of languages decidable by a non-deterministic Turing machine (NDTM) of space-complexity $\mathcal{O}(s(n))$. That is,

$$NSPACE(n) = \{L : L \text{ is a language decided by}$$
$$\text{a NDTM of } \mathcal{O}(t(n)) \text{ space}\}. \quad (3.54)$$

The details of the relation between time and space complexity are as yet unknown. However, in most cases, space is a more powerful resource than time.

Proposition 3.3.1 Let DTIME(n), NTIME(n), DSPACE(n) and NSPACE(n) be as defined in Definition 3.3.3, then

$$DTIME(n) \subseteq NTIME(n). \quad (3.55)$$

$$DSPACE(n) \subseteq NSPACE(n). \quad (3.56)$$

$$DTIME(n) \subseteq DSPACE(n). \quad (3.57)$$

$$NTIME(n) \subseteq NSPACE(n). \quad (3.58)$$

$$NTIME(n) \subseteq DSPACE(n). \quad (3.59)$$

$$DTIME(n) \subseteq NTIME(n) \subseteq DSPACE(n) \subseteq NSPACE(n). \quad (3.60)$$

Theorem 3.3.1 *(Savitch)* For any function $f : \mathbb{N} \to \mathbb{N}$ with $f(n) \geq n$,

$$NSPACE(n) \subseteq DSPACE(n^2). \quad (3.61)$$

3.3.2 Complexity Classification of Formal Languages

In Chomsky hierarchy, formal languages are classified with classes of automata. That is, regular languages are languages accepted by finite automata, context-free languages are languages accepted by push-down automata, context-sensitive languages are languages accepted by linear-bounded automata, recursively-enumerable languages are languages accepted by Turing machines. There is, yet, another way to classify formal languages via the complexity classes defined in the previous subsection.

Proposition 3.3.2 Complexity Classification and the Chomsky Hierarchy:

(i) Every regular language can be accepted by a DFA in time proportional to the length of the input. Therefore,

$$L_{\text{REG}} \subseteq \text{DTIME}(n). \tag{3.62}$$

(ii) Every context-free language can be accepted by a DPDA in $\mathcal{O}(n^3)$ time, thus $L_{\text{CF}} \in \text{DTIME}(n^3)$, but there are languages in $\text{DTIME}(n^3)$ that are not context-free. Therefore,

$$L_{\text{CF}} \subset \text{DTIME}(n^3). \tag{3.63}$$

If nondeterminism is allowed, then

$$L_{\text{CF}} \subseteq \text{NTIME}(n). \tag{3.64}$$

(iii) A context-sensitive language can be accepted by a linear-bounded automata LBA. Therefore,

$$L_{\text{CS}} \subseteq \text{NSPACE}(n). \tag{3.65}$$

But by Savitch theorem, we have

$$L_{\text{CS}} \subseteq \text{DSPACE}(n^2). \tag{3.66}$$

(iv) There is no total Turing computable function $f(n)$ such that every recursive language L_{RE} is in $\text{NSPACE}(f(n))$. That is,

$$L_{\text{RE}} \supset \text{NSPACE}(n). \tag{3.67}$$

This implies that $\text{NSPACE}(n)$ cannot cover recursive and recursively enumerable languages.

3.3.3 Polynomial-Bounded Complexity Classes

As mentioned previously, the theory of computational complexity is mainly concerned with the set of problems that can be solved by a Turing machine, but within this set, the theory attempts to classify problems according to their computational difficulty measured by the amount of time or space their solution would take. There are many *complexity classes* [64] within this solvable set; we shall only introduce some of the important classes of computational complexity in this section.

Definition 3.3.4 ($\mathcal{L}$ and $\mathcal{NL}$)

(i) $\mathcal{L}$ is the class of languages that are decidable in logarithmic space on a deterministic Turing machine (DTM). That is,

$$\mathcal{L} = \text{DSPAC}(\log n). \qquad (3.68)$$

(ii) $\mathcal{NL}$ is the class of languages that are decidable in *logarithmic space* on a nondeterministic Turing machine (NDTM). That is,

$$\mathcal{NL} = \text{NSPAC}(\log n). \qquad (3.69)$$

Definition 3.3.5 ($\mathcal{P}$, $\mathcal{NP}$, co-$\mathcal{NP}$, $\mathcal{NP}$-hard and $\mathcal{NP}$-complete)

(i) $\mathcal{P}$ is the class of languages that are decidable in *polynomial time* by a *deterministic* (single-tape) Turing machine (DTM). That is,

$$\mathcal{P} = \bigcup_{i \geq 1} \text{DTIME}(n^i). \qquad (3.70)$$

(ii) $\mathcal{NP}$ is the class of languages that are decidable in *polynomial* time by a *nondeterministic* Turing machine (NDTM). That is,

$$\mathcal{NP} = \bigcup_{i \geq 1} \text{NTIME}(n^i). \qquad (3.71)$$

(iii) co-$\mathcal{NP}$ is the class of languages such that $\overline{L}$ belongs to $\mathcal{NP}$. That is, co-$\mathcal{NP}$ consists of the complements of all $\mathcal{NP}$ languages:

$$\text{co-}\mathcal{NP} = \{L : \overline{L} \in \mathcal{NP}\}. \qquad (3.72)$$

(iv) A language L is called $\mathcal{NP}$-hard if for every language $Q \in \mathcal{NP}$, Q is reduced to L in polynomial time.

(v) An $\mathcal{NP}$-hard problem that is also in $\mathcal{NP}$ is called $\mathcal{NP}$-complete.

Example 3.3.1 The class of all regular languages L_{REG} is in $\mathcal{P}$; that is, $L_{\text{REG}} \subset \mathcal{P}$. The class of all context-free languages L_{CF} also belongs to $\mathcal{P}$; that is, $L_{\text{CF}} \subset \mathcal{P}$.

Example 3.3.2 The class $\mathcal{NP}$ is a subset of recursive languages, that is, $\mathcal{NP} \subset L_{\text{REC}}$; the polynomial bound on the number of transitions ensures that all computations of a Turing machine M will eventually terminate. Since every deterministic machine is also nondeterministic, then, $\mathcal{P} \subseteq \mathcal{NP}$. But it is not known whether $\mathcal{NP} \subseteq \mathcal{P}$.

It is widely believed, although no proof has been given, that problems in $\mathcal{P}$ are tractable (feasible) and problems in $\mathcal{NP}$ are intractable (infeasible) by computers. More formally, we have the following so-called *Cook-Karp Thesis*[5] :

Cook-Karp Thesis: The problem of determining membership of strings in a given language L is tractable if and only if $L \in \mathcal{P}$.

Definition 3.3.6 Let Q and L be languages over alphabets Σ_1 and Σ_2, respectively. We say that Q is reducible to L in polynomial time if there is a polynomial time computable function $f : \Sigma_1^* \to \Sigma_2^*$ such that $u \in Q$ if $f(u) \in L$.

---

[5] In honor of Stephen Cook and Richard Karp — Stephen A. Cook (1939–) is currently professor in Computer Science and Mathematics at the University of Toronto. He received his PhD in mathematics from Harvard University in 1966. He is author of over 50 research papers, including his famous 1971 paper "The Complexity of Theorem Proving Procedures" which introduced the theory of $\mathcal{NP}$-completeness. Dr. Cook is the 1982 recipient of the Turing award, a fellow of the Royal Society of Canada, and a member of the National Academy of Sciences (United States) and the American Academy of Arts and Sciences. Richard M. Karp (1935–) earned his PhD in applied mathematics from Harvard University in 1959. He had been a researcher at the IBM Thomas J. Watson Research Center in New York and a professor of Computer Science, Operations Research and Mathematics at the University of California at Berkeley. He is now professor of Computer Science & Engineering and adjunct professor of Molecular Biotechnology at the University of Washington, Seattle. Dr. Karp was the 1985 Turing award winner for his fundamental contributions to complexity theory, which extended the earlier work of Stephen Cook in $\mathcal{NP}$-completeness theory. He was elected to membership in the National Academy of Sciences and National Academy of Engineering (United States).

Theorem 3.3.2 Let Q be reducible to L in polynomial time and $L \in \mathcal{P}$. Then $Q \in \mathcal{P}$.

Theorem 3.3.3 If there is an $\mathcal{NP}$-complete language that is also in $\mathcal{P}$, then $\mathcal{P} = \mathcal{NP}$.

Definition 3.3.7 The language L is co-$\mathcal{NP}$-hard if every problem in co-$\mathcal{NP}$ reduces in polynomial time to L. It is co-$\mathcal{NP}$-complete if in addition it is also in co-$\mathcal{NP}$.

It is not known whether $\mathcal{NP} = $ co-$\mathcal{NP}$; this is also arguably one of the most important problems in theoretical computer science.

Definition 3.3.8 ($\mathcal{P}$-SPACE and $\mathcal{NP}$-SPACE)

(i) $\mathcal{P}$-SPACE is the class of languages that are decidable in *polynomial space* by a *deterministic* Turing machine (DTM). That is,

$$\mathcal{P}\text{-SPACE} = \bigcup_{i \geq 1} \text{DSPACE}(n^i). \tag{3.73}$$

(ii) $\mathcal{NP}$-SPACE is the class of languages that are decidable in *polynomial space* by a *non-deterministic* Turing machine (NDTM). That is,

$$\mathcal{NP}\text{-SPACE} = \bigcup_{i \geq 1} \text{NSPACE}(n^i). \tag{3.74}$$

Example 3.3.3 The class of context-sensitive languages L_{CS} belongs to $\mathcal{P}$-SPACE. That is, $L_{\text{CS}} \in \mathcal{P}$-SPACE.

The following are some conjectured relations between these classes:
Let $\mathcal{L}$, $\mathcal{NL}$, $\mathcal{P}$, $\mathcal{NP}$, $\mathcal{P}$-SPACE and $\mathcal{NP}$-SPACE be defined above. Then

$$\mathcal{L} \subseteq \mathcal{NL} \subseteq \mathcal{P}. \tag{3.75}$$

$$\mathcal{P} \subseteq \mathcal{NP} \subseteq \mathcal{NP}\text{-SPACE}. \tag{3.76}$$

$$\mathcal{P}\text{-SPACE} = \mathcal{NP}\text{-SPACE}. \tag{3.77}$$

Thus we have:

$$\mathcal{L} \subseteq \mathcal{NL} \subseteq \mathcal{P} \subseteq \mathcal{NP} \subseteq \mathcal{P}\text{-SPACE} = \mathcal{NP}\text{-SPACE} \tag{3.78}$$

although it is not known whether or not these containments are proper.

Definition 3.3.9 (EXPTIME and NEXPTIME)

(i) EXPTIME, denoted sometimes by $\mathcal{EXP}$, is the class of languages decidable in time bounded by 2^{n^i}. That is,

$$\text{EXPTIME} = \bigcup_{i \geq 1} \text{DTIME}(2^{n^i}). \tag{3.79}$$

(ii) NEXPTIME is the nondeterministic analog of EXPTIME, which is defined by

$$\text{NEXPTIME} = \bigcup_{i \geq 1} \text{NTIME}(2^{n^i}). \tag{3.80}$$

It is not difficult to see that if $\mathcal{P} = \mathcal{NP}$, then EXPTIME = NEXPTIME. At present, it is safe to conjecture that EXPTIME $\subseteq$ NEXPTIME. Note also that we know $\mathcal{P} \neq \mathcal{EXP}$, but we do not know whether or not $\mathcal{P} = \mathcal{NP}$; $\mathcal{P} \stackrel{?}{=} \mathcal{NP}$ is an important open problem.

Example 3.3.4 The compositeness problem, i.e., determining if a given number $N \in \mathbb{N}_{>1}$ is composite, is in EXPTIME – try all *possible* divisors of N (because the complexity is measured in terms of bit operations, rather than arithmetic operations, so to perform $\sqrt{N}$ number of arithmetic operations, we will need approximately $2^{0.5 \log N}$ bit operations; see Example 3.4.4 in subsection 3.4.2 for more information). Note that the compositeness problem can also be determined in $\mathcal{NP}$ – simply guess a nontrivial pair of integers whose product is N. However, it is not known whether or not the set of composite numbers is in $\mathcal{P}$.

Any problem in $\mathcal{NP}$ can be solved by brute-force in EXPTIME.

Theorem 3.3.4 If $L \in \mathcal{NP}$, then L can be recognised deterministically in time $2^{n^{\mathcal{O}(1)}}$.

Figure 3.4 gives a diagrammatical illustration of some common and important computational complexity classes [51].

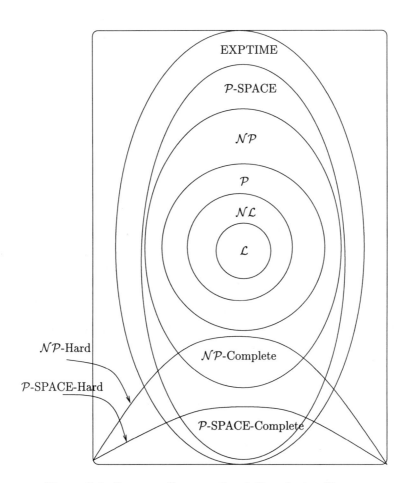

Figure 3.4: Common Computational Complexity Classes

3.3.4   Parallel and Probabilistic Complexity Classes

The computational models we have studied so far are actually sequential in nature. The following statement, often called the *Sequential Computation Thesis*, claims that all sequential universal models of computation, including those that have not yet been invented, have polynomially related time and space behaviour, so that the class of functions computable in reasonable time and space is precisely the same for all models. That is, all sequential computers and Turing machines are *polynomially equivalent* or *related*.

> **Sequential Computation Thesis**:   If, giving an input n, a number-theoretic function f is computable in time-space complexity $(\mathcal{O}(t(f(n))), \mathcal{O}(s(f(n))))$ on a fast computer, then there is an equivalent Turing machine to compute the function f that will take no more than then time-space complexity $(\mathcal{O}(p(t(f(n)))), \mathcal{O}(p(s(f(n)))))$, for some fixed polynomial p.

This thesis actually indicates that not only the class of computable functions is robust, but also the tractable functions. For example, giving a function computable in time $\mathcal{O}(n)$ on a fast computer, the time may grow, if it is computed on a slow computer, from $\mathcal{O}(n)$ to $\mathcal{O}(n^2)$ or $\mathcal{O}(n^{23})$, but not to $\mathcal{O}(2^n)$.

The sequential computation thesis may, however, not hold for models of parallel and probabilistic computations. In this subsection, we shall study the computational complexity of parallel and probabilistic computations.

There are many general models of parallel computation. Among the most popular are:

(i) Parallel Random Access Machines (PRAM): A PRAM consists of a set of processors that have access to a common shared memory. Each processor may have its own registers and local memory. A PRAM can exclusive (resp. concurrent) read (resp. write), giving four versions of PRAM; CRCW, CREW, ERCW, EREW.

(ii) Vector Machines (VM): This model can be SIMD (Single Instruction Multiple Data) or MIMD (Multiple Instruction Multiple Data). The processors are arranged in an array and all execute synchronously.

(iii) Boolean/Arithmetic Circuits (or Circuits for short): These are a common parallel computation model; the *size* of the circuit (i.e., the number of nodes) corresponds roughly to the number of processors in PRAM, and the *depth* of the circuit (i.e., the length of the longest path from an input to an output) corresponds to time, so we can define the *processor complexity* of a circuit to be its *size*, denoted by $z(n)$, and *parallel time complexity* a circuit to be its *depth*, denoted by $d(n)$.

Many interesting problems have size-depth complexity

$$
\begin{aligned}
\{z(n),\ d(n)\} &= \{p(n),\ p/\log(n)\} \\
&= \left\{ \bigcup_{i \geq 1} \mathcal{O}(n^i),\ \bigcup_{i \geq 1} \mathcal{O}(\log^i n) \right\}.
\end{aligned} \tag{3.81}
$$

That is, the size complexity $z(n)$ is bounded by some polynomial in n, denoted by $p(n)$, and the depth complexity $d(n)$ by some polynomial in the logarithm of n, denoted by $p/\log(n)$. This prompts the formal definition of the parallel complexity class $\mathcal{NC}$:

Definition 3.3.10 The parallel complexity class of $\mathcal{NC}$ is defined as follows:

$$
\mathcal{NC} = \bigcup_{i \geq 1} \mathcal{NC}^i \tag{3.82}
$$

where $\mathcal{NC}^i$ is the set of all problems solvable by a uniform circuit family $C = \{C_n\}$, $n = 1, 2, \cdots$; the size $z(n)$ and the depth $d(n)$ of the n-th circuit C_n are bounded by a polynomial in n and a polynomial in the logarithm of n, respectively, where n is the size of the problem. That is,

$$
\{z(n),\ d(n)\} = \{p(n),\ p/\log(n)\}. \tag{3.83}
$$

Functions that are computed by such circuit families are called $\mathcal{NC}^i$ *computable* or $\mathcal{NC}$ *computable*.

Note that the name $\mathcal{NC}$ stands for *Nicks's Class*, in honor of Nicholas Pippenger, the first researcher to study the class seriously. Note also that the above definition critically depends on the notion of uniform family of Boolean circuits. The uniformity condition says essentially that the n-th circuit in this family is easily constructed, and is a technical condition that allows circuits and PRAMs to simulate each other efficiently.

The relationship between the Boolean circuit model and other different parallel models (e.g., PRAM, vector machine VM and alternating Turing machine

ATM [67]) and connectionist machines (CM) can be summarized as follows:

$$
\begin{aligned}
\mathcal{NC} &= \text{PRAM}\,(p(n),\ p/\log(n)) & (3.84)\\
&= \text{VM}\,(p(n),\ p/\log(n)) & (3.85)\\
&= \text{ATM}\,(\log(n),\ p/\log(n)) & (3.86)\\
&= \text{CM}\,(\log(n),\ p/\log(n))\,. & (3.87)
\end{aligned}
$$

Example 3.3.5 Consider the multiplication of two $m \times m$ Boolean matrices $A = \{a_{ik}\}$ and $B = \{b_{ik}\}$: The input has $2m^2 = n$ entries to represent A and B, the output has m^2 entries representing the matrix $C = \{c_{ik}\}$:

$$
c_{ik} = \bigvee_j (a_{ij} \wedge b_{jk}).
$$

The circuit for the matrix multiplication has gates g_{ijk} that compute $a_{ij} \wedge b_{jk}$ for each i, j, k. In addition, for each i and k the circuit contains a binary tree of $\vee$ gates to compute $\bigvee_j g_{ijk}$. each such tree contains $m - 1$ OR-gates and has $\log m$ depth. So the size-depth complexity of the Boolean matrix multiplication is $\left(\mathcal{O}(n^{3/2}),\ \mathcal{O}(\log n)\right)$.

It is no doubt that parallelism can be used to improve the efficiency of the sequence computation. But can we devise a parallel computer or algorithm for noncomputable functions or undecidable problems? The answer is "no", since every parallel computer/algorithm can be simulated by a single processor. In this sense, the Church-Turing Thesis is applied to parallel computation too: the class of unsolvable problems is insensitive even to the addition of parallelism. This leads to a refinement of the Church-Turing Thesis for parallel computations:

Parallel Computation Thesis I (in terms of uniform families of circuits): A number-theoretic function can be mechanically computed by a parallel computer of the size-depth complexity

$$
(\mathcal{O}(z(n)),\ \mathcal{O}(d(n)))
$$

only if there is a uniform family of circuits of the size-depth complexity

$$
(\mathcal{O}(p(z(n)d(n))),\ \mathcal{O}(p(d(n)))),
$$

for some fixed polynomial p.

This thesis can, of course, be restated in terms of other parallel models of computation, such as PRAM or vector machines (VM). In fact, all parallel models of computation can be shown to be polynomially equivalent.

The second question about parallel computations to ask is that if parallelism can turn *intractable* problems into *tractable* ones? The answer to this question is just simply that "we don't know". But one thing we certainly know about parallel computations is that

> **Parallel Computation Thesis II**: Parallel time is polynomially equivalent to sequence space. That is, up to polynomial differences, parallel polynomial time $\mathcal{P}$ is the same as sequential polynomial memory space $\mathcal{P}$-SPACE:

$$\text{sequential-}\mathcal{P}\text{-SPACE} = \text{parallel-}\mathcal{P}\text{ time} \qquad (3.88)$$

It is clear that the question whether or not parallel algorithms can solve intractable problems is the same as that whether or not the sequential complexity class $\mathcal{P}-$SPACE contains intractable problems. This question is still open, and probably is as difficult as the famous $\mathcal{P} \overset{?}{=} \mathcal{NP}$ problem.

Now we move on to the study of the computational complexity classes of probabilistic computations.

Definition 3.3.11 A probabilistic Turing machine M is said of time complexity $t(n)$ if M halts within $t(n)$ time in each computation on each input of length n. If $t(n)$ is a polynomial, then M is said to have polynomial time complexity. M is said of *expected time complexity* $t(n)$ if for each input x of M the function $t(n)$ satisfies

$$t(|w|) \geq \sum_{j=0}^{\infty} \text{Prob}(M \text{ accepts } w \text{ with } j \text{ moves}) \cdot j \qquad (3.89)$$

If $t(n)$ is a polynomial, then M is said to have *expected* polynomial time complexity.

Definition 3.3.12 (Probabilistic Computational Complexity Classes $\mathcal{RP}$, $\mathcal{ZPP}$, and $\mathcal{BPP}$)

(i) $\mathcal{RP}$ is the class of languages L that are accepted by a *probabilistic Turing machine* (PTM) in expected polynomial time, where inputs in L are accepted with a probability of at least $\frac{1}{2}$ and inputs not in L are *rejected* with a probability of 1. That is;

$$\mathcal{RP} = \{L : \exists \text{ a set } S \in \mathcal{P} \text{ and a polynomial } p \text{ such that} \qquad (3.90)$$
$$x \in L \Longrightarrow \text{for at least } \frac{1}{2} \text{ of all } y, \ \log y \leq p(\log x),$$
$$\text{we have } (x, y) \in S;$$
$$x \notin L \Longrightarrow \text{for all } y, \ \log y \leq p(\log x),$$
$$\text{we have } (x, y) \notin S\}.$$

L is sometimes said to be accepted with *one-sided errors*. (Note that $\mathcal{RP}$ stands for "Random Polynomial").

(ii) $\mathcal{BPP}$ is the class of languages L that are accepted by a *probabilistic Turing machine* (PTM) in expected polynomial time, where inputs in L are *accepted* with a probability of at least $\frac{3}{4}$ and inputs not in L are *rejected* with a probability of also $\frac{3}{4}$. That is;

$$\mathcal{RP} = \{L : \exists \text{ a set } S \in \mathcal{P} \text{ and a polynomial } p \text{ such that} \qquad (3.91)$$
$$x \in L \Longrightarrow \text{for at least } \frac{3}{4} \text{ of all } y, \ \log y \leq p(\log x),$$
$$\text{we have } (x, y) \in S;$$
$$x \notin L \Longrightarrow \text{for at least } \frac{3}{4} \text{ of all } y, \ \log y \leq p(\log x),$$
$$\text{we have } (x, y) \notin S.$$

L is sometimes said to be accepted with *two-sided errors*. (Note that the "$\mathcal{B}$" in $\mathcal{BPP}$ stands for "Bounded away the error probability from $\frac{1}{2}$", whereas $\mathcal{PP}$ stands for "Probabilistic Polynomial").

(iii) $\mathcal{ZPP}$ is the languages L that are accepted by a *probabilistic Turing machine* (PTM) in expected polynomial time with *zero error*. That is,

$$\mathcal{RP} = \{L : \exists \text{ a set } S \in \mathcal{P} \text{ and a polynomial } p \text{ such that} \qquad (3.92)$$
$$x \in L \Longrightarrow \text{for all } y, \ \log y \leq p(\log x), \text{ we have } (x, y) \in S;$$
$$x \notin L \Longrightarrow \text{for all } y, \ \log y \leq p(\log x), \text{ we have } (x, y) \notin S\}.$$

L is sometimes said to be accepted with *zero error*. Note that by "zero error" we mean that the machine will answer "yes" when the answer is "yes" (with zero probability of error), and will answer "no" when the answer is "no" (with zero probability of error). But note that the machine may also answer "?", which means that the machine does not know the answer is "yes" or "no". However, it is guaranteed that at most half of simulation cases the machine will answer "?". Another way to define $\mathcal{ZPP}$ is via $\mathcal{RP}$:

$$\mathcal{ZPP} = \mathcal{RP} \cap \text{co}-\mathcal{RP}, \tag{3.93}$$

where co-$\mathcal{RP}$ is defined to be the complementary language of $\mathcal{RP}$, i.e., co-$\mathcal{RP} = \{L : \overline{L} \in \mathcal{RP}\}$. (Note that $\mathcal{ZPP}$ stands for "Zero Probabilistic Polynomial").

The following is a simple comparison of the probabilistic errors in $\mathcal{ZPP}$, $\mathcal{RP}$ and $\mathcal{BPP}$:

Complexity Classes		Error of Acceptance	Error of Rejection
$\mathcal{ZPP}$	(zero error)	0	0
$\mathcal{RP}$	(one-sided error)	1/2	0
$\mathcal{BPP}$	(two-sided error)	3/4	3/4

Example 3.3.6 The following are some examples of $\mathcal{RP}$ and $\mathcal{BPP}$:

$$\text{Composite Numbers} \in \mathcal{RP},$$
$$\text{Perfect Numbers} \in \mathcal{BPP},$$
$$\text{Amicable Numbers} \in \mathcal{BPP},$$
$$\text{Prime Numbers} \in \mathcal{ZPP}.$$

The complexity classes for probabilistic computation can be diagrammatically shown in Figure 3.5.

As far as universal algorithmic powers concerned, the Church-Turing Thesis extends to probabilistic computation as well. Probabilism, like parallelism, cannot be used to solve the noncomputable or the undecidable, at least not under the following assumptions: it is possible to simulate every probabilistic Turing machine by a standard Turing machine, and moreover, this can be done with at most an exponential loss of time.

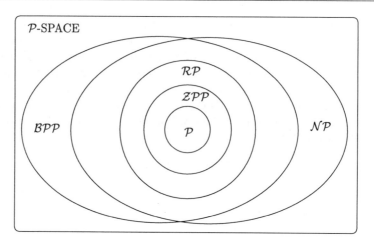

Figure 3.5: Common Probabilistic Complexity Classes

Probabilistic Computation Thesis: A number-theoretic function that can be algorithmically computed with the aid of probabilistic choices can also be computed by a probabilistic Turing machine of polynomially related time complexity and polynomially related, expected time complexity.

Just as the complexity class $\mathcal{P}$ has its zero-error and one-sided-error probabilistic counterparts $\mathcal{ZPP}$ and $\mathcal{RP}$, the parallel complexity class $\mathcal{NC}$ also has its zero-error and one-sided-error probabilistic counterparts: $\mathcal{ZNC}$ and $\mathcal{RNC}$. It is reasonable to conjecture that

$$\mathcal{NC} \subseteq \mathcal{ZNC} \subseteq \mathcal{RNC} \subseteq \mathcal{P} \tag{3.94}$$

As a conclusion, we summarize the hierarchy of all the common computational complexity classes as follows:

$$\mathcal{L} \subseteq \mathcal{NL} \subseteq \mathcal{NC} \subseteq \mathcal{ZNC} \subseteq \mathcal{RNC} \subseteq \mathcal{P} \subseteq \mathcal{ZPP} \subseteq \mathcal{RP} \subseteq \begin{pmatrix} \mathcal{BPP} \\ \mathcal{NP} \end{pmatrix}$$

$$\subseteq \mathcal{P}\text{-SPACE} = \mathcal{NP}\text{-SPACE} \subseteq \text{EXPTIME} \subseteq \text{NEXPTIME}. \tag{3.95}$$

Besides the proper inclusions $\mathcal{NL} \subset \mathcal{P}\text{-SPACE}$ and $\mathcal{P} \subset \text{EXPTIME}$, it is not known whether any of the other inclusions in the above hierarchy is proper.

Up to now, we have studied various "computation theses" for various abstract machines, such as Turing machines, sequential machines, parallel machines, and probabilistic machines. Remarkably enough, all these machines are actually *similar* [59]:

> **Similarity Thesis for Different (Sequential and Parallel) Computation Models**: All (at least all the existing) computation models are similar; that is, in order to solve the same class of problems, they need essentially the same parallel time, essentially the same space, and essentially the same sequential time. By "essentially the same", we mean "polynomially related", and in fact polynomially related by low degree of polynomials.

3.3.5 $\mathcal{NP}$-Complete Problems

The $\mathcal{NP}$-complete problems are the hardest problems in $\mathcal{NP}$, in the sense that if Q' is any decision problem in $\mathcal{NP}$, and Q is an $\mathcal{NP}$-complete problem, then every instance of Q' is polynomially reducible to an instance of Q. $\mathcal{NP}$-completeness has many nice properties, which computer scientists are interested in. For example, if one can find an efficient algorithm for one $\mathcal{NP}$-complete problem, one can find an efficient algorithm for all of the $\mathcal{NP}$-complete problems. That is, finding a polynomial-time algorithm for any one of the $\mathcal{NP}$-complete problems would imply the existence of a polynomial-time algorithm for all of them.

The first $\mathcal{NP}$-complete problem, the so-called *satisfiability problem* (SAT), was found by Stephen Cook of the University of Toronto in 1971. His work is now very well understood as Cook's Theorem, which actually tries to identify a language in the class $\mathcal{NP}$ which any other language in $\mathcal{NP}$ can be reduced to by some polynomial reductions. Thus, if this language is in $\mathcal{P}$, then all the languages in $\mathcal{NP}$ must be in $\mathcal{P}$. Here is the satisfiability problem.

Definition 3.3.13 Let $X = \{x_1, x_2, \cdots, x_n\}$ be a set of Boolean *variables*. A *literal* is defined to be either one of the variables x_i or the negation of x_i, denoted by $\neg x_i$. (Clearly, there are $2n$ possible literals). A *truth assignment* for X is a function $t : X \to \{T, F\}$. If $t(x) = T$ we say that x is "true" under t; otherwise, we say that x is "false". The literal x is true under t if and only if the variable x is true; the literal $\neg x$ is true under t if and only if the variable x is false. A *clause* over X is a set of literals over X, such as $\{x_1, \neg x_4, x_9\}$. It represents the disjunction of those literals and is *satisfied* by a truth assignment if and only if at least one of its members is true under that assignment. For example, the clause $\{x_1, \neg x_4, x_9\}$ will be satisfied by t unless $t(x_1) = F$, $t(x_4) = T$, $t(x_9) = F$. A set of clauses C over X is satisfied if and only if there exists some truth assignment for X that simultaneously satisfies all clauses in C. Such a truth assignment is called a *satisfying truth assignment* for C.

Definition 3.3.14 (The satisfiability problem (SAT)) Given a set of variables X and a set of clauses C over X. Does there exist a satisfying truth assignment for C? Formally, the language of SAT can be defined as follows:

$$L_{\text{SAT}} = \{C : \text{ every clause in C is satisfiable}\}.$$

What Cook proved about SAT is that

Theorem 3.3.5 (Cook's Theorem) $L_{\text{SAT}} \in \mathcal{NP}$-complete. That is, SAT is $\mathcal{NP}$-complete.

The proof of this theorem is very lengthy, we shall just present an idea, and leave the detailed proof to the reader as an exercise. To prove that SAT is $\mathcal{NP}$-complete, we have to prove first that SAT is in $\mathcal{NP}$. This can be done by constructing an nondeterministic Turing machine that accepts L_{SAT}. Then we need to prove that SAT is $\mathcal{NP}$-hard, that is, we need to prove that every problem in $\mathcal{NP}$ is polynomially reducible to an instance of SAT.

Cook's theorem opened the way to the identification of a large number of $\mathcal{NP}$-complete problems: to prove that some problem P is $\mathcal{NP}$-complete, we need only to show that SAT reduces to problem p, For if that is so, then every problem in $\mathcal{NP}$ can be reduced to problem P by first reduced to an instance of SAT and then to an instance of P. So, life after Cook's theorem is much easier; we do not need to go all the way back to Turing machine computations any more. More formally, to prove a language L is $\mathcal{NP}$-complete, what we need to do is just to follow the following five steps:

(i) Prove L is in $\mathcal{NP}$.

(ii) Select a language L' that is known to be $\mathcal{NP}$-complete (e.g., L_{SAT}).

(iii) Design a function f that maps members of L' to members of L.

(iv) Show that $x \in L'$ if and only if $f(x) \in L$.

(v) Show that f can be computed in polynomial time.

Now we list some $\mathcal{NP}$-complete problems, which readers may wish to prove by follow the above five steps.

(i) Travelling Salesman Problem (TSP): TSP can be best described as a language as follows:

$$L_{\text{TSP}} = \{\langle G, c, k \rangle : \ G = \langle V, E \rangle \text{ is a complete graph},$$
$$c : V \times V \to \mathbb{Z},$$
$$k \in \mathbb{Z},$$
$$G \text{ has a travelling-salesman tour with cost}$$
$$\text{at most } k\}$$

A travelling-salesman tour (or a Hamiltonian cycle; see the following item (ii)) is a route that visits each city in G exactly once and finishes at the starting city.

(ii) Hamiltonian Cycle Problem[6] (HCP): Given a network of cities and roads linking them, is there a route that starts and finishes at the same city and visits every other cities exactly once?

(iii) Directed Hamiltonian Path Problem (HPP): Given a directed graph G and two distinct vertices u and v, does there exist a directed path in the graph from u to v that visits every vertex exactly once?

(iv) The Clique Problem (CLIQUE): A *clique* in an undirected graph $G = (V, E)$ is a subset $V' \in V$ of vertices, each pair of which is connected by an edge in E. The size m of a clique is the number of vertices it contains. The clique problem is then: given a finite graph $G = (V, E)$ and positive integer $m \leq |v|$, does G have a clique of size m? The formal definition of CLIQUE is

$$L_{\text{CLIQUE}} = \{\langle G, m \rangle : \ G \text{ is a graph with clique of size } m\}$$

---

[6]After Ireland's greatest mathematician William R. Hamilton (1805–1865), who became Professor of Astronomy at Trinity College Dublin and Royal Astronomer of Ireland at the age of 22. Hamilton is perhaps best known in pure mathematics for his algebraic theory of complex numbers, the invention of *quaternions* and the exploitation of noncommutative algebra. The terms *Hamiltonian graph*, *Hamiltonian cycle*, and *Hamiltonian path* arise from his letter describing a mathematical game on the dodecahedron.

(v) The Map-colouring problem: Given a map, is it possible to colour that map using three colours in such a way that no two countries with a common border are coloured the same?

(vi) The Quadratic Residue Problem: Given positive integers a, b, c with $a < b$, is there a positive integer $x < c$ such that

$$x^2 \bmod b = a?$$

(vii) The Quadratic Diophantine Equation Problem: Given positive integers a, b, c, do there exist positive integers x and y such that

$$ax^2 \bmod by = c?$$

(viii) The Subset-Sum Problem: Given a finite set $S \subseteq \mathbb{N}$ and a target $t \in \mathbb{N}$, is there a subset $S' \subseteq S$ whose elements sum to t? The subset-sum problem can be defined as a language:

$$L_{\text{SS}} = \left\{ \langle S, t \rangle : \text{ there exsists a subset } S' \subseteq S \text{ such that } t = \sum_{s \in S'} s \right\}.$$

For example, if $S = \{1, 4, 16, 64, 256, 1040, 1041, 1093, 1284, 1344\}$ and $t = 3754$, then the subset $S' = \{1, 16, 64, 256, 1040, 1093, 1284\}$ is a solution.

3.4   Design and Analysis of Algorithms

In this section, we move our study of algorithms a bit further on to the more practical aspects: design and analysis of algorithms. More specifically, we shall study the asymptotic behaviours of some common complexity functions, the complexity analysis of algorithms measured in both arithmetic operations and bits operations, with an emphasis on bit operations, and the design of efficient algorithms for modular exponentiations and point additions on elliptic curves.

3.4.1 Asymptotic Complexity

Definition 3.4.1 Let f and g be functions from $\mathbb{N}$ to $\mathbb{R}$. Then

(i) f is $\mathcal{O}(g(n))$ if

$$\exists c \in \mathbb{N} \; \overset{\infty}{\forall} \; f(n) \leq c \cdot g(n). \tag{3.96}$$

The notation $\overset{\infty}{\forall}$ means "for almost all" or "for all but finitely many". Intuitively, f grows no faster asymptotically than g within a constant multiple.

(ii) f is $o(g(n))$ if

$$\exists c \in \mathbb{N} \; \overset{\infty}{\forall} \; f(n) \leq \frac{1}{c} \cdot g(n). \tag{3.97}$$

Intuitively, f grows strictly more slowly than any arbitrary small positive constant multiple of g.

(iii) f is $\Omega(g(n))$ if $g(n) = \mathcal{O}(f(n))$. That is,

$$\exists c \in \mathbb{N} \; \overset{\infty}{\forall} \; f(n) \geq \frac{1}{c} \cdot g(n). \tag{3.98}$$

(iv) f is $\omega(g(n))$ if

$$g(n) = o(f(n)). \tag{3.99}$$

(v) f is $\Theta(g(n))$ if

$$f(n) = \mathcal{O}(g(n)) \quad \text{and} \quad f(n) = \Omega(g(n)). \tag{3.100}$$

Remark 3.4.1 The $\mathcal{O}$-notation provides an asymptotic *upper* bound on a function. For example, in $f(n) = \mathcal{O}(g(n))$, $g(n)$ is an asymptotic upper bound for $f(n)$. But the asymptotic upper bound provided by $\mathcal{O}$-notation may or may not be asymptotically tight. For example, the bound $13n^5 = \mathcal{O}(n^5)$ is asymptotically tight, whereas $2n^3 = \mathcal{O}(n^5)$ is not. The o-notation provides an upper bound that is not asymptotically tight. For example, $12n = o(n^2)$, but $12n^2 \neq o(n^2)$. asymptotic upper bound for $f(n)$. Just as $\mathcal{O}$-notation provides an asymptotic *upper* bound on a function and o-notation provides an upper bound that is not asymptotically tight, Ω-notation provides an asymptotic *lower* bound on a function and ω-notation provides a *lower* bound that is not asymptotically tight. For example, $n^2/2 = \omega(n)$, but $n^2/2 \neq \omega(n^2)$. It should be kept in mind that

always use $\mathcal{O}$ and o for upper bounds, and Ω and ω for lower bounds.
Never use $\mathcal{O}$ for lower bounds.

The Θ-notation provides an asymptotic *tight* bound on a function. For example, $\frac{1}{3}n^5 + \frac{1}{3}n^2 - 3 = \Theta(n^5)$, but $\frac{1}{3}n^5 + \frac{1}{3}n^2 - 3 \neq \Theta(n^4)$.

In this book, we shall be mainly concerned with the asymptotic complexity described by $\mathcal{O}$-notation. That is, we shall mainly consider the asymptotic *upper* bounds of algorithms. The asymptotic complexity of an algorithm can often be characterised in one of the following classes:

Definition 3.4.2 Let f be a function $f : \mathbb{N} \to \mathbb{R}$. Then f can be characterised in one of the following classes:

(i) f is $\mathcal{O}(1)$: For any algorithm with complexity $\mathcal{O}(1)$, there exists some $k \in \mathbb{N}$ such that execution of the algorithm will cost $r \leq k$ regardless of the input value of n. Any function which is $\mathcal{O}(c)$, where $c \in \mathbb{R}$, is $\mathcal{O}(1)$. An algorithm of $\mathcal{O}(1)$ complexity is said to have *constant complexity*.

(ii) f is $\mathcal{O}(\log n)$: An algorithm of $\mathcal{O}(\log n)$ complexity is said to have *logarithmic complexity*. For algorithms of logarithmic complexity, the cost of applying the algorithm to problems of sufficiently large size n can be bounded by a function of the form $k \log n$, where $k \in \mathbb{R}$.

(iii) f is $\mathcal{O}(\log^c n)$, where $c \geq 1$: An algorithm of $\mathcal{O}(\log^c n)$ complexity is said to have *polylogarithmic complexity*.

(iv) f is $\mathcal{O}(n^r)$, where $0 < r < 1$: An algorithm of $\mathcal{O}(n^r)$ with $0 < r < 1$ is said to have *sublinear complexity*.

(v) f is $\mathcal{O}(n)$: An algorithm of $\mathcal{O}(n)$ complexity is said to have *linear complexity*. For any such algorithm, there will exist $k \in \mathbb{Z}_{\geq 0}$ such that executing the algorithm to a problem of sufficiently large size n will cost no more than kn.

(vi) f is $\mathcal{O}(n \log n)$: For any algorithm of $\mathcal{O}(n \log n)$ complexity, there will exist $k \in \mathbb{N}$ such that applying the algorithm to a problem of sufficiently large size n will cost no more than $kn \log n$. Such an algorithm is said to have $n \log n$ complexity.

(vii) f is $\mathcal{O}(n^2)$: An algorithm of $\mathcal{O}(n^2)$ complexity is said to have *quadratic complexity*.

(viii) f is $\mathcal{O}(n^r)$, where $1 < r < 2$: An algorithm of $\mathcal{O}(n^r)$ with $1 < r < 2$ is said to have *subquadratic complexity*.

(ix) f is $\mathcal{O}(n^3)$: An algorithm of $\mathcal{O}(n^3)$ complexity is said to have *cubic complexity*.

(x) f is $\mathcal{O}(n^c)$, where $c \geq 1$: An algorithm of $\mathcal{O}(n^c)$ with $c \geq 1$ is said to have *polynomial complexity*.

(xi) f is $\mathcal{O}(r^n)$ where $r > 1$: An algorithm of complexity $\mathcal{O}(r^n)$ where $r > 1$ is said to have *exponential complexity*.

(xii) f is $\mathcal{O}(n!)$: An algorithm of complexity $\mathcal{O}(n!)$ is said to have *factorial complexity*.

The hierarchy of the above mentioned complexity functions can be shown as follows:

$$
\begin{aligned}
&\mathcal{O}(1) \subset \mathcal{O}(\log n) \subset \mathcal{O}(\log^c n) \\
&\quad \subset \mathcal{O}(n) \subset \mathcal{O}(n \log n) \subset \mathcal{O}(n^2) \subset \mathcal{O}(n^3) \subset \cdots \subset \mathcal{O}(n^c) \\
&\quad \subset \mathcal{O}(2^n) \subset \mathcal{O}(3^n) \subset \cdots \mathcal{O}(r^n) \\
&\quad \subset \mathcal{O}(n!) \\
&\quad \subset \mathcal{O}(n^n).
\end{aligned}
$$

Complexity functions involved in various powers of n often occur in the analysis of algorithms. In fact, we are more interested in polynomial complexity $p(n) = \mathcal{O}(n^c)$, where $p(n)$ is a polynomial in n of order c.

The asymptotic behaviour of a complexity function provides important information about whether it will be feasible to execute an algorithm for moderate or large values of input n. (Of course, any algorithm can be executed on small problems). This point is illustrated in the following two tables. In Tables 3.1, the growth of the various complexity functions has been illustrated by some selected *problem size* (i.e., *input size*) n. By examining this table, one can see that exponential and factorial complexity functions grow faster than any polynomial functions when n is large. Table 3.2 gives a comparison of the maximum sizes of problems that can be solved in one second, one minute, and one hour by each of the algorithms with eight common complexity functions. We assume that one unit of time equals one microsecond (10^{-6} second), i.e., an average execution time of one operation per microsecond is assumed. Proportional values hold for other speeds. In Table 3.3 we gives a comparison of execution

Input	Complexity Functions					
Size n	$\log n$	n	$n \log n$	n^2	2^n	$n!$
5	2	5	12	25	32	120
10	3	10	33	100	1024	3.6×10^6
5×10	6	50	282	2500	1.1×10^{15}	3×10^{64}
10^2	7	100	664	10^4	1.3×10^{30}	9.3×10^{157}
5×10^2	9	500	4483	25×10^4	3.3×10^{150}	1.2×10^{1134}
10^3	10	10^3	9966	10^6	1.1×10^{301}	4.0×10^{2567}
10^4	13	10^4	132877	10^8	1.9×10^{3010}	2.8×10^{35659}
10^5	17	10^5	1.6×10^6	10^8	(too large)	(too large)

Table 3.1: Comparison of Growth Rates of Complexity Functions with Input Sizes

Complexity	Maximum Number of Operations Can Be Done in				
Function	1 second	1 hour	1 day	1 year	1 century
$\log n$	2^{10^6}	$2^{3.6 \times 10^9}$	$2^{8.6 \times 10^{10}}$	$2^{3.1 \times 10^{13}}$	$2^{3.1 \times 10^{15}}$
$\sqrt{n}$	10^{12}	1.3×10^{19}	7.4×10^{21}	9.6×10^{26}	9.6×10^{30}
n	10^6	3.6×10^9	8.6×10^{10}	3.1×10^{13}	3.1×10^{15}
$n \log n$	6.3×10^4	1.3×10^8	2.8×10^9	7.9×10^{11}	6.8×10^{13}
n^2	10^3	6×10^4	2.9×10^5	5.6×10^6	5.6×10^7
n^3	10^2	1.5×10^3	4.4×10^3	3.1×10^4	1.5×10^5
n^5	15.8	81.5	1.5×10^2	5×10^2	1.3×10^3
2^n	19.9	31.7	36.3	44.8	51.5
3^n	12.6	20	22.9	28.3	32.5
5^n	8.6	13.7	15.6	19.3	22.2
$n!$	9.5	11.9	14	16.2	17.8

Table 3.2: Comparison of Maximum Input Sizes with Times and Complexity Functions

| Complexity | Execution Time on Input Size n | | | | | |
Function	10	20	30	40	50	60
n	0.00001 seconds	0.00002 seconds	0.00003 seconds	0.00004 seconds	0.00005 seconds	0.00006 seconds
n^2	0.0001 seconds	0.0004 seconds	0.0009 seconds	0.0014 seconds	0.0025 seconds	0.0036 seconds
n^3	0.001 seconds	0.008 seconds	0.027 seconds	0.064 seconds	0.125 seconds	0.216 seconds
n^5	0.1 seconds	3.2 seconds	24.3 seconds	1.7 minutes	5.2 minutes	13.0 minutes
2^n	0.01 seconds	1.0 seconds	17.9 minutes	12.7 days	35.7 years	366 centuries
3^n	0.59 seconds	58 minutes	6.5 years	3855 centuries	2.3×10^8 centuries	1.3×10^{13} centuries
5^n	9.8 seconds	3 years	3×10^5 centuries	2.9×10^{12} centuries	2.8×10^{19} centuries	2.8×10^{26} centuries
$n!$	3.6 seconds	7.7×10^4 years	8.4×10^{14} centuries	2.6×10^{32} centuries	9.6×10^{48} centuries	2.6×10^{66} centuries

Table 3.3: Comparison of Execution times of Complexity Functions

times for algorithms with some common complexity functions. For example, when $n = 50$, the computation of algorithms with $n \log n$ complexity will take 6×10^{-6} second, while for algorithms with exponential complexity 2^n will need 35.7 years.

One may think that improved computer technology (e.g., fast-speed and highly parallel computer architectures) might change the asymptotic complexity. Unfortunately, it is not true. In Figure 3.4, we show how the largest problem instance solvable in one hour would change if we had a computer 100 or 1000 times faster than our present computer. Observe that with the 3^n algorithm a thousand times faster computer only adds about 7 to the size of the largest problem instance we can solve in an hour, whereas with the n^5 algorithm this size increases nearly four times.

Complexity	Present and Improved Computer Technology		
Function	With Present Computer	With Computers 100 Times Faster	With Computers 1000 Times Faster
n	N_1	$100N_1$	$1000N_1$
n^2	N_2	$10N_2$	$31.6N_2$
n^3	N_3	$4.64N_3$	$10N_3$
n^5	N_4	$2.5N_4$	$3.98N_4$
2^n	N_5	$6.44 + N_5$	$9.97 + N_5$
3^n	N_6	$4.19 + N_6$	$6.29 + N_6$

Table 3.4: Effect of Improved Technology on Complexity Functions

3.4.2   Analysis of Algorithms

Analysing even a simple algorithm can be a challenge, because it often requires deep mathematical knowledge from many areas of mathematics, including combinatorics, probability theory, number theory, and real/complex analysis. As we have already seen, the complexity we are interested in is measured by the *input size* n of a problem. There are two conceptually different interpretations of the input size n of a problem:

(i) n is the *number of items in the input* – for many problems such as sorting and computing discrete Fourier transforms, the most natural measure is the number of items in the input; for example, a list of integers for sorting, so in this case, a *"large input"* means an input containing "many integers".

(ii) n is the *number of bits in the input* – for many other problems such as multiplying two integers, the best measure of input size is the *total number of bits* needed to represent the input in ordinary binary notation. For example, to factor an integer into its prime decomposition form; in this case, a *"large input"* means an input containing "large integers" rather than "many integers".

In general, the computation time taken by an algorithm grows with the size of the input, so it is traditional to describe the running time of an algorithm as a function of the size of the its input. The *running time* of an algorithm on a particular input is the number of *primitive operations* or *steps* executed. Of course, such primitive operations or steps should be as machine-independent as possible. In this book, the primitive operations or steps we consider will either be *arithmetic operations* or *bit operations* or may be both. But in Chapter 4, we shall mainly consider bit operations rather than arithmetic operations as our primitive operations in the analysis of number-theoretic algorithms.

(I) Complexity Measured by Arithmetic Operations

We are all familiar with matrix operations from high-school. Suppose we wish to multiply two $n \times n$ matrices A and B. Using the usual method, we get

$$C = AB = \left\{ \sum_{k=1}^{n} a_{ik}b_{kj}, \quad i, j = 1, 2, \cdots, n \right\}. \tag{3.101}$$

It is clear that this method will need $\mathcal{O}(n^3)$ arithmetic operations. However, the number of arithmetic operations can be reduced by a simple divide-and-conquer method. Let us first examine the multiplication of two 2×2 matrices:

$$\left(\begin{array}{c|c} A_1 & A_2 \\ \hline A_3 & A_4 \end{array} \right) = \left(\begin{array}{c|c} B_1 & B_2 \\ \hline B_3 & B_4 \end{array} \right) \left(\begin{array}{c|c} C_1 & C_2 \\ \hline C_3 & C_4 \end{array} \right) \tag{3.102}$$

Using Strassen's idea, we can get:

$$\left\{ \begin{array}{l} A_1 = P_0 + P_5 - P_4 + P_6 \\ A_2 = P_2 + P_4 \\ A_3 = P_1 + P_5 \\ A_4 = P_0 - P_1 + P_2 + P_3 \end{array} \right. \quad \text{where} \quad \left\{ \begin{array}{l} P_0 = (B_1 + B_4)(C_1 + C_4) \\ P_1 = (B_3 + B_4)C_1 \\ P_2 = B_1(C_2 - C_4) \\ P_3 = (-B_1 + B_3)(C_1 + C_2) \\ P_4 = (B_1 + B_2)C_4 \\ P_5 = B_4(-C_1 + C_3) \\ P_6 = (B_2 - B_4)(C_3 + C_4) \end{array} \right. \tag{3.103}$$

It is plain that we will need 7 multiplications and 18 additions/subtractions. But the advantage of this algorithm is that we can recursively apply it to a $2^k \times 2^k$ matrix product:

$$A \Rightarrow \left(\begin{array}{c|c} A_1 & A_2 \\ \hline A_3 & A_4 \end{array} \right) \Rightarrow \left(\begin{array}{cc|cc} A_{11} & A_{12} & A_{21} & A_{22} \\ A_{13} & A_{14} & A_{23} & A_{24} \\ \hline A_{31} & A_{32} & A_{41} & A_{42} \\ A_{33} & A_{34} & A_{43} & A_{44} \end{array} \right) \Rightarrow \cdots \cdots \tag{3.104}$$

which will only need $M(n) = 7^k$ multiplications (by induction), rather than the upper bound $(2^k)^3 = 8^k$. So the complexity is $M(n) = \mathcal{O}(n^{\log_2 7}) = \mathcal{O}(n^{2.807354922})$ rather than the upper bound $M(n) = \mathcal{O}(n^3)$. That is,

Theorem 3.4.1 Two $n \times n$ matrices over an arbitrary ring can be multiplied in $\mathcal{O}(n^{\log_2 7})$ arithmetic operations.

By using some complicated mathematical ideas, this low bound can be reduced to $M(n) = \mathcal{O}(n^{\log_{70} 143640}) = \mathcal{O}(n^{2.795122690})$. Based on Strassen's work, Pan [105] has improved the lower bound to $M(n) = \mathcal{O}(n^{2.496})$. Note that the above idea can be extended to any $n \times n$ matrix by padding the matrix with 0 if it is not in the form 2^k, which is so because

$$\left(\begin{array}{c|c} A & 0 \\ \hline 0 & 0 \end{array}\right) = \left(\begin{array}{c|c} B & 0 \\ \hline 0 & 0 \end{array}\right)\left(\begin{array}{c|c} AB & 0 \\ \hline 0 & 0 \end{array}\right). \tag{3.105}$$

Of course, every algorithm must be $\Omega(n^2)$, since it has to look at all the entries of the matrices; no better lower bound is possible.

(II) Complexity Measured by Bit Operations

Any integer $n \in \mathbb{N}$ written to the base b is a notation of the form

$$\begin{aligned} n &= (d_{\beta-1} d_{\beta-2} \cdots d_1 d_0)_b \\ &= d_{\beta-1} b^{\beta-1} + d_{\beta-2} b^{\beta-2} + \cdots + d_1 b + d_0 \end{aligned} \tag{3.106}$$

where d_i $(i = \beta - 1, \beta - 2, \cdots, 1, 0)$ are digits. If $d_{\beta-1} \neq 0$, we call n a β-digit base-b number. Clearly, any number $b^{\beta-1} \leq n < b^\beta$ is a β-digit number to the base b. For example, $10^5 \leq 780214 < 10^6$ is a 6-digit number to the base 10. By the definition of logarithms, this gives the following formula for the number of base-b digits for n:

$$\begin{aligned} \text{number of digits of } n &= \lfloor \log_b n \rfloor + 1 \\ &= \left\lfloor \frac{\ln n}{\ln b} \right\rfloor + 1 \\ &= \mathcal{O}(\log n) \end{aligned} \tag{3.107}$$

For example, let $n = 999$, then

$$
\begin{aligned}
\text{the number of digits of } 999 &= \lfloor \log_{10} 999 \rfloor + 1 \\
&= \left\lfloor \frac{\ln 999}{\ln 10} \right\rfloor + 1 \\
&= \lfloor 2.999565488 \rfloor + 1 \\
&= 2 + 1 = 3 \\
\text{the number of bits of } 999 &= \lfloor \log_2 999 \rfloor + 1 \\
&= \left\lfloor \frac{\ln 999}{\ln 2} \right\rfloor + 1 \\
&= \lfloor 9.964340868 \rfloor + 1 \\
&= 9 + 1 = 10
\end{aligned}
$$

It is easy to verify that 999 has 10 bits, since $999 = 1111100111$. Note that the word "bits" is short for "binary digits", and usually refer to Shannon[7] bits. In Chapter 5, we shall introduce quantum bits (qubits for short), in which one bit can have infinitely many states, rather than just two states as the Shannon bits. Now let us observe a simple arithmetic problem, the addition of two β-bit binary integers (if one of the two integers has fewer bits than the other, we just fill in zeros in the left), for example;

$$
\begin{array}{r}
11101011000 \\
+ \quad 01000110101 \\
\hline
100110001101
\end{array}
$$

Clearly, we must repeat the following steps β times:

(i) look at the top and bottom bits, and also at whether there is a carry above the top bit.

(ii) if both bits are 0 and there is no carry, then put down 0 and move on.

(iii) if either one of the following occurs

---

[7]Claude E. Shannon (1916–) received his PhD in mathematics from MIT in 1940. He joined MIT as Professor in Electrical Engineering in 1958, and has been Professor Emeritus there since 1980. Shannon is the inventor of information theory, the first to apply Boolean algebra to the design of circuits, and the first to use "bits" to represent information.

- both bits are 0 and there is a carry
- one of the bits is 0 and there is no carry

then put down 1 and move on.

(iv) if either one of the following occurs

- both bits are 1 and there is no carry
- one of the bits is 0 and there is a carry

then put down 0, put a carry on the next column, and move on.

(v) if both bits are 1 and there is a carry, then put down 1, put a carry on the next column, and move on.

Doing this procedure *once* is called a *bit operation*. So adding two β-bit numbers requires β bit operations. That is,

$$
\begin{aligned}
T(\beta\text{-bits} + \beta\text{-bits}) &= \mathcal{O}(\beta) \\
&= \mathcal{O}(\log n)
\end{aligned}
$$

To multiply two β-bit integer, for example

$$
\begin{array}{r}
11101011001 \\
\times \quad 01000110101 \\
\hline
11101011001 \\
11101011001 \\
11101011001 \\
11101011001 \\
+ \ 11101011001 \\
\hline
100000011011101101101
\end{array}
$$

that is,

$$11101011001 + 1110101100100 = 10010010111101$$
$$10010010111101 + 111010110010000 = 1001101010010011101$$
$$1001101010010011101 + 1110101100100000 = 1001101001001101$$
$$11000010101101101 + 11101011001000000000 = 100000011011101101101.$$

and hence,

$$11101011001 \cdot 1000110101 = 100000011011101101101.$$

The result can easily be verified to be correct, since

$$11101011001 = 1881, \quad 1000110101 = 565$$
$$1881 \cdot 565 = 1062765 = 100000011011101101101_2.$$

The above example tells us that multiplying two β-bit integers requires at most β^2 bit operations. That is,

$$\begin{aligned} T(\beta\text{-bits} \times \beta\text{-bits}) &= \mathcal{O}(\beta^2) \\ &= \mathcal{O}(\log n)^2 \end{aligned}$$

How fast can we multiply two integers? Earlier attempts at improvements employed simple algebraic identities and resulted with a reduction to

Theorem 3.4.2 There is an algorithm that can multiply two β-bit integers in

$$\begin{aligned} T(\beta\text{-bits} \times \beta\text{-bits}) &= \mathcal{O}(\beta^{\log_2 3}) \\ &= \mathcal{O}(\beta^{1.584962501}) \\ &= \mathcal{O}(\log n)^{1.584962501} \end{aligned}$$

bit operations.

Schönhage and Strassen in 1971 utilized some number-theoretic ideas and FFT to obtain the following theorem:

Theorem 3.4.3 There is an algorithm that can multiply two β-bit integers in

$$\begin{aligned} T(\beta\text{-bits} \times \beta\text{-bits}) &= \mathcal{O}(\beta \log \beta \log \log \beta) \\ &= \mathcal{O}(\log n \log \log n \log \log \log n) \end{aligned}$$

bit operations. (Assuming $\beta = \mathcal{O}(n)$).

As far as for the lower bounds for complexity of integer multiplication, Paterson, Fisher and Meyer obtained the following result:

Theorem 3.4.4 At least

$$\begin{aligned} T(\beta\text{-bits} \times \beta\text{-bits}) &= \mathcal{O}\left(\beta \frac{\log \beta}{\log \log \beta}\right) \\ &= \mathcal{O}\left(\log n \frac{\log \log n}{\log \log \log n}\right) \end{aligned}$$

bit operations are necessary for multiplying two β-bit integers.

3.4.3   Design of Efficient Algorithms

In computer science, we are not just interested in design an algorithm, but an efficient algorithm. In this subsection we shall study two efficient algorithms for modular exponentiations and point additions on elliptic curves, the two most important operations in modern number-theoretic computations.

(I) Fast Algorithm for Modular Exponentiations

A frequently occuring operation in computer science is raising one number to a power modulo another number $x^e \bmod m$, also knows as *modular exponentiation*. The conventional method of *repeated multiplication* would take $\Theta(e \log^2 n)$ bit operations, which is too slow when e is large. Fortunately, the method of *repeated squaring* will solve this problem efficiently using the binary representation of b. The idea of the repeated squaring method is as as follows:

Theorem 3.4.5 Suppose we want to compute $x^e \bmod m$ with $x, e, m \in \mathbb{Z}^+$. Suppose moreover that the binary form of e is as follows:

$$e = \beta_k 2^k + \beta_{k-1} 2^{k-1} + \cdots + \beta_1 2^1 + \beta_0 2^0 \tag{3.108}$$

where each β_i $(i = 0, 1, 2, \cdots k)$ is either 0 or 1. Then we have

$$x^e = x^{\beta_k 2^k + \beta_{k-1} 2^{k-1} + \cdots + \beta_1 2^1 + \beta_0 2^0} \tag{3.109}$$

$$= \prod_{i=0}^{k} x^{\beta_i 2^i} \tag{3.110}$$

$$= \prod_{i=0}^{k} \left(x^{2^i} \right)^{\beta_i} \tag{3.111}$$

Further, by the exponentiation law

$$x^{2^{i+1}} = (x^{2^i})^2$$

the final value of the exponentiation can be obtained by *repeated squaring* operations.

Example 3.4.1 Suppose we wish to compute a^{100}, we first write $100_{10} = 1100100_2$, and then compute

$$a^{100} = (((((((a)^2 \cdot a)^2)^2)^2 \cdot a)^2)^2 \tag{3.112}$$
$$\Rightarrow a, a^3, a^6, a^{12}, a^{24}, a^{25}, a^{50}, a^{100}.$$

Note that for each "1" in the binary form, we perform a squaring and a multiplication (except the first "1", for which we just write down a, indicated in the first bracket); whereas for each "0" in the binary form, we perform only a squaring.

Exercise 3.4.1 Write down the similar expressions as in (3.112) for computing x^{931} and x^{6501}. (Hints: $931_{10} = 1110100011_2$ and $6501_{10} = 1100101100101$).

Now we are in a position to introduce a fast algorithm for modular exponentiations (note that we can just simply remove the "mod m" operation if we only wish to compute the exponentiation $c = x^e$):

Algorithm 3.4.1 This algorithm will compute the modular exponentiation

$$c = x^e \bmod m,$$

where $x, m \in \mathbb{N}$ and $e \in \mathbb{Z}^+$.

[1] (Precomputation) Write the power e in the following binary form

$$e_{10} = (e_1 e_2 \cdots e_r)_2,$$

where each e_i is either 1 or 0. For example, $5 = 101$ with $r = 3$ and $562 = 1000110010$ with $r = 10$. Note that we write the binary form in the order $e_1 e_2 \cdots e_r$, rather than the conventional order $e_r e_{r-1} \cdots e_1 e_0$.

[2] (Initialization) Set $c \leftarrow 1$.

[3] (Modular Exponentiation) Compute $c = x^e \bmod m$ in the following way:

> for i from 1 to r do
> $\quad c \leftarrow c^2 \bmod m$
> $\quad$ if $e_i = 1$ then
> $\quad\quad c \leftarrow c \cdot x \bmod m$

[4] Print c and terminate the algorithm.

Theorem 3.4.6 Suppose e is a β-bit number, or more generally, x, e and m are β-bit numbers, then the total number of arithmetic operations required for Algorithm 3.4.1 to compute $x^e \bmod m$ is $\mathcal{O}(\beta)$, or $\mathcal{O}(\log e)$. That is,

$$T(\text{compute } x^e \bmod m) = \mathcal{O}(\beta) = \mathcal{O}(\log e). \qquad (3.113)$$

Example 3.4.2 Use the above algorithm to compute $7^{9007} \bmod 561$ (here $x = 7$, $e = 9007$ and $m = 561$). By writing e in the binary form $e = e_1 e_2 \cdots e_r$, we have $9007 = 10001100101111$ with $k = 14$. Now we just perform the following computations as described in Algorithm 3.4.1:

$$c \leftarrow 1$$
$$x \leftarrow 7$$
$$m \leftarrow 561$$
for i from 1 to 14 do
$\qquad c \leftarrow c^2 \bmod m$
$\qquad$ if $e_i = 1$ then $c \leftarrow c \cdot x \bmod m$
print c; (now $c = x^e \bmod m$)

The results at each loop for i from 1 to 14 are as follows:

i	1	2	3	4	5	6	7	8	9	10	11	12	13	14
e_i	1	0	0	0	1	1	0	0	1	0	1	1	1	1
c	7	49	157	526	160	241	298	166	469	49	538	337	46	226

So, at the end of the computation, the final result $c = 7^{9007} \bmod 561 = 226$ will be returned. It is clear that at most $2 \log_2 9007$ multiplications and $2 \log_2 9007$ divisions will be needed to complete the computation. In fact, only 22 multiplications and 22 divisions will be needed to complete this computation task.

Exercise 3.4.2 Use the above fast exponentiation method to compute $F = 3^{4294967296} \bmod 4294967297$ by completing the F items in following table (note that $4294967296 = 1 \underbrace{000 \cdots 00}_{32 \text{ zeroes}}$ in binary):

i	1	2	3	4	5	6	$\cdots\cdots$	31	32	33
e_i	1	0	0	0	0	0	$\cdots\cdots$	0	0	0
F	3	9	81	6561	43046721	3793201458				

What is the final value for F? (Historical Note: Fermat in 1640 conjectured that all the numbers of the form $F_n = 2^{2^n} + 1$, for $n \geq 0$, were prime, after he had verified the five numbers 3, 5, 17, 257, 65537 corresponding to $n = 0, 1, 2, 3, 4$ of F_n. Of course, Fermat was wrong, since $F_5 = 2^{2^5} + 1 = 4294967297$ is composite. Fermat should not really have made such a false conjecture, since F_5 can be relatively easily verified to be composite, by using the fact that "if $\gcd(a, n) = 1$ and $a^{n-1} \not\equiv 1 \pmod{n}$, then n is composite"; see Corollary 1.3.2

in Chapter 1. So, once you have correctly completed the above table, you shall be able to deduce that F_5 is composite).

Remark 3.4.2 The above introduced fast exponentiation algorithm is about half as good as the best; more efficient algorithms are known. For example, Brickell, et. al. [20] developed a more efficient algorithm, using precomputed values to reduce the number of multiplications needed. Their algorithm allows the computation of g^n for $n < N$ in time $\mathcal{O}(\log N / \log \log N)$. They also showed that their method can be parallelized, to compute powers in time $\mathcal{O}(\log \log N)$ with $\mathcal{O}(\log N / \log \log N)$ processors.

(II) Fast Algorithm for Point Additions on Elliptic Curves

A similar problem to exponentiations discussed above is to compute

$$kP = \underbrace{P + P + \cdots + P}_{k \text{ times}} \qquad (3.114)$$

or

$$kP = \underbrace{P + P + \cdots + P}_{k \text{ times}} \mod N, \qquad (3.115)$$

for large k, where P is a point on an elliptic curve $E : y^2 = x^3 + ax + b$ over $\mathbb{Z}/N\mathbb{Z}$ (see subsection 1.4.4 in Chapter 1 for more information about elliptic curves). A straightforward method will need $\mathcal{O}(k)$ group operations. For example, to find the multiples of the point $P = (3, 2)$ on the elliptic curve $E : y^2 = x^3 - 2x - 3$ over $\mathbb{Z}/7\mathbb{Z}$, we use the following formula:

$$\begin{cases} x_3 = \left(\dfrac{3x_1^2 + a}{2y_1}\right)^2 - 2x_1 \pmod{N}, \\[2mm] y_3 = \left(\dfrac{3x_1^2 + a}{2y_1}\right)(x_1 - x_3) - y_1 \pmod{N} \end{cases} \qquad (3.116)$$

to compute $2P = P + P$ (here $x_1 = 3$, $y_1 = 2$, $a = -2$, $N = 7$):

$$\begin{cases} x_3 = \left(\dfrac{3 \cdot 3^2 - 2}{2 \cdot 2}\right)^2 - 2 \cdot 3 \pmod{7} = 2 \\[2mm] y_3 = \left(\dfrac{3 \cdot 3^2 - 2}{2 \cdot 2}\right)(3 - 2) - 2 \pmod{7} = 6 \end{cases}$$

So, $2P = (2,6)$. It is easy to verify that the point $(2,6)$ is on the elliptic curve $E : y^2 = x^3 - 2x - 3$ over $\mathbb{Z}/7\mathbb{Z}$, since $6^2 \pmod{7} = 2^3 - 2 \cdot 2 - 3 \pmod{7} = 1$. Next we use the following formula:

$$
\begin{cases}
x_3 = \left(\dfrac{y_2 - y_1}{x_2 - x_1}\right)^2 - x_1 - x_2 \pmod{N}, \\[2mm]
y_3 = \left(\dfrac{y_2 - y_1}{x_2 - x_1}\right)(x_1 - x_3) - y_1 \pmod{N}
\end{cases}
\tag{3.117}
$$

to compute $3P = P + 2P$ (now $x_1 = 3$, $y_1 = 2$, $x_2 = 2$, $y_2 = 6$, $a = -2$, $N = 7$):

$$
\begin{cases}
x_3 = \left(\dfrac{6-3}{2-3}\right)^2 - 3 - 2 \pmod{7} = 4, \\[2mm]
y_3 = \left(\dfrac{6-3}{2-3}\right)(3 - 4) - 2 \pmod{7} = 2.
\end{cases}
$$

So, $3P = (4,2)$. Similarly, we have

$$
\begin{aligned}
4P &= P + 3P = (3,2) + (4,2) = (0,5), & 5P &= P + 4P = (3,2) + (0,5) = (5,0), \\
6P &= P + 5P = (3,2) + (5,0) = (0,2), & 7P &= P + 6P = (3,2) + (0,2) = (4,5), \\
8P &= P + 7P = (3,2) + (4,5) = (2,1), & 9P &= P + 8P = (3,2) + (2,1) = (3,5), \\
& & 10P &= P + 9P = (3,2) + (3,5) = \mathcal{O}_E.
\end{aligned}
$$

That is, there are 10 points on the elliptic curve $y^2 = x^3 - 2x - 3$ over $\mathbb{Z}/7\mathbb{Z}$. This can be verified by using point counting formula (1.181) in Chapter 1:

$$
1 + 7 + \sum_{x \in \mathbb{Z}/7\mathbb{Z}} \left(\frac{x^3 + ax + b}{7}\right) = 10.
$$

This method is indeed "straightforward" and also "easy" to implement, but its main drawback is that it is slow and requires $\mathcal{O}(k)$ arithmetic operations. Since we intend to perform many such doublings and additions in elliptic curve related computations, it is important that these basic operations be performed as efficient as possible. Remarkably enough, the idea of *repeated squaring* for fast exponentiations can be almost directly used to fast point addition of elliptic curves. The idea of the fast point addition is just as follows:

[1] compute P, $2P$, $\cdots$, $2^r P$, with $2^r \leq k < 2^{r+1}$, by repeated doubling,

[2] add together suitable multiples of P, determined by the binary expansion of k, to get kP.

For example, to compute $109P$, we just perform the following computation steps:

[1] determine r: since we require $2^r \leq k < 2^{r+1}$, then $2^6 < 109 < 2^7$, so $r = 6$.

[2] repeated double P, $2P$, $4P$, $4P$, $8P$, $16P$, $32P$, $64P$, since $r = 6$ and $2^6 = 64$.

[3] Now $k = 109_{10} = 1101101_2$. Then we have:

$64P$		$32P$		$16P$		$8P$		$4P$		$2P$		P
$64P$	$+$	$32P$				$8P$	$+$	$4P$				P
1		1		0		1		1		0		1

So we finally have $109P = 64P + 32P + 8P + 4P + P$. Note that additions on the suitable multiples of P is determined by the binary expansion of k.

It is clear that using the repeated doubling method to compute $109P$, only 6 doublings and 4 additions will be needed; it is much faster than the straightforward method that would need 108 additions. In what follows, we shall present a systematic procedure to implement the idea of the *repeated doubling* for computing kP.

Algorithm 3.4.2 This algorithm computes kP, where k is a large integer, and P is assumed to be a point on an elliptic curve $E : y^2 = x^3 + ax + b$ (note that we will not actually do the additions for the coordinates of P in this algorithm).

[1] Write k in the binary expansion form $k = e_1 e_2 \cdots e_\beta$, where each e_i is either 1 or 0. (Assume k has β bits).

[2] Set $c \leftarrow 0$

[3] Compute kP:

$$\text{for } i \text{ from 1 to } \beta \text{ do}$$
$$c \leftarrow 2c$$
$$\text{if } e_i = 1 \text{ then } c \leftarrow c + P$$

[4] Print c; (now $c = kP$)

Example 3.4.3 Use Algorithm 3.4.2 to compute $100P$. Let

$$k = 105 = 1101001 := e_1 e_2 e_3 e_4 e_5 e_6 e_7.$$

At the initial stage of the algorithm, we set $c = 0$. Now we perform the following computation steps according to Algorithm 3.4.2:

$$
\begin{array}{lllll}
e_1 = 1: & c \leftarrow P + 2c & \Longrightarrow & c \leftarrow P & \Longrightarrow & c = P \\
e_2 = 1: & c \leftarrow P + 2c & \Longrightarrow & c \leftarrow P + 2P & \Longrightarrow & c = 3P \\
e_3 = 0: & c \leftarrow 2c & \Longrightarrow & c \leftarrow 2(P + 2P) & \Longrightarrow & c = 6P \\
e_4 = 1: & c \leftarrow P + 2c & \Longrightarrow & c \leftarrow P + 2(2(P + 2P)) & \Longrightarrow & c = 13P \\
e_5 = 0: & c \leftarrow 2c & \Longrightarrow & c \leftarrow 2(P + 2(2(P + 2P))) & \Longrightarrow & c = 26P \\
e_6 = 0: & c \leftarrow 2c & \Longrightarrow & c \leftarrow 2(2(P + 2(2(P + 2P)))) & \Longrightarrow & c = 52P \\
e_7 = 1: & c \leftarrow P + 2c & \Longrightarrow & c \leftarrow P + 2(2(2(P + 2(2(P + 2P))))) & \Longrightarrow & c = 105P.
\end{array}
$$

That is, $P + 2(2(2(P + 2(2(P + 2P))))) = 105P$.

Example 3.4.4 Suppose we want to compute $kP \bmod 1997$, where $k = 9007 = 100011001011112$. The computation can be summarized in the following table that shows the values for each execution of the **"for"** loop in Algorithm 3.4.2 (plus an additional modular operation $\bmod 1997$ at the end of each loop):

i	13	12	11	10	9	8	7	6	5	4	3	2	1	0
e_i	1	0	0	0	1	1	0	0	1	0	1	1	1	1
c	P	2P	4P	8P	17 P	35 P	70 P	140 P	281 P	562 P	1125 P	254 P	509 P	1019 P

The final result of the computation is $c \equiv 1019P \pmod{1997}$. It is clear that the above computation will need at most $\mathcal{O}(\log 9007)$ arithmetic operations.

Note that Algorithm 3.4.2 does not actually calculate the coordinates (x, y) of kP on an elliptic curve over $\mathbb{Q}$ or over $\mathbb{Z}/N\mathbb{Z}$. To make Algorithm 3.4.2 a practically useful algorithm for point additions on an elliptic curve E, we must however incorporate the actual coordinate addition $P_3(x_3, y_3) = P_1(x_1, y_1) + P_2(x_2, y_2)$ on E into the algorithm. To do this, we use the following formulas to compute x_3 and y_3 for P_3:

$$(x_3, y_3) = (\lambda^2 - x_1 - x_2, \ \lambda(x_1 - x_3) - y_1)$$

where

$$\lambda = \begin{cases} \dfrac{3x_1^2 + a}{2y_1} & \text{if } P_1 = P_2 \\[2mm] \dfrac{y_1 - y_2}{x_1 - x_2} & \text{otherwise.} \end{cases}$$

Algorithm 3.4.3 This algorithm will compute the point kP, where $k \in \mathbb{Z}^+$ and P is an initial point (x, y) on an elliptic curve $E : y^2 = x^3 + ax + b$ over $\mathbb{Z}/N\mathbb{Z}$; if we require E over $\mathbb{Q}$, just compute kP, rather than $kP \bmod N$. Let the initial point $P = (x_1, y_1)$, and the result point $P = (x_c, y_c)$.

[1] (Precomputation) Write k in the following binary expansion form $k = e_1 e_2 \cdots, e_\beta$. (Suppose k has β bits).

[2] (Initialization) Initialize the values for a, x_1 and y_1. Let $(x_c, y_c) = (x_1, y_1)$; this is exactly the computation task for e_1 (e_1 always equals 1).

[3] (Doublings and Additions) Computing $kP \bmod N$. Note that the computation loop will start from 2, 3, 4, upto β (β is the total number of bits in k).

$$
\begin{aligned}
&\text{for } i \text{ from 2 to } \beta \text{ do} \\
&\quad m_1 \leftarrow 3x_c^2 + a \bmod N \\
&\quad m_2 \leftarrow 2y_c \bmod N \\
&\quad M \leftarrow m_1/m_2 \bmod N \\
&\quad x_3 \leftarrow M^2 - 2x_c \bmod N \\
&\quad x_3 \leftarrow M^2 - 2x_c \bmod N \\
&\quad y_3 \leftarrow M(x_c - x_3) - y_c \bmod N \\
&\quad x_c \leftarrow x_3 \\
&\quad y_c \leftarrow y_3 \\
&\quad \text{if } e_i = 1 \\
&\quad\quad \text{then } c \leftarrow 2c + P \\
&\quad\quad\quad m_1 \leftarrow y_c - y_1 \bmod N \\
&\quad\quad\quad m_2 \leftarrow x_c - x_1 \bmod N \\
&\quad\quad\quad M \leftarrow m_1/m_2 \bmod N \\
&\quad\quad\quad x_3 \leftarrow M^2 - x_1 - x_c \bmod N \\
&\quad\quad\quad y_3 \leftarrow M(x_1 - x_3) - y_1 \bmod N \\
&\quad\quad\quad x_c \leftarrow x_3 \\
&\quad\quad\quad y_c \leftarrow y_3 \\
&\quad\quad \text{else } c \leftarrow 2c
\end{aligned}
$$

[4] Print c (now $c = kP \bmod N$) and terminate the algorithm. (Note that this algorithm will stop whenever $m_1/m_2 \equiv \mathcal{O}_E \pmod{N}$, that is, it will stop whenever a modular inverse does not exit at any step of the computation).

Theorem 3.4.7 Assume k is a β-bit number, then Algorithm 3.4.3 requires only $\mathcal{O}(\beta)$ arithmetic operations. That is,

$$T(\text{computing } kP \bmod N) = \mathcal{O}(\beta) = \mathcal{O}(\log k). \tag{3.118}$$

Exercise 3.4.3 Let

$$E : y^2 = x^3 - x - 1$$

be an elliptic curve over $\mathbb{Z}/1098413\mathbb{Z}$ and $P = (0, 1)$ be a point on E. Use Algorithm 3.4.3 to compute the coordinates (x, y) of the following points kP on E modulo $\mathbb{Z}/1098413\mathbb{Z}$:

$$8P, \qquad 31P, \qquad 92P, \qquad 261P, \qquad 320P, \qquad 7892P, \qquad 10319P.$$

3.4.4 Relations between Arithmetic and Bit Operations

Both arithmetic and bit operations are important measures of complexity in the design and analysis of algorithms (to avoid any possible confusion, we may use $\mathcal{O}_A(\cdot)$ and $\mathcal{O}_B(\cdot)$ to denote the arithmetic and bit operation measures of complexity, respectively). In certain cases, it is sufficient to only consider the cost of arithmetic operations. For example, the computation of the product of two polynomials in one variable:

$$p(x)q(x) = \left(\sum_{i=0}^{n} a_i x^i \right) \left(\sum_{i=0}^{n} a_i x^i \right) = \sum_{k=0}^{2n-2} c_k x^k, \quad \text{where} \quad c_k = \sum_{m=0}^{n-1} a_m b_{k-m}$$

will require $\mathcal{O}(n \log n)$ arithmetic operations by use of the Fast Fourier Transform (FFT). That is, two n-th degree polynomials with real coefficients can be multiplied in time (we use e.g., $T(n)$ to denote the time complexity):

$$T(\text{evaluate } p(x)q(x)) = \mathcal{O}_A(n \log n) \tag{3.119}$$

by use of the FFT. The arithmetic operation measure of complexity is reasonable, since in practice, we would represent polynomials by their coefficients to fixed precision and implement the various polynomial operations by arithmetic operations on the coefficients. However, we claim that for number-theoretic computations, as we shall discuss in detail in the next chapter, bit operations are the only measure of interest. To see this, let us observe the following algorithm for primality testing (see also the compositeness test in Example 3.3.4):

Algorithm 3.4.4 (Trial-Division for Primality Testing)

> Input $N \in \mathbb{N}_{>1}$
> for i from 2 to $\sqrt{N}$ do
>   if $\forall (i \bmod N) \neq 0$ then N is prime
>   otherwise, N is composite

This algorithm of course requires "only" $\mathcal{O}_A(\sqrt{N})$ arithmetic operations. It seems to be a very efficient algorithm, but unfortunately it is not, since in terms of bit operations, this algorithm requires $\mathcal{O}_B(2^{0.5 \log N})$ bit operations (recall that if n is encoded in binary using β bits, then, $\beta = \lfloor \log N \rfloor + 1$, so $\sqrt{N} = \mathcal{O}(2^{\beta/2})$), which grows faster than any polynomial complexity function. In short, the algorithm is of complexity

$$T(\text{primality test for } N) = \mathcal{O}_A(N^{1/2}) \qquad (3.120)$$
$$= \mathcal{O}_B(2^{0.5 \log N}) \qquad (3.121)$$

and hence it is very inefficient and more specifically hopeless to test numbers larger than e.g., 40 digits. An algorithm is said to be *efficient* if its worst case execution time is bounded by a polynomial function of the *length* of the input. For example, if we want to test N for primality, whose length is $\mathcal{O}(\log N)$, then an $\mathcal{O}(\log N)^{10}$ algorithm would be polynomial time, whereas $\mathcal{O}(N)^{0.1}$ is not. More generally, we have the following definition for efficient algorithms:

Definition 3.4.3 An algorithm to perform a computation involving integers $N_1, N_2, \cdots, N_r$ of $\beta_1, \beta_2, \cdots, \beta_r$ bits, respectively, is said to be polynomial time algorithm, if there exist integers $k_1, k_2, \cdots, k_r$ such that the number of bit operations (denoted by $T(N)$) required to perform the computation is

$$T(N) = \mathcal{O}_B\left(\beta_1^{k_1}, \beta_2^{k_2}, \cdots, \beta_r^{k_r}\right) \qquad (3.122)$$
$$= \mathcal{O}_B\left((\log N_1)^{k_1}, (\log N_2)^{k_2}, \cdots, (\log N_r)^{k_r}\right). \qquad (3.123)$$

This leads us to the following important conclusion:

> *The complexity measured in terms of arithmetic operations is often not very natural and realistic for number-theoretic computations such as primality testing and integer factorization, since e.g., a polynomial-time factoring algorithm measured in arithmetic operations would be essentially useless (see the next exercise). For these type of computations, it is best to consider the complexity measured in terms of bit operations, rather than in arithmetic operations.*

Exercise 3.4.4 In 1979, Adi Shamir proposed a deterministic algorithm that can factor integer N in $\mathcal{O}_A(\log N)$ arithmetic operations [128]. Is this algorithm practically useful? What is the complexity of this algorithm if it is measured in terms of bit operations? (Hints: the result that N can be factored in $\mathcal{O}(\log N)$ arithmetic operations is useless since the operations are on numbers as large as 2^n).

In practice, we can reclassify the computational complexity classes, in terms of *bit operations*, into the following four categories:

$\mathcal{O}\left((\log N)^c\right)$	Polynomial Complexity	Tractable	fast
$\mathcal{O}\left((\log N)^{c\log\log\log N}\right)$	Superpolynomial Complexity		$\uparrow$
$\mathcal{O}\left(N^{c\sqrt{\log\log N/\log N}}\right)$	Subexponential Complexity	Intractable	$\downarrow$
$\mathcal{O}\left(N^{\epsilon}\right)$	Exponential Complexity		slow

Generally speaking, we only think of problems that are solvable by polynomial-time algorithms as being *tractable*, and problems that require superpolynomial-time or more as being *intractable*.

To see why $\mathcal{O}\left((\log N)^c\right)$ is of polynomial order, whereas $\mathcal{O}\left(N^{\epsilon}\right)$ is not, let us compare the complexity functions

$$f(N) = \left\{(\log N)^{10},\ (\log N)^{\log\log N},\ N^{\sqrt{\log\log N/\log N}},\ N^{0.1}\right\}$$

for

$$N = \left\{10^{100},\ 10^{1000},\ 10^{10000},\ 10^{100000}\right\}.$$

Note first that the log here is natural log; and secondly we express

$$N^{0.1} = 2^{0.1\log N}$$

and

$$N^{\sqrt{\log\log N/\log N}} = 2^{\log N\sqrt{\log\log N/\log N}}.$$

Then we have:

$f(N)$	$N = 10^{100}$	$N = 10^{1000}$	$N = 10^{10000}$	$N = 10^{100000}$
$(\log N)^{10}$	$4.2 \cdot 10^{23}$	$4.2 \cdot 10^{33}$	$4.2 \cdot 10^{43}$	$4.2 \cdot 10^{53}$
$(\log N)^{\log \log N}$	$7.1 \cdot 10^{12}$	$1.1 \cdot 10^{26}$	$6.5 \cdot 10^{43}$	$1.6 \cdot 10^{66}$
$N^{\sqrt{\log \log N / \log N}}$	$4.5 \cdot 10^{10}$	$1.5 \cdot 10^{40}$	$5.9 \cdot 10^{144}$	$3.7 \cdot 10^{507}$
$N^{0.1}$	$8.5 \cdot 10^{6}$	$2.1 \cdot 10^{69}$	$1.4 \cdot 10^{693}$	$3.0 \cdot 10^{6931}$

By now we know the reason that algorithms with complexity $\mathcal{O}(N^{\epsilon})$ e.g., when $\epsilon = 0.1$ are not polynomial-time algorithms, whereas algorithms with complexity $\mathcal{O}((\log N)^c)$ e.g., when $c = 10$ are polynomial-time algorithms (although they may be high order polynomials).

Although algorithms whose complexity is measured in terms of arithmetic operations may be not very useful for number-theoretic computations, it however often forms a basis for estimating the bit operations needed to perform arithmetic-related computations. To see this, let us review some results that show the relationships between arithmetic and bit operations needed to perform some common arithmetic-related computations:

(i) Let a and b are both β-bit (or less) numbers, then $\gcd(a, b)$ can be computed by Euclid's algorithm in time

$$
\begin{align}
T(\gcd(a,b)) &= \mathcal{O}_A(\beta) & (3.124) \\
&= \mathcal{O}_B(\beta^3) & (3.125) \\
&= \mathcal{O}_B(\beta^2 \log \beta \log \log \beta). & (3.126)
\end{align}
$$

Note that a much faster algorithm for computing $\gcd(a, b)$ is known [8], which has running time of $\mathcal{O}_B(\beta \log^2 \beta \log \log \beta)$.

(ii) Let $M(n)$ be the time required to compute the product of two $n \times n$ Boolean matrices, and all the entries in the two matrices are β-bits (or less), then by using Victor Pan's method [105], we can have

$$
\begin{align}
M(n) &= \mathcal{O}_A(n^{2.496}) & (3.127) \\
&= \mathcal{O}_B(n^{2.496} \beta^2) & (3.128) \\
&= \mathcal{O}_B(n^{2.496} \beta \log \beta \log \log \beta). & (3.129)
\end{align}
$$

(iii) Let x, e and n are all β-bit numbers (or less), then $x^e \bmod n$, the modular exponentiations, can be computed by the *repeated squaring method* in time

$$
\begin{align}
T(\text{compute } x^e \bmod n) &= \mathcal{O}_A(\beta) && (3.130) \\
&= \mathcal{O}_B(\beta^3) && (3.131) \\
&= \mathcal{O}_B(\beta^2 \log \beta \log \log \beta). && (3.132)
\end{align}
$$

(iv) The computation of

$$
kP \bmod n = \underbrace{P + P + \cdots + P}_{k \text{ times}} \bmod n
$$

where $P = (x, y)$ is a point on an elliptic curve over a field, can be computed by the *repeated doubling method* in time

$$
\begin{align}
T(\text{compute } kP \bmod n) &= \mathcal{O}_A(\beta) && (3.133) \\
&= \mathcal{O}_B(\beta^3) && (3.134) \\
&= \mathcal{O}_B(\beta^2 \log \beta \log \log \beta) && (3.135)
\end{align}
$$

assuming k, n, x and y are all β-bit (or less) numbers.

The proof of the above results is easy; since the multiplication of two β-bit (or less) integers has running time of $\mathcal{O}_B(\beta^2)$. The best known method for multiplying two β-bit (or less) numbers has running time of $\mathcal{O}_B(\beta \log \beta \log \log \beta)$, so the results follow from these two facts immediately. It is worthwhile pointing out that for all practical purposes, the $\mathcal{O}_B(\beta^2)$ is often best, so we could just use this bound as a basis for our analyses. In other words, the term $\mathcal{O}_B(\beta^2)$ is a generous allowance for the cost of performing arithmetic operations on numbers which are $\mathcal{O}(\beta)$-bits. Note that if you wish to express the complexity in terms of e.g., n, the actual number rather than β, its number of bits, then since $\beta = \mathcal{O}(\log n)$, you can write for example,

$$
\begin{align}
T(\text{compute } kP \bmod n) &= \mathcal{O}_A(\log n) && (3.136) \\
&= \mathcal{O}_B(\log^3 n) && (3.137) \\
&= \mathcal{O}_B(\log^2 n \log \log n \log \log \log n). && (3.138)
\end{align}
$$

Exercise 3.4.5 Use both the $\mathcal{O}_A$ and $\mathcal{O}_B$ notations to estimate the number of arithmetic operations and the number of bit operations required to compute

(i) the factorial $n!$, and n^n,

(ii) the binomial coefficient $\begin{pmatrix} n \\ m \end{pmatrix}$,

(iii) the Jacobi symbol $\left(\frac{a}{n}\right)$,

(iv) $n(n+1)(2n+1)/6$,

(v) multiplication of an $k \times s$ matrix by an $s \times t$ matrix, where all matrix entries are $\leq n$,

(vi) Euclid's algorithm for computing $\gcd(a,b)$, where $a, b \in \mathbb{Z}$, but a and b cannot be both zero.

Exercise 3.4.6 Using the Prime Number Theory (i.e., Theorem 1.3.4 in Chapter 1),

(i) estimate the number of bit operations required to compute the product of all prime numbers less than n.

(ii) estimate the number of bit operations required to test a large odd number n; suppose you have a list of prime numbers up to $\sqrt{n}$, and you test the primality of n by trial division by these primes numbers.

Further Reading

Computability and complexity, or theoretical computer science in general, is the central topic as well as a very active area in computer science. The books edited by Davis [29] and Herken [58] provide interesting sources of the development of theoretical computer science over the last 60 years or so, whereas the two-volume handbook edited by van Leeuwen [76] is an important source for working theoretical computer scientists. For general reading in computability and complexity, we strongly recommend the following well-written textbooks; [8], [21], [26], [31], [32], [42], [54], [53], [57], [61], [59], [64], [68], [69], [73], [82], [86], [87], [88], [96], [97], [136], [138], and [141].

Chapter 4

Number-Theoretic Computations and Applications

To the layman, a lot of math (like primality testing and factoring large numbers) may seem a frivolous waste of time. However, this research often pays off unexpectedly years later. Factoring and primality testing have become important because the need to make electronic communications secure. ··· So what used to be an esoteric playground for mathematicians has become applicable research.

– DAVID GRIES (1939–)

In this chapter, we shall introduce some methods for large number-theoretic computation from a complexity-theoretic point of view. By a "large number", we mean that the number is several hundred or several thousand digits long. For large number-theoretic computation, say, for example, to factor a two-hundred digit number into its prime decomposition form, we are not interested in just finding an algorithm, rather, we are interested in finding an *efficient* algorithm, i.e., an algorithm of polynomial complexity, or at least of subexponential complexity. Three fundamental number-theoretic computation problems (i.e., primality testing, integer factorization and discrete logarithms) and their applications in computing science, particularly in cryptography and computer security will be discussed. In the last section of this chapter, we shall also introduce the basic concepts of residue number systems based on the Chinese Remainder Theorem, and discuss their applications in high-speed computation and special-purpose computer architecture design.

4.1   Primality Testing

4.1.1   Basic Algorithm

Primality testing of large numbers is *very* important in many areas of mathematics, computer science and cryptography. For example, in public-key cryptography, if we can find two large primes p and q, each with 100 digits or more, then we can get a composite

$$n = p \cdot q$$

with 200 digits or more. This composite n can be used to encrypt a message securely even when n is made public. The message cannot be decrypted without knowledge of the prime factors of n. Of course, we can try to use a modern integer factorization method, e.g., the Elliptic Curve Method (ECM) to factor n and to get its prime factors p and q, but it might take about 20 million years to complete the job even on a supercomputer. Thus, it is impossible in practice to decrypt the message. Another good example is searching for amicable numbers. In the following algebraic method for generating amicable numbers, if we can make sure that the following four integers p, q, r, s:

$$\begin{cases} p = 2^x \cdot g - 1 \\ q = 2^y + (2^{n+1} - 1) \cdot g \\ r = 2^{n-y} \cdot g \cdot q - 1 \\ s = 2^{n-y+x} \cdot g^2 \cdot q - 1 \end{cases} \quad \text{where} \quad \begin{cases} 0 < x < n \\ g = 2^{n-x} + 1 \\ 0 < y < n \end{cases}$$

are all primes, then the pair

$$(m, n) = (2^n qpr, \ 2^n qs)$$

is an amicable pair. Thus, searching for amicable numbers is often the same as primality testing of some related integers.

The search for efficient primality tests is not just an open problem, but one of the oldest unfinished projects in mathematics, with a root that possibly goes back to the ancient Greeks about 2000 years ago. It is a typical *decision problem*:

$$\begin{cases} \text{Input}: & n \in \mathbb{N} \text{ with } n > 1. \\ \text{Output}: & \begin{cases} \text{Yes}, & \text{if } n \in \text{Primes}, \\ \text{No}, & \text{otherwise.} \end{cases} \end{cases} \quad (4.1)$$

Unfortunately, it is not a simple matter to determine whether or not a random integer n is prime, particularly when n is very large. An efficient algorithm for primality testing from the complexity point of view would have to run in $\mathcal{O}(\log^k n)$ steps, for some fixed k. But unfortunately, no such *deterministic* algorithm exists for a random integer n.

According to Dickson (page 59 in [34]) and more recently to Ribenboim [117], the (ancient) Chinese mathematicians seem to have known at least before Fermat that

$$\text{for } p \text{ prime } 2^p - 2 \text{ is divisible by } p. \tag{4.2}$$

This fact was rediscovered by Fermat in June 1640 while investigating perfect numbers. Shortly afterwards (18 October 1640), Fermat stated that he had a proof of the more general fact now known as Fermat's little theorem. By using the Chinese idea, a simple but practical primality testing algorithm (called the Chinese test) for all numbers between i and j can be designed as follows:

Algorithm 4.1.1 (Chinese Primality Testing)

[1] Initialize the values $i > 3$ and $j > i$. Set $n \leftarrow i$.

[2] If $2^n \pmod{n} = 2$, then n is a *probable prime*, else n is composite.

[3] $n \leftarrow n + 1$. If $n \leq j$ goto [2], else goto [4].

[4] Terminate the execution of the algorithm.

Among the numbers below 2000 that pass the Chinese test, only seven are composites: 341, 561, 645, 1105, 1387, 1729 and 1905; all the rest are indeed primes. Further computation shows that such composite numbers seem to be rare and so we define a composite number n to be a *base 2 pseudoprime* if n divides $2^n - 2$. To exhibit quite how rare these are, note that up to 10^{10} there are around 450 million primes, but only about 15000 base 2 pseudoprimes, while up to $2 \cdot 5 \times 10^{10}$ there are over a billion primes, but fewer than 22000 base 2 pseudoprimes. So, if we were to choose a random number $n < 2 \cdot 5 \times 10^{10}$ for which n divides $2^n - 2$, then there would be a less than a 1-in-50000 chance that our number would be composite. We quote the following comments on the importance of the Chinese test from [122]:

> *Because most composite integers are not pseudoprimes, it is possible to develop primality tests based on the original Chinese idea, together with extra observations.*

4.1.2   Strong Pseudoprimality Test

In this section we shall introduce some basic concepts and ideas of probable primes, pseudoprimes and pseudoprimality tests, which will be used throughout the book.

Theorem 4.1.1 (Fermat's Theorem) If b is a positive integer, p is a prime and $\gcd(b, p) = 1$, then

$$b^{p-1} \equiv 1 \pmod{p}. \tag{4.3}$$

Most modern primality testing algorithms depend in some way on the converse (an immediate corollary) of Fermat's Theorem:

Corollary 4.1.1 (Converse of Fermat's Theorem – Fermat test) Let n be an odd positive integer. If $\gcd(b, n) = 1$ and

$$b^{n-1} \not\equiv 1 \pmod{n}, \tag{4.4}$$

then n is composite.

By Corollary 4.1.1 we know that if there exists a b with $1 < b < n$, $\gcd(b, n) = 1$ and $b^{n-1} \not\equiv 1 \pmod{n}$, then n must be composite. What happens if we find a number n such that $b^{n-1} \equiv 1 \pmod{n}$? Can we conclude that n is certainly a prime? The answer is unfortunately not, because n usually is indeed a prime, but sometimes is not! This leads to the following important concepts of probable primes and pseudoprimes.

Definition 4.1.1 If $b^{n-1} \equiv 1 \pmod{n}$, then we call n a (Fermat) *probable prime* to the base b. A (Fermat) probable prime n to the base b is called a (Fermat) *pseudoprime* to the base b if n is composite. A composite integer n that satisfies $b^{n-1} \equiv 1 \pmod{n}$ for every positive integer b with $\gcd(b, n) = 1$ is called a *Carmichael number*, after the American mathematician Robert D. Carmichael.

Example 4.1.1 For the integer 1387, we have $2^{1387-1} \equiv 1 \pmod{1387}$. Thus 1387 is a Fermat probable prime to the base 2. But since $1387 = 19 \cdot 73$ is composite, it is in fact a Fermat pseudoprime to the base 2.

It is usually much harder to show that a given large integer n is a Carmichael number than to show that it is a Fermat pseudoprime. The most obvious method is to factor n and then use the following two properties:

Proposition 4.1.1 Every Carmichael number is square free.

Proposition 4.1.2 If the prime p divides the Carmichael number n, then $n \equiv 1 \pmod{p-1}$.

Example 4.1.2 To show 561 is a Carmichael number, we first note that it is square free. Now since $561 = 3 \cdot 11 \cdot 17$, then we have:

$$561 \equiv 1 \pmod{2},$$
$$561 \equiv 1 \pmod{10},$$
$$561 \equiv 1 \pmod{16}.$$

Thus, 561 is a Carmichael number.

A further and immediate improvement over the Fermat test is the strong pseudoprimality test (or just strong test, for short), which is based on the following idea:

Theorem 4.1.2 If $n = 1 + 2^j d$ with d odd is prime and $b^d \not\equiv 1 \pmod{n}$, then the following Miller-Rabin sequence

$$b^d, b^{2d}, b^{4d}, b^{8d}, \cdots, b^{2^{j-1}d}, b^{2^j d} \tag{4.5}$$

will end with 1, and the value just preceding the first appearance of 1 will be $n - 1$. (The only solutions to $x^2 \equiv 1 \pmod{p}$ are $x \equiv \pm 1$ when p is prime.) That is, if n is prime and $n = 1 + 2^j d$, then the sequence (4.5) has the form

$$(1, 1, \cdots, 1) \tag{4.6}$$

or

$$(?, \cdots, ? - 1, 1, \cdots, 1). \tag{4.7}$$

when considered mod n (for any b, $1 < b < n$).

The above theorem can easily be implemented by the following algorithm:

Algorithm 4.1.2 (Strong Pseudoprimality Test):

[1] Let n be an odd number, and the base b a random number in the range $1 < b < n$. Find j and d with d odd, so that $n - 1 = 2^j d$.

[2] Set $i \leftarrow 0$ and $y \leftarrow b^d \pmod{n}$.

[3] If $i = 0$ and $y = 1$, or $y = n - 1$, then terminate the algorithm and output "n is probably prime". If $i > 0$ and $y = 1$ goto [5].

[4] $i \leftarrow i + 1$. If $i < j$, set $y \equiv y^2 \pmod{n}$ and return to [3].

[5] Terminate the algorithm and output "n is definitely not prime".

Definition 4.1.2 A positive integer n with $n - 1 = d \cdot 2^j$ and d odd, is called a *strong probable prime* to the base b if it passes the strong pseudoprimality test described above (i.e., the last term in the Rabin-Miller sequence is 1, and the first occurrence of 1 is either the first term or is preceded by -1). A strong probable prime to the base b is called a *strong pseudoprime* to the base b if it is composite.

Theorem 4.1.3 Let $n > 1$ be an odd composite integer. Then n passes the strong test for at most $(n - 1)/4$ bases b with $1 \le b < n$.

PROOF. (Sketch – for a more detailed proof, see [70] or [122]) We first assert that if p is an odd prime, and α and q are positive integers, then the number of incongruent solutions of the congruence

$$x^{q-1} \equiv 1 \pmod{p^\alpha}$$

is $\gcd(q - 1, p^{\alpha-1}(p - 1))$.

Let $n - 1 = d \cdot 2^j$, where d is an odd positive integer and j is a positive integer. For n to be a strong pseudoprime to the base b, either

$$b^d \equiv 1 \pmod{n}$$

or

$$b^{2^i d} \equiv -1 \pmod{n}$$

for some integer i with $0 < i < j - 1$. In either case, we have

$$b^{n-1} \equiv 1 \pmod{n}.$$

Let the standard prime decomposition form of n be

$$n = p_1^{\alpha_1} p_2^{\alpha_2} \cdots p_k^{\alpha_k}.$$

By the assertion made at the beginning of the proof, we know that there are

$$\gcd\left(n - 1, \; p_i^{\alpha_i}(p_i - 1)\right) = \gcd(n - 1, \; p_i - 1)$$

incongruent solutions to the congruence

$$x^{n-1} \equiv 1 \pmod{p_i^{\alpha_i}}, \; i = 1, 2, \cdots, k.$$

Further, by the Chinese remainder theorem, we know that there are exactly

$$\prod_{i=1}^{k} \gcd(n - 1, \; p_i - 1)$$

incongruent solutions to the congruence

$$x^{n-1} \equiv 1 \pmod{n}.$$

To prove the theorem, there are three cases to consider:

[i] the standard prime decomposition form of n contains a prime power $p_r^{\alpha_r}$ with exponent $\alpha_r \geq 2$;

[ii] $n = pq$, with p and q distinct odd primes.

[iii] $n = p_1 p_2 \cdots p_k$, with $p_1, p_2, \cdots, p_k$ distinct odd primes.

The second case can actually be included into the third case. We consider here only the first case. Since

$$\frac{p_r - 1}{p_r^{\alpha_r}} = \frac{1}{p_r^{\alpha_r - 1}} - \frac{1}{p_r^{\alpha_r}}$$

$$\leq \frac{2}{9}$$

we have

$$\prod_{i=1}^{k} \gcd(n-1,\ p_i - 1) \ \le\ \prod_{i=1}^{k}(p_i - 1)$$

$$\le\ \left(\prod_{\substack{i=1 \\ i \ne r}}^{k} p_i\right)\left(\frac{2}{9}p_r^{\alpha_r}\right)$$

$$\le\ \frac{2}{9}n$$

$$\le\ \frac{n-1}{4} \qquad \text{for } n \ge 9.$$

So, there are at most $(n-1)/4$ integers b, $1 < b < n-1$, for which n is a strong pseudoprime to the base b, and for which n can pass the strong test. ∎

A probabilistic interpretation of Theorem 4.1.3 is as follows:

Corollary 4.1.2 Let $n > 1$ be an odd composite integer and b be chosen randomly from $\{2, 3, \cdots, n-1\}$. Then the probability that n passes the strong test is less than $1/4$.

From Corollary 4.1.2, we can construct a simple, general purpose, polynomial time primality test which has a positive (but arbitrarily small) probability of giving the wrong answer. Suppose an error probability of ϵ is acceptable. Choose k such that $4^{-k} < \epsilon$, and select $b_1, b_2, \cdots, b_k$ randomly and independently from $\{2, 3, \cdots, n-1\}$. If n fails the strong test on a_i, $i = 1, 2, \cdots, k$, then n is strong probable prime.

Theorem 4.1.4 The strong test (Algorithm 4.1.2) requires, for $n - 1 = 2^j d$ with d odd and for k randomly selected bases, at most $k(2+j)\log n$ steps. If n is prime, then the result is always correct. If n is composite, then the probability that n passes all k tests is at most $1/4^k$.

PROOF. The first two statements are obvious, only the last statement requires proof. An error will occur only when the n to be tested is composite and the bases $a_1, a_2, \cdots, a_k$ chosen in this particular run of the algorithm are all nonwitnesses. (An integer a is a *witness* to the compositeness of n if it is possible using a to prove that n is composite, otherwise it is a *nonwitness*.)

Since the probability of randomly selecting a nonwitness is smaller than $1/4$ (by Corollary 4.1.2), the probability of independently selecting k nonwitnesses is smaller than $1/4^k$. Thus the probability that with any given number n, a particular run of the algorithm will produce an erroneous answer is smaller than $1/4^k$ [113]. ∎

In the following table, we give some values of k and $1/4^k$ for the purposes of comparison:

k	10	25	30	50	100	168	1000
$1/4^k$	$< 10^{-6}$	$< 10^{-15}$	$< 10^{-18}$	$< 10^{-30}$	$< 10^{-60}$	$< 10^{-101}$	$< 10^{-602}$

Let n be a composite positive integer. Using the strong test, if we pick 100 different integers between 1 and n at random and perform the strong test for each of these 100 bases, then the probability that n passes all the tests is less than $4^{-100} < 10^{-60}$, an extremely small number. In fact, it may be more likely that a computer error was made than that a composite integer passes all the 100 tests. We conclude that for all *practical* purposes, we can test primality in polynomial time.

There is a famous conjecture in analytic number theory called the Generalized Riemann Hypothesis (GRH) about the Dirichlet L-series. A consequence of this hypothesis is the following conjecture (see [102] or [122]):

Conjecture 4.1.1 For every composite positive integer n, there is a base b with $1 < b \le 2(\log n)^2$, such that n fails the strong test for the base b.

If this conjecture is true (many number theorists do believe it to be true), the following result provides a rapid primality test:

Theorem 4.1.5 There is an algorithm to determine whether or not a positive integer n is prime using $\mathcal{O}(\log^5 n)$ bit operations.

PROOF. Let b be a positive integer less than n. To perform the strong test for the base b on n takes $\mathcal{O}(\log^3 n)$ bit operations, because this test requires that we perform no more than $\log^3 n$ modular exponentiations, each using $\mathcal{O}(\log^2 b)$ bit operations. Assume that the GRH is true. If n is composite, then, by Conjecture 4.1.1, there is a base b with $1 < b \le 2(\log n)^2$ such that n fails the strong test for b. To discover this b requires less than $\mathcal{O}(\log^3 n) \cdot \mathcal{O}(\log^2 n) = \mathcal{O}(\log^5 n)$ bit operations. Hence, using $\mathcal{O}(\log^5 n)$ bit operations, we can determine whether n is composite or prime. ∎

Although very few composites can pass the strong pseudoprimality test, such numbers do exist. For example, the composite $n = 2047 = 23 \cdot 89$ can pass the strong pseudoprimality test base 2, because $n - 1 = 2^1 \cdot 1023$, $d = 1023$ and the Miller–Rabin sequence is $2^{1023} \equiv 1 \pmod{2047}$, $2^{2046} \equiv 1 \pmod{2047}$. So $n = 2047$ is a *strong pseudoprime to the base 2*. Thus we cannot conclude with certainty that n is prime just by a strong primality test, we will need some other tests as well. One of the other tests is the Lucas (pseudoprimality) test, which we shall discuss in the next subsection.

4.1.3   Lucas Pseudoprimality Test

In this subsection, we shall study the Lucas sequences and their applications in pseudoprimality testing.

Let a, b and $D = a^2 - 4b$ be non-zero integers. Consider the equation $x^2 - ax + b = 0$; its discriminant is $D = a^2 - 4b$, and α and β are the two roots:

$$\begin{cases} \alpha = \dfrac{a + \sqrt{D}}{2} \\ \beta = \dfrac{a - \sqrt{D}}{2}. \end{cases} \tag{4.8}$$

So

$$\begin{cases} \alpha + \beta = a \\ \alpha - \beta = \sqrt{D} \\ \alpha\beta = b. \end{cases} \tag{4.9}$$

Define the sequences of U_k and V_k by

$$\begin{cases} U_k(a, b) = \dfrac{\alpha^k - \beta^k}{\alpha - \beta} \\ V_k(a, b) = \alpha^k + \beta^k. \end{cases} \tag{4.10}$$

In particular, $U_0(a, b) = 0, U_1(a, b) = 1$, while $V_0(a, b) = 2, V_1(a, b) = a$. We also have the recursions

$$\begin{cases} U_k(a, b) = aU_{k-1} - bU_{k-2} \\ V_k(a, b) = aV_{k-1} - bV_{k-2} \end{cases} \tag{4.11}$$

for $k \geq 2$. The sequences

$$
\begin{cases}
U(a,b) = (U_k(a,b))_{k \geq 0} \\
V(a,b) = (V_k(a,b))_{k \geq 0}
\end{cases}
\tag{4.12}
$$

are called the *Lucas[1] sequences associated with the pair (a,b), or just Lucas sequences*. Special cases had been considered by Fibonacci, Fermat, and Pell, among others. For example, the sequence $U_k(1,-1)$ for $k = 0, 1, \cdots$ was first considered by Fibonacci[2], and it begins as follows:

$0, 1, 1, 2, 3, 5, 8, 13, 21, 34, 55, 89, 144, 233, 377, 610, 987, 1597, 2584, 4181, 6765, \cdots$

These are called *Fibonacci numbers*.

The companion sequence $V_k(1,-1)$ to the Fibonacci numbers is the sequence of *Lucas numbers*. It begins as follows:

$2, 1, 3, 4, 7, 11, 18, 29, 47, 76, 123, 199, 322, 521, 843, 1364, 2207, 3571,$

$$5778, 9349, 15127, \cdots$$

If $a = 3$, $b = 2$, then the sequences obtained are

$U_k(3,2) = 2^k - 1$:

$$0, 1, 3, 7, 15, 31, 63, 127, 255, 511, 1023, 2047, 4095, 8191, 16383, \cdots$$

$V_k(3,2) = 2^k + 1$:

$$2, 3, 5, 9, 17, 33, 65, 129, 257, 513, 1025, 2049, 4097, 8193, 16385, \cdots$$

for $n \geq 0$.

---

[1]Francois Édouard Lucas (1842–1891) was born in Amiens, France, and was educated at the Ecole Normale, one of the two most prestigious French institutions of the time. After finishing his studies, he worked as an assistant at the Paris Observatory, and later on, became a mathematics teacher at three Paris secondary schools. He was the last "largest prime record" holder in the pre-computer age, by discovering the 12th Mersenne prime in 1876, though it was confirmed only in 1914. In the world mathematics community, Lucas is perhaps best known for his work on Lucas numbers, the Lucas test and the Tower of Hanoi problem.

[2]Leonardo Fibonacci (about 1170–1250), pseudonym of one of the first European mathematicians to emerge after the Dark Ages. An Italian merchant by the name of Leonardo da Pisa or Leonardo Pisano, he was one of those who introduced the Hindu-Arabic number system to Europe. He strongly advocated this system in *Liber abaci*, published in 1202, which also contained problems including one that gives rise to the *Fibonacci numbers*. Fibonacci's other writings dealt with non-Euclidean geometry and Diophantine analysis.

The sequences associated with $a = 2$, $b = -1$ are called *Pell sequences*; they begin as follows:

$$U_k(2,-1): \ 0,1,2,5,12,29,70,169,408,985,2378,5741,13860,33461,80782,\cdots$$
$$V_k(2,-1): \ 2,2,6,14,34,82,198,478,1154,2786,6726,16238,39202,94642,\cdots$$

Now we are in a position to study some analogues of pseudoprimes in which $a^{n-1} - 1$ is replaced by a Lucas sequence. Recall that odd composite numbers n for which

$$a^{n-1} \equiv 1 \ (\text{mod } n)$$

are called pseudoprimes to base a.

Theorem 4.1.6 (Lucas Theorem) Let a and b be integers and put $D = a^2 - 4b \neq 0$. Define the Lucas sequence $\{U_k\}$ with the parameters D, a, b by

$$U_k = \frac{\alpha^k - \beta^k}{\alpha - \beta}, \quad k \geq 0 \tag{4.13}$$

where α and β are the two roots of $x^2 - ax + b = 0$. If p is an odd prime, $p \nmid b$ and $\left(\frac{D}{p}\right) = -1$, where $\left(\frac{D}{p}\right)$ is the Jacobi symbol, then $p \mid U_{p+1}$.

The above theorem can be used directly to construct a primality test, often called the *Lucas test*:

Corollary 4.1.3 (Converse of the Lucas Theorem – Lucas Test) Let n be an odd positive integer. If $n \nmid U_{n+1}$, then n is composite.

Just as there are Fermat probable primes and Fermat pseudoprimes, we also have the concepts of Lucas probable primes and Lucas pseudoprimes.

Definition 4.1.3 An odd positive integer n is called a *Lucas probable prime* for D, a and b, if $n \nmid b$, $\left(\frac{D}{n}\right) = -1$ and $n \mid U_{n+1}$. A Lucas probable prime n is called a *Lucas pseudoprime* if n is composite.

Another different but equivalent presentation of Theorem 4.1.6 is as follows:

Theorem 4.1.7 Let n be an odd positive integer, $\varepsilon(n)$ the Jacobi symbol $\left(\frac{D}{n}\right)$, and $\delta(n) = n - \varepsilon(n)$. If n is prime and $\gcd(n, b) = 1$, then

$$U_{\delta(n)} \equiv 0 \pmod{n}. \tag{4.14}$$

If n is composite, but (4.14) still holds, then n is called a *Lucas pseudoprime with parameters a and b*.

Remark 4.1.1 Although Theorem 4.1.7 is true when $\left(\frac{D}{n}\right) = 1$, it is best to avoid this situation. If $D \equiv w^2 \pmod{n}$, then modulo n the roots of $x^2 - ax + b = 0$ are

$$\alpha = (a + w)/2$$
$$\beta = (a - w)/2.$$

Then

$$U_{n-1} = \frac{\alpha^{n-1} - \beta^{n-1}}{\alpha - \beta} = \frac{[(a+w)/2]^{n-1} - [(a-w)/2]^{n-1}}{w}.$$

If for example $a = w + 2$, then

$$U_{n-1} = \frac{(w+1)^{n-1} - 1}{w}.$$

So in that case,

$$U_{n-1} \equiv 0 \pmod{n}$$

implies

$$(w+1)^{n-1} \equiv 1 \pmod{n}.$$

But this is the same as the ordinary Fermat probable prime test which we do not want to just repeat in the Lucas test.

A good way to avoid this situation is to select a suitable D such that $\left(\frac{D}{n}\right) = -1$. Two methods have been proposed [11]:

(i) Let D be the first element of the sequence $5, -7, 9, -11, 13, \cdots$ for which $\left(\frac{D}{n}\right) = -1$. Let $a = 1$ and $b = (1 - D)/4$.

(ii) Let D be the first element of the sequence $5, 9, 13, 17, 21, \cdots$ for which $\left(\frac{D}{n}\right) = -1$. Let a be the least odd number exceeding $\sqrt{D}$ and $b = (a^2 - D)/4$.

The first 10 Lucas pseudoprimes by the first method are: 323, 377, 1159, 1829, 3827, 5459, 5777, 9071, and 10877. The first 10 Lucas pseudoprimes by the second method are: 323, 377, 1349, 2033, 2651, 3569, 3599, 3653, 3827, and 4991. The most interesting thing about the Lucas test is that if we choose the parameters D, a and b as described in the second method, then the first 50 Carmichael numbers and several other Fermat pseudoprimes base 2 will never be Lucas pseudoprimes [11]. This leads to the general belief that a combination of a strong pseudoprimality test and a Lucas pseudoprimality test (called a combined test, for short) might be an infallible test for primality. Since up to now, no composites have been found to pass such a combined test, it is thus reasonable to conjecture that:

Conjecture 4.1.2 Let n be a positive integer greater than 1. If n can pass the combination of a strong pseudoprimality test and a Lucas test, then n is prime.

The advantages of the combination of a strong test and a Lucas test seem to be that the two pseudoprimality tests are independent. That is, n being a probable prime of the first type does not affect the probability of n being a probable prime of the second type. In fact, if n is a strong pseudoprime (to a certain base), then n is less likely than a typical composite to be a Lucas pseudoprime (to the parameters a and b), provided a and b are chosen properly, and vice versa. If n passes both a strong test and a Lucas test, we can be more certain that it is prime than if it merely passes several strong tests, or several Lucas tests. Pomerance et al in [111] issued a challenge (with a total prize now $620) for an example of a composite number which passes both a strong pseudoprimality test base 2 and a Lucas test, or a proof that no such number exists. At the moment, the prize is unclaimed; no counter-example has yet been found.

There is also a very special Lucas test suitable for Mersenne primes (recall that a positive integer $M_n = 2^n - 1$ is called a Mersenne prime whenever n is a prime), known as the *Lucas-Lehmer test* after Lucas the French mathematician Francois Lucas who discovered the basic idea in 1876 and the American mathematician Derrick Lehmer[3] who refined the method in 1930 based on the following theorem:

---

[3]Derrick H. Lehmer (1905–1991), the father of computational number theory and perhaps best known for his sharp and definitive form of Lucas test for Mersenne primes, was born in Berkeley, California. He received his bachelor's degree in physics from the University of California, Berkeley, whereupon he went to the University of Chicago for graduate studies in number theory with L. E. Dickson. But since he didn't like working with Dickson, he later went to Brown University to study for a PhD. He served as a faculty member in California

Theorem 4.1.8 *(Lucas-Lehmer Theorem for Mersenne Primes M_n)* Let $a = 2$, $b = -2$, and consider the associated Lucas sequences $(U_k)_{k \geq 0}$, $(V_k)_{k \geq 0}$, which have discriminant $D = 12$. Then $N = M_n$ is prime if and only if

$$N \mid V_{(N+1)/2}. \tag{4.15}$$

Example 4.1.3 First we notice that the Lucas sequence $V_k(2, -2)$ begins as follows:

$$2, 2, 8, 20, 56, 152, 416, 1136, 3104, 8480, 23168, 63296, 172928,$$
$$472448, 1290752, 3526400, 9634304, \cdots.$$

Now suppose we wish to test the primality of $N = 2^7 - 1$. Compute $V_{(N+1)/2}$ for $N = 2^7 - 1$:

$$
\begin{aligned}
V_{(N+1)/2} &= V_{2^7/2} \\
&= V_{64} \\
&= 8615517765800787268541087744 \\
&\equiv 0 \pmod{(2^7 - 1)},
\end{aligned}
$$

then by Theorem 4.1.8, $N = 2^7 - 1$ is a prime.

For the purpose of computation, it is convenient to replace the Lucas sequence $(V_k)_{k \geq 0}$ by the following Lucas-Lehmer sequence $(L_k)_{k \geq 1}$, defined recursively as follows:

$$
\begin{cases}
L_0 = 4 \\
L_{k+1} = L_k^2 - 2.
\end{cases}
\tag{4.16}
$$

The Lucas-Lehmer sequence begins with

$$4, 14, 194, 37634, 1416317954, 2005956546822746114,$$
$$4023861667741036022825635656102100994, \cdots.$$

Institute of Technology, Lehigh University, and the University of Cambridge before joining the Mathematics Department at Berkeley in 1940. Lehmer made many significant contributions to number theory, and also invented many special purpose devices for number theoretic computations, some with his father who was also a mathematician at Berkeley. Lehmer was interested in primality testing throughout his life, and he was also involved throughout his life with the theory and practice of integer factorization.

The reason that we can replace the Lucas sequence $V_k(2, -2)$ by the Lucas-Lehmer sequence L_k is based on the following observations:

$$\begin{cases} L_0 = V_2/2 \\ L_{k-1} = V_{2^k}/2^{2^{k-1}}. \end{cases} \qquad (4.17)$$

Note that (4.17) follows immediately from (4.10), and can also be easily verified to be correct:

$$L_0 = V_2/4 = 8/2 = 4$$

$$L_1 = V_{2^2}/2^{2^{2-1}} = V_4/2^2 = 56/4 = 14$$

$$L_2 = V_{2^3}/2^{2^{3-1}} = V_8/2^4 = 3104/16 = 194$$

$$L_3 = V_{2^4}/2^{2^{4-1}} = V_{16}/2^8 = 9634304/256 = 37634$$

$$L_4 = V_{2^5}/2^{2^{5-1}} = V_{32}/2^{16} = 92819813433344/65536 = 1416317954$$

$$\cdots\cdots\cdots .$$

So, Theorem 4.1.8 can be rewritten as follows:

Theorem 4.1.9 (Lucas-Lehmer Test for Mersenne Primes M_n) Let n be an odd prime. Then $M_n = 2^n - 1$ is prime if and only if

$$M_n \mid L_{n-2}, \qquad (4.18)$$

or equivalently

$$L_{n-2} \equiv 0 \pmod{(2^n - 1)}. \qquad (4.19)$$

Example 4.1.4 Suppose we wish to test the primality of $2^7 - 1$; we first compute the Lucas-Lehmer sequence $\{L_k\}$ for $2^7 - 1$ ($k = 0, 1, \cdots, p - 2 = 5$):

$$L_0 = 4$$

$$L_1 \equiv 14$$

$$L_2 \equiv 67$$

$$L_3 \equiv 42$$

$$L_4 \equiv 111$$

$$L_5 \equiv 0 \pmod{127}.$$

Since $L_{p-2} \equiv 0 \pmod{(2^p - 1)}$, then $2^7 - 1$ is a prime.

So a practical algorithm for Mersenne primes can now be derived as follows:

Algorithm 4.1.3 (Lucas-Lehmer Test for Mersenne Primes)

> Initialize the value for $p \in$ Primes
>
> $L \leftarrow 4$
>
> for i from 1 to $p - 2$ do
>
> $\qquad L \leftarrow L^2 - 2 \pmod{2^p - 1}$
>
> if $L = 0$ then $2^p - 1$ is prime
>
> $\qquad$ else $2^p - 1$ is composite

Remark 4.1.2 The above Lucas-Lehmer test for Mersenne primes is very efficient since the major step in the algorithm is to compute

$$L = L^2 - 2 \pmod{2^p - 1}$$

which can be performed in polynomial time. But still, the computation required to test a single Mersenne prime M_p has order $\mathcal{O}(p^3)$. Thus, to test M_{2n+1} would take approximately eight times as long as to test M_n with the same algorithm [135]. Historically, it has required about four times as much computation to discover the next Mersenne prime as it would to re-discover all previously known Mersenne primes. The search for Mersenne primes has been an accurate measure of computing power for the past two hundred years, and even in the modern era, it has been an accurate measure of computing power for new super-computers.

4.1.4 Elliptic Curve Test

In this subsection, we introduce a novel application of elliptic curves to primality testing. Although the test based on elliptic curves is still probabilistic, its answer will always be correct, only the running time is random; in practice, the expected running time is finite, it is possible that the algorithm does not terminate, but with probability zero.

First let us introduce one of the very useful converses of Fermat's little theorem:

Theorem 4.1.10 (Pocklington's theorem) Let s be a divisor of $N - 1$. Let a be an integer prime to N such that

$$\begin{cases} a^{N-1} \equiv 1 \ (\text{mod } N) \\ \gcd(a^{(N-1)/q}, N) = 1 \end{cases} \tag{4.20}$$

for each prime divisor q of s. Then each prime divisor p of N satisfies

$$p \equiv 1 \ (\text{mod } s). \tag{4.21}$$

Corollary 4.1.4 If $s > \sqrt{N} - 1$, then N is prime.

A similar theorem can be stated for elliptic curves.

Theorem 4.1.11 Let N be an integer greater than 1 and prime to 6, E an elliptic curve over $\mathbb{Z}/N\mathbb{Z}$, P a point on E, m and s two integers with $s \mid m$. Suppose we have found a point P on E that satisfies $mP = \mathcal{O}_E$, and that for each prime factor q of s, we have verified that $(m/q)P \neq \mathcal{O}_E$. Then if p is a prime divisor of N, $\#E(\mathbb{Z}/p\mathbb{Z}) \equiv 0 \ (\text{mod } s)$.

Corollary 4.1.5 If $s > (\sqrt[4]{N} + 1)^2$, then N is prime.

Combining the above theorem with Schoof's algorithm [126] that computes $\#E(\mathbb{Z}/p\mathbb{Z})$ in time $\mathcal{O}\left((\log p)^{8+\epsilon}\right)$, we can obtain the following so-called Goldwasser-Kilian (GK) algorithm [47]:

Algorithm 4.1.4 (Procedure GK(N), N is a probable prime)

[1] choose an elliptic curve E over $\mathbb{Z}/N\mathbb{Z}$, for which the number of points m (computed with Schoof's algorithm) satisfies $m = 2q$, with q a probable prime;

[2] if (E, m) satisfies the conditions of Theorem 4.1.11 with $s = m$, then N is prime, otherwise it is composite;

[3] the primality q is proved in the same way;

[4] Exit.

The running time of the GK algorithm is analyzed in the following two theorems [9]:

Theorem 4.1.12 Suppose that there exist two positive constants c_1 and c_2 such that the number of primes in the interval $[x, x + \sqrt{2x}]$ where $(x \geq 2)$ is greater than $c_1\sqrt{x}(\log x)^{-c_2}$, then GK proves the primality of N in expected time $\mathcal{O}\left((\log N)^{10+c_2}\right)$.

Theorem 4.1.13 There exist two positive constants c_3 and c_4 such that for all $k \geq 2$, the proportion of prime numbers N of k bits for which the expected time of GK is bounded by $c_3(\log N)^{11}$ is at least

$$1 - c_4 2^{-k^{1/\log\log k}}.$$

A serious problem with the GK algorithm is that Schoof's algorithm seems almost impossible to implement. In order to avoid the use of Schoof's algorithm, Atkin and Morain in 1991 developed a new implementation method called ECPP (Elliptic Curve Primality Proving) [9], which uses the properties of elliptic curves over finite fields related to complex multiplication. We summarize the principal properties as follows:

Theorem 4.1.14 Let p be a rational prime number that splits as the product of two principal ideals in a field $\mathbb{K}$: $p = \pi\pi'$ with π, π' integers of $\mathbb{K}$. Then there exists an elliptic curve E defined over $\mathbb{Z}/p\mathbb{Z}$ having complex multiplication by the ring of integers of $\mathbb{K}$, whose cardinality is

$$m = N_{\mathbb{K}}(\pi - 1) = (\pi - 1)(\pi' - 1) = p + 1 - t$$

with $|t| \leq 2\sqrt{p}$ (Hasse's Theorem) and whose invariant is a root of a fixed polynomial $H_D(X)$ (depending only upon D) modulo p.

For more information on the computation of the polynomials H_D, see [100]. We are now in a position to present an *algorithmic* description of the ECPP method:

Algorithm 4.1.5 (Procedure ECPP(N), N is a probable prime)

[1] [Initialization]. Set $i \leftarrow 0$ and $N_0 \leftarrow N$.

[2] [Building the sequence].

　　While $N_i > N_{small}$

　　　　[2.1] Find a D_i such that $N_i = \pi_i\pi_i'$ *in* $\mathbb{K} = \mathcal{Q}(\sqrt{-D_i})$;

[2.2] If one of the $w(-D_i)$ numbers $m_1, ..., m_w$ ($m_r = N_K(\omega_r - 1)$ where ω_r is a conjugate of π) is probably factored goto step [2.3] else goto [2.1];

[2.3] Store $\{i, N_i, D_i, \omega_r, m_r, F_i\}$ where $m_r = F_i N_{i+1}$. Here F_i is a completely factored integer and N_{i+1} a probable prime; set $i \leftarrow i + 1$ and goto step [2.1].

[3] [Proving].

For i from k downto 0

[3.1] Compute a root j of $H_{D_i}(X) \equiv 0 \pmod{N_i}$;

[3.2] Compute the equation of the curve E_i of the invariant j and whose cardinality modulo N_i is m_i;

[3.3] Find a point P_i on the curve E_i;

[3.4] Check the conditions of Theorem 4.1.14 with $s = N_{i+1}$ and $m = m_i$.

[4] [Exit]. Terminate the execution of the algorithm.

For ECPP, only the following heuristic analysis is known [100]:

Theorem 4.1.15 The expected running time of the ECPP algorithm is roughly proportional to $\mathcal{O}\left((\log N)^{6+\epsilon}\right)$ for some $\epsilon > 0$.

One of the largest primes verified so far with the ECPP algorithm is

$$391587 \times 2^{216193} - 1$$

which has 65087 digits.

Normally, probabilistic primality test (i.e., the Maple primality test [146], which is a combined use of a strong pseudoprimality test and a Lucas test) is always far more efficient than ECPP, especially for large numbers. But of course, we usually need to use an elliptic curve test or some other deterministic (or probabilistic but with zero error) test such as APR test [6] to confirm the results obtained by a probabilistic test. So we are finally approaching a more practical and realistic primality test for large numbers:

Algorithm 4.1.6 (Practical primality testing for an odd integer n)

[1] [Primality Testing – Probabilistic Method] Using a combination of Rabin-Miller's test and a Lucas test to determine if n is a probable prime. If it is, goto [2], else report that n is composite and goto [3].

[2] [Primality Proving – Elliptic Curve Method] Using the elliptic curve method to verify whether or not n is indeed a prime. If it is, then report that n is prime, or otherwise report that n is composite.

[3] [Exit] Terminate the algorithm.

4.1.5 Complexity of Primality Testing

In this subsection, we summarize some computational complexity results of primality testing, most of the results have been explicitly discussed in the previous subsections.

Determining whether a given integer $N \in \mathbb{N}$ is composite is easily seen to be in $\mathcal{NP}$ – simply multiply a non-trivial pair of integers whose product is N. In 1975 Pratt showed that determining primality is also in $\mathcal{NP}$ by exhibiting polynomial-time verifiable certificates of primality. (The basic idea is to prove N prime by showing that the multiplicative group of integers modulo N is cyclic of order $N - 1$; this requires a generator for the group, as well as a primality proof for each prime factor of $N - 1$, which can be found recursively.) Finding these certificates, however, requires the ability to factor, a problem much harder than primality testing.

Miller in 1976 [95] showed that primality could be tested in polynomial time (see Theorem 4.1.5) if the following conjecture (a version of the Generalized Riemann Hypothesis GRH) holds: For all $a, N \in \mathbb{N}$ and $\gcd(a, N) = 1$, we have the following estimate as $x \to \infty$:

$$\# \{p \in \text{Primes}, \ p \leq x, \ p \equiv a \ (\text{mod } N)\}$$
$$= \frac{li(x)}{\phi(N)} + \mathcal{O}\left(\sqrt{x} \ \log x\right). \tag{4.22}$$

Here $\phi(N) := \#(\mathbb{Z}/N\mathbb{Z})^*$, the number of units in $\mathbb{Z}/N\mathbb{Z}$, and $li(x)$ denotes the *logarithm integral* $\int_2^x dt/\log t$.

Miller's result can be re-stated as follows. For $l \in \mathbb{N}$, let $v_2(l) = \max\{e : 2^e \mid l\}$. Also, for $a \in \mathbb{N}$, let $L(a)$ be the Boolean expression ($L(a)$ says that N is a strong probable prime to the base a):

$$L(a) = [a^{N-1} \equiv 1 \ (\text{mod } N) \text{ and } \forall k < v_2(N - 1),$$
$$a^{\frac{N-1}{2^k}} \equiv 1 \Longrightarrow a^{\frac{N-1}{2^{k+1}}} \equiv \pm 1 \ (\text{mod } N)]. \tag{4.23}$$

Now assume that N is odd, and restrict a to be nonzero modulo N. Miller showed that if the GRH holds, then there is some $c > 0$ such that N is prime if and only if for all such a with $1 \leq a \leq c \log^2 N$, $L(a)$ holds. Because $L(a)$ can be checked in $\mathcal{O}(\log^3 N)$ steps, this shows that the set of primes is in $\mathcal{P}$ (assuming GRH).

Shortly after Miller published his results, Solovay and Strassen [137] noted that N is prime if and only if every a with $\gcd(a, N) = 1$ satisfies

$$M(a) := \left[a^{(N-1)/2} \equiv \left(\frac{a}{N} \right) \ (\bmod\ N) \right]. \tag{4.24}$$

Furthermore, they observed that if N is composite, then $M(a)$ holds for at most half of the residues modulo N. One could therefore obtain statistical evidence of primality by choosing a from $\{1, \cdots, N-1\}$ at random, and testing $M(a)$. Because $M(a)$ will always hold when N is prime, and can be checked in $\mathcal{O}\left((\log N)^3 \right)$ steps, this shows that the set of composite numbers belongs to $\mathcal{RP}$. Rabin [113] showed that Miller's predicate $L(a)$ has the same property, and is somewhat more reliable. He proved that if a is chosen at random from $\{1, \cdots, N-1\}$ and N is composite, then the probability that $L(a)$ holds is at most $1/4$ (see Theorem 4.1.3).

Recent work has the goal of proving primality quickly, unconditionally, and without error. The fastest known *deterministic* algorithm, called the APRCL test, originally invented by Adleman, Pomerance, Rumely (known as the APR test) in 1980 [6], but further simplified and improved by Cohen and Lenstra in 1981 using the idea of Jacobi sums [27], can determine the primality of $N \in \mathbb{N}_{>1}$ in time

$$\mathcal{O}\left((\log N)^{c(\log \log \log N)} \right), \qquad \text{for some suitable constant } c > 0. \tag{4.25}$$

The exponent $c(\log \log \log N)$ is an extremely slow growing function; for example, if N has a million decimal digits, then $\log \log \log N$ is only about 2.68. Riesel in 1985 reported an algorithm based on this method, which, when implemented on the CDC Cyber 170-750 Computer is able to deal with 100-digit numbers in about 30 seconds and 200-digit numbers in about 8 minutes. It is now possible to prove the primality of numbers with 1000 digits in a not too unreasonable amount of time. Of course, the APRCL test does not run in *polynomial* time, nor does it provide polynomial-length certificates of primality. Note that although $\mathcal{O}\left((\log N)^{c(\log \log \log N)} \right)$ is not a polynomial time complexity, we shall also not regard it as an exponential time complexity; the most suitable name for this type of complexity is *superpolynomial complexity*.

In 1986, another modern primality testing algorithm, based on the theory of elliptic curves was invented, first for theoretical purposes by Goldwasser and Killain [47], and then considerably modified by Atkin and implemented by Atkin and Morain [9]; we normally call Atkin and Morain's version the ECPP (Elliptic Curves and Primality Proving) test. The ECPP test is also practical for numbers with 1000 digits and possibly with several thousand digits within a reasonable amount of time. The *expected* running time for ECPP is $\mathcal{O}\left((\log N)^6\right)$, hence, is polynomial time, but this is only on average since for some numbers the running time of ECPP could be much larger. A totally non-practical version (i.e., only of theoretical interest and very difficult to implement at present) based on Abelian varieties (higher dimensional analog of elliptic curves) over finite fields has been given by Adleman and Huang [7] in 1992; they proved that without any hypothesis, primality testing can be done in random polynomial time. In other words, they proved (note that both the ECPP and the Adleman-Huang tests belong to the probabilistic complexity class $\mathcal{ZPP}$):

Theorem 4.1.16 There exists a random polynomial time algorithm that can determine whether or not a given number $N \in \mathbb{N}_{>1}$ is prime.

The techniques used in their proof of the theorem are from algebraic geometry, algebraic number theory and analytic number theory. The theorem is one of the major achievements of theoretical algorithmic number theory, and complements the Solovay and Strassen's result that there exists a random polynomial time algorithm for the set of composite numbers.

More recently, Konyagin and Pomerance [71] proposed several algorithms that can find proofs of primality in deterministic polynomial time for *some* primes and that do not rely on any unproved assertions such as the Riemann Hypothesis. In particular they have a deterministic polynomial time algorithm that will prove prime more than $x^{1-\epsilon}$ primes up to x. The key tool they use is the idea of a smooth number, that is, a number with only small prime factors.

Finally, we summarize some of the main complexity results in primality testing as follows:

(i) Primes/Composites $\in \mathcal{EXP}$, just try all the possible divisors

(ii) Composites $\in \mathcal{NP}$, guess a divisor

(iii) Primes $\in \mathcal{NP}$, Pratt (1975)

(iv) Primes $\in \mathcal{P}$, Miller (1976); assuming Extended Riemann Hypothesis

(v) Composites $\in \mathcal{RP}$, Rabin (1976); using Miller's randomised algorithm

(vi) Primes $\in$ super-$\mathcal{P}$, the APRCL test, due to Adleman, Pomerance and Rumely (1980), and Cohen and Lenstra (1981)

(vii) Primes $\in \mathcal{ZPP}$, Elliptic Curve Test, due to Goldwasser and Killian (1985), and Atkin and Morain (1991); not yet proved to work on all primes

(viii) Primes $\in \mathcal{ZPP}$, Hyperelliptic Curve Test, due to Adleman and Huang (1992); does not rely on any hypothesis, but totally non-practical.

(ix) Primes $\in \mathcal{P}$, Konyagin and Pomerance (1997); only for some primes.

4.2 Integer Factorization

Of all problems in the theory of numbers to which computers have been applied, and of all problems in computing to which number theory has been applied, probably none has been more influential than the problem of factoring. Because it is basic to the theory of numbers and easily understood, it has always, since Euclid's time, had an immense appeal. In this section, we shall introduce some widely used algorithms for factoring large integers.

4.2.1 Complexity of Integer Factorization

According to the Fundamental Theorem of Arithmetic, any integer bigger than one can be written in a unique prime product form. The problem of integer factorization is to factor an integer into its prime product form. More formally, the integer factorization problem can also be described as:

$$\begin{cases} \text{Input}: & n \in \mathbb{N} \text{ with } n > 1 \\ \\ \text{Output}: & \begin{cases} p_1, p_2, \cdots, p_k \in \text{ Primes} \\ \alpha_1, \alpha_2, \cdots, \alpha_k \in \mathbb{N} \\ \text{such that } n = p_1^{\alpha_1} p_2^{\alpha_2} \cdots p_k^{\alpha_k}. \end{cases} \end{cases} \qquad (4.26)$$

Primality testing and integer factorization are very important in mathematics. In 1801 Gauss wrote the following famous statement in his most profound

publication *Disquistiones Arithmeticae* (translated into English by A. A. Clarke
and published by Yale University Press in 1966):

> *The problem of distinguishing prime numbers from composite num-*
> *bers and of resolving the latter into their prime factors is known to*
> *be one of the most important and useful in arithmetic.* ⋯ *the dig-*
> *nity of science itself seems to require that every possible means be*
> *explored for the solution of a problem so elegant and so celebrated.*

But unfortunately, the primality testing problem, and particularly the the
integer factorization problem are computationally intractable; no-one at present
has found an efficient algorithm to test for primality or to factor an integer into
its prime product form, nor has anyone proved that no such algorithm exists [2].
Donald Knuth[4] thus wrote the following statement in his encyclopaedic work
[69]:

> *It is unfortunately not a simple matter to find this prime factoriza-*
> *tion of n, or to determine whether or not n is prime.* ⋯ *therefore*
> *we should avoid factoring large numbers whenever possible.*

Despite this, remarkable progress has been made in recent years, and mathematicians (at least some mathematicians) believe that efficient primality testing
and integer factorization algorithms are somewhere around the corner waiting
for discovery, although it is very hard to find such an algorithm. Generally
speaking, the most useful factoring algorithms fall into one of following two
main classes [18] (note that it is sufficient to have an algorithm for finding a
nontrivial factor p of N, because this can be applied recursively to p and N/p to
obtain the complete prime power decomposition of N, thus, the timings in the
following two classes refer to finding a nontrivial factor of N, not the complete
prime power decomposition of N):

(I) The running time depends mainly on the size of N, the number to be
factored, and is not strongly dependent on the size of the factor p found.
Examples are:

---

[4]Donald E. Knuth (1938–), perhaps best known for his series of books *The Art of Com-*
puter Programming (TAOCP). He studied mathematics as an undergraduate at Case Institute
of Technology, and received a PhD in Mathematics in 1963 from California Institute of Technology. Knuth joined Stanford University as Professor of Computer Science in 1968, and is
now Professor Emeritus there. Knuth received in 1974 the prestigious Turing Award from
the Association for Computing Machinery for his work in analysis of algorithms and for his
series books TAOCP.

(i) *Lehman's method* [77], which has a rigorous worst-case running time bound $\mathcal{O}\left(N^{1/3+\epsilon}\right)$.

(ii) *Shanks' SQUare FOrm Factorization method* SQUFOF, which has expected running time $\mathcal{O}\left(N^{1/4}\right)$.

(iii) *Shanks' class group method*, which has running time $\mathcal{O}\left(N^{1/5+\epsilon}\right)$.

(iv) *The Continued FRACtion (CFRAC) method*, which under plausible assumptions has expected running time

$$\mathcal{O}\left(\exp\left(c\sqrt{\log N \log\log N}\right)\right) = \mathcal{O}\left(N^{c\sqrt{\log\log N/\log N}}\right),$$

where c is a constant depending on the details of the algorithm (usually $c = \sqrt{2} \approx 1.414213562$).

(v) *The Multiple Polynomial Quadratic Sieve* (MPQS), which under plausible assumptions has expected running time

$$\mathcal{O}\left(\exp\left(c\sqrt{\log N \log\log N}\right)\right) = \mathcal{O}\left(N^{c\sqrt{\log\log N/\log N}}\right),$$

where c is a constant depending on the details of the algorithm (usually $c = \frac{3}{2\sqrt{2}} \approx 1.060660172$).

(vi) *The Number Field Sieve* (NFS), which under plausible assumptions has expected running time

$$\mathcal{O}\left(\exp\left(c(\log N)^{1/3}(\log\log N)^{2/3}\right)\right)$$

where $c = (64/9)^{1/3} \approx 1.922999427$ if GNFS (a general version of NFS) is used to factor an arbitrary integer N, whereas $c = (32/9)^{1/3} \approx 1.526285657$ if SNFS (a special version of NFS) is used to factor a special integer N such as $N = r^e \pm s$, where r and s are small, r and e is large. This is substantially asymptotically faster than any other currently known factoring method.

(II) The running time depends mainly on the size of p, the factor found of N. (We can assume that $p \leq \sqrt{N}$.) Examples are:

(i) *The Trial division algorithm*, which has running time $\mathcal{O}\left(p(\log N)^2\right)$.

(ii) *The Pollard's "rho" method* (also known as Pollard's ρ-algorithm), which under plausible assumptions has expected running time $\mathcal{O}\left(p^{1/2}(\log N)^2\right)$.

(iii) *Lenstra's Elliptic Curve Method (ECM)*, which under plausible assumptions has expected running time

$$\mathcal{O}\left(\exp\left(c\sqrt{\log p \log\log p}\right)\cdot(\log N)^2\right),$$

where $c \approx 2$ is a constant (depending on the details of the algorithm).

The term $\mathcal{O}\left((\log N)^2\right)$ is a generous allowance for the cost of performing arithmetic operations on numbers which are $\mathcal{O}(\log N)$ or $\mathcal{O}\left((\log N)^2\right)$ bits long; these could theoretically be replaced by $\mathcal{O}\left((\log N)^{1+\epsilon}\right)$ for any $\epsilon > 0$.

In practice, algorithms in both categories are important. It is often very difficult to say whether one method is better than another, but it is generally worth attempting to find small factors with algorithms in the second class before using the algorithms in the first class. That is, we could first try the *trial division algorithm*, then use some other methods such as NFS. So the trial division algorithm is still useful in integer factorization even though it is simple. In the subsections that follow, we shall introduce some of the most useful and widely used factoring algorithms.

Remark 4.2.1 An algorithm is said to be of *exponential complexity*, measured in terms of bit operations, if its required running time is

$$\mathcal{O}\left(N^\epsilon\right), \tag{4.27}$$

where a typical value for ϵ would be between 0.1 and 0.5. But note that we usually do not regard the type of complexity

$$\mathcal{O}\left(N^{\epsilon(N)}\right) = \mathcal{O}\left(N^{c\sqrt{\log\log N/\log N}}\right) = \mathcal{O}\left(\exp\left(c\sqrt{\log N \log\log N}\right)\right) \tag{4.28}$$

as an exponential complexity; it is not really an exponential, the most suitable name for it is *"subexponential complexity"*. The relationship between the polynomial, superpolynomial, subexponential, and exponential complexities, and some examples, are as follows:

$$\mathcal{O}\left((\log N)^k\right) \subset \mathcal{O}\left((\log N)^{c\log\log\log N}\right) \subset \mathcal{O}\left(N^{c\sqrt{\log\log N/\log N}}\right) \subset \mathcal{O}\left(N^\epsilon\right)$$

$\updownarrow$	$\updownarrow$	$\updownarrow$	$\updownarrow$
polynomial	superpolynomial	subexponential	exponential
$\Uparrow$	$\Uparrow$	$\Uparrow$	$\Uparrow$
Euclid's Algorithm	APRCL Test	MPQS Factoring	Trial Division

Remark 4.2.2 It is sometimes convenient to use abbreviations to denote the subexponential complexity; one such abbreviation is as follows:

$$L_N(\gamma, c) \stackrel{\text{def}}{=} \exp\left(c(\log N)^\gamma (\log \log N)^{1-\gamma}\right). \qquad (4.29)$$

So, in this notation, we could write for example

$$T(\text{CFRAC}) = \mathcal{O}\left(L_N\left(1/2, \sqrt{2}\right)\right), \qquad (4.30)$$

$$T(\text{ECM}) = \mathcal{O}\left(L_p\left(1/2, \sqrt{2}\right) \cdot (\log N)^2\right), \qquad (4.31)$$

$$T(\text{MPQS}) = \mathcal{O}\left(L_N\left(1/2, 3/(2\sqrt{2})\right)\right), \qquad (4.32)$$

$$T(\text{GNFS}) = \mathcal{O}\left(L_N\left(1/3, (64/9)^{1/3}\right)\right). \qquad (4.33)$$

Note also that some authors prefer to use the following abbreviation

$$L(N) \stackrel{\text{def}}{=} e^{\sqrt{\log N \log \log N}}. \qquad (4.34)$$

In this notation, we could write for example

$$T(\text{CFRAC}) = \mathcal{O}\left(L(N)^{2+o(1)}\right), \qquad (4.35)$$

$$T(\text{ECM}) = \mathcal{O}\left(L(p)^{2+o(1)} \cdot (\log N)^2\right), \qquad (4.36)$$

$$T(\text{MPQS}) = \mathcal{O}\left(L(N)^{1+o(1)}\right). \qquad (4.37)$$

$$(4.38)$$

To avoid confusion we shall stick to the ordinary full notation in this book.

4.2.2 Trial Division and Fermat Method

(I) Factoring by Trial Division

The simplest factoring algorithm is the trial division method, which tries all possible divisors of n to obtain its complete prime factorization:

$$n = p_1 p_2 \cdots p_t, \qquad p_1 \leq p_2 \leq \cdots \leq p_t. \qquad (4.39)$$

The following is the algorithm:

Algorithm 4.2.1 (Factoring by Trial Division)

[1] Input n and set $t \leftarrow 0$, $k \leftarrow 2$.

[2] If $n = 1$, then goto [5].

[3] $q \leftarrow n/k$ and $r \leftarrow n \pmod{k}$.

    If $r \neq 0$, goto [4].

    $t \leftarrow t + 1$, $p_t \leftarrow k$, $n \leftarrow q$, goto [2].

[4] If $q > k$, then $k \leftarrow k + 1$, and goto [3].

    $t \leftarrow t + 1$; $p_t \leftarrow n$.

[5] Exit: terminate the algorithm.

Exercise 4.2.1 Use Algorithm 4.2.1 to factor $n = 2759$.

An immediate improvement on Algorithm 4.2.1 is to make use of an auxiliary sequence of *trial divisors*:

$$2 = d_0 < d_1 < d_2 < d_3 < \cdots \tag{4.40}$$

which includes all primes $\leq \sqrt{n}$ and at least one value $d_k \geq \sqrt{n}$. The algorithm can be described as follows:

Algorithm 4.2.2 (Factoring by Trial Division – an improvement)

[1] Input n and set $t \leftarrow 0$, $k \leftarrow 0$.

[2] If $n = 1$, then goto [5].

[3] $q \leftarrow n/d_k$ and $r \leftarrow n \pmod{d_k}$.

    If $r \neq 0$, goto [4].

    $t \leftarrow t + 1$, $p_t \leftarrow d_k$, $n \leftarrow q$, goto [2].

[4] If $q > d_k$, then $k \leftarrow k + 1$, and goto [3].

    $t \leftarrow t + 1$; $p_t \leftarrow n$.

[5] Exit: terminate the algorithm.

Exercise 4.2.2 Use Algorithm 4.2.2 to factor $n = 2759$; assume that we have the list L of all primes $\leq \lfloor\sqrt{2759}\rfloor = 52$ and at least one $\geq \sqrt{n}$, that is, $L = \{2, 3, 5, 7, 11, 13, 17, 19, 23, 29, 31, 37, 41, 43, 47, 53\}$.

Theorem 4.2.1 Algorithm 4.2.2 requires a running time in $\mathcal{O}\left(\max\left(p_{t-1}, \sqrt{p_t}\right)\right)$, where p_t is the largest prime factor of n. If a primality test is inserted between steps [2] and [3] the running time would be in $\mathcal{O}(p_{t-1})$ (or $\mathcal{O}(p_{t-1}/\ln p_{t-1})$ if one does trial division only by primes), the second largest prime factor of n, rather than $\mathcal{O}\left(\max\left(p_{t-1}, \sqrt{p_t}\right)\right)$.

The trial division algorithm is very useful for removing small factors, but should not be used for complete factoring, except when n is very small, say $n < 10^8$.

(II) Fermat Factoring Method

Now suppose n is an odd integer (if n were even we could repeatedly divide by 2 until an odd integer results). If $n = pq$, where $p \leq q$ are both odd, then on setting $x = (p+q)/2$ and $y = (q-p)/2$ we find that $n = x^2 - y^2$, or $y^2 = x^2 - n$. The following algorithm tries to find $n = pq$ using the above idea.

Algorithm 4.2.3 (Fermat's Algorithm)

[1] Input n and set $k \leftarrow \lfloor\sqrt{n}\rfloor + 1$, $y \leftarrow k \cdot k - n$, $d \leftarrow 1$

[2] If $\lfloor\sqrt{y}\rfloor = \sqrt{y}$ goto step [4] else $y \leftarrow y + 2 \cdot k + d$ and $d \leftarrow d + 2$

[3] If $\lfloor\sqrt{y}\rfloor < n/2$ goto step [2] else print "No Factor Found" and goto [5]

[4] $x \leftarrow \sqrt{n+y}$, $y \leftarrow \sqrt{y}$, print $x - y$ and $x + y$, the nontrivial factors of n

[5] Exit: terminate the algorithm.

Exercise 4.2.3 Use the Fermat method to factor $n = 278153$.

Theorem 4.2.2 (The Complexity of Fermat's Method): The Fermat method will try as many as $(n+1)/2 - \sqrt{n}$ arithmetic steps to factor n. That is, it is of complexity $\mathcal{O}((n+1)/2 - \sqrt{n})$.

4.2.3   Legendre's Congruence

In the next two subsections, we shall introduce three widely used *general purpose* integer factorization methods, namely, the continued fraction method (abbreviated CFRAC), the quadratic sieve (abbreviated QS) and the number field sieve (abbreviated NFS). By a general purpose factoring method, we mean one that will factor *any* integer of a given size in about the same time as any other of that size. The method will take as long, for example, to split a 100-digit number into the product of a 1-digit and a 99-digit prime, as it will to split a different number into the product of two 50-digit primes. These methods do not depend upon any special properties of the number or its factors [133].

The CFRAC method, as well as other powerful general purpose factoring methods such as Quadratic Sieve (QS) and Number Filed Sieve (NFS), makes use of the simple but important observation that if we have two integers x and y such that

$$x^2 \equiv y^2 \pmod{N},\ 0 < x < y < N,\ x \neq y,\ x + y \neq N \tag{4.41}$$

then $\gcd(x - y,\ N)$ and $\gcd(x + y,\ N)$ are possible nontrivial factors of N, because $N \mid (x + y)(x - y)$, but $N \nmid (x + y)$ and $N \nmid (x - y)$. The congruence (4.41) is often called Legendre's congruence. So, to use Legendre's congruence for factorization, we just simply perform the following two steps:

[1] Find a nontrivial solution to the congruence $x^2 \equiv y^2 \pmod{N}$.

[2] Compute the factors d_1 and d_2 of N by using the Euclid's algorithm:

$$(d_1, d_2) = (\gcd(x + y,\ N),\ \gcd(x - y,\ N)).$$

Example 4.2.1 Let $N = 119$. Since $12^2 \bmod 119 = 5^2 \bmod 119 = 0$, then $12^2 \equiv 5^2 \pmod{119}$. So, $(d_1, d_2) = (\gcd(12 + 5, 119),\ \gcd(12 - 5, 119)) = (17, 7)$. In fact, $119 = 7 \cdot 17$.

The best methods for constructing congruences of the form (4.41) start by accumulating several congruences of the form

$$\left(A_i = \prod p_k^{e_k} \right) \equiv \left(B_i = \prod p_j^{e_j} \right) \pmod{N} \tag{4.42}$$

and multiplying some of these congruences in order to generate squares on both sides [99]. We illustrate this idea in the following example.

Example 4.2.2 Let $N = 77$. Then, in the left hand side of the following table, we collect eight congruences of the form (4.42) over the prime factor base FB $= \{-1, 2, 3, 5\}$ (note that we include -1 as a "prime" factor), and the right hand side of the table contains the exponent vector information of $v(A_i)$ and $v(B_i)$ modulo 2.

$45 = 3^2 \cdot 5$	$\equiv$	$-32 = -2^5$	$\Longleftrightarrow (0\ 0\ 0\ 1)$	$\equiv (1\ 1\ 0\ 0)$
$50 = 2 \cdot 5^2$	$\equiv$	$-27 = -3^3$	$\Longleftrightarrow (0\ 1\ 0\ 0)$	$\equiv (1\ 0\ 1\ 0)$
$72 = 2^3 \cdot 3^2$	$\equiv$	-5	$\Longleftrightarrow (0\ 1\ 0\ 0)$	$\equiv (1\ 0\ 0\ 1)$
$75 = 3 \cdot 5^2$	$\equiv$	-2	$\Longleftrightarrow (0\ 0\ 1\ 0)$	$\equiv (1\ 1\ 0\ 0)$
$80 = 2^4 \cdot 5$	$\equiv$	3	$\Longleftrightarrow (0\ 0\ 0\ 1)$	$\equiv (0\ 0\ 1\ 0)$
$125 = 5^3$	$\equiv$	$48 = 2^4 \cdot 3$	$\Longleftrightarrow (0\ 0\ 0\ 1)$	$\equiv (0\ 0\ 1\ 0)$
$320 = 2^6 \cdot 5$	$\equiv$	$243 = 3^5$	$\Longleftrightarrow (0\ 0\ 0\ 1)$	$\equiv (0\ 0\ 1\ 0)$
$384 = 2^7 \cdot 3$	$\equiv$	-1	$\Longleftrightarrow (0\ 1\ 1\ 0)$	$\equiv (1\ 0\ 0\ 0)$

Now we multiply some of these congruences in order to generate squares on both sides; both sides will be squares precisely when the sum of the exponent vectors is the zero vector modulo 2. We first multiply the sixth and seventh congruences and get:

$125 = 5^3$	$\equiv$	$48 = 2^4 \cdot 3$	$\Longleftrightarrow (0\ 0\ 0\ 1)$	$\equiv (0\ 0\ 1\ 0)$
$320 = 2^6 \cdot 5$	$\equiv$	$243 = 3^5$	$\Longleftrightarrow (0\ 0\ 0\ 1)$	$\equiv (0\ 0\ 1\ 0)$
			$\downarrow \downarrow \downarrow \downarrow$	$\downarrow \downarrow \downarrow \downarrow$
			$(0\ 0\ 0\ 0)$	$(0\ 0\ 0\ 0)$

Since the sum of the exponent vectors is the zero vector modulo 2, then we find squares on both sides:

$$5^3 \cdot 2^6 \cdot 5 \equiv 2^4 \cdot 3 \cdot 3^5 \Longleftrightarrow (5^2 \cdot 2^3)^2 \equiv (2^2 \cdot 3^3)^2$$

and hence we have $\gcd(5^2 \cdot 2^3 \pm 2^2 \cdot 3^3, 77) = (77, 1)$, but this does not split 77, we try to multiply some other congruences, e.g., the fifth and the seventh, and get:

$80 = 2^4 \cdot 5$	$\equiv$	3	$\Longleftrightarrow (0\ 0\ 0\ 1)$	$\equiv (0\ 0\ 1\ 0)$
$320 = 2^6 \cdot 5$	$\equiv$	$243 = 3^5$	$\Longleftrightarrow (0\ 0\ 0\ 1)$	$\equiv (0\ 0\ 1\ 0)$
			$\downarrow \downarrow \downarrow \downarrow$	$\downarrow \downarrow \downarrow \downarrow$
			$(0\ 0\ 0\ 0)$	$(0\ 0\ 0\ 0)$

the sum of the exponent vectors is the zero vector modulo 2, so we find

$$2^4 \cdot 5 \cdot 2^6 \cdot 5 \equiv 3 \cdot 3^5 \Longleftrightarrow (2^5 \cdot 5)^2 \equiv (3^3)^2$$

and compute $\gcd(2^5 \cdot 5 \pm 3^3, 77) = (11, 7)$. This time, it splits 77. Once we split N, we stop the process. But as an example, we just try one more instance, which will also split N.

$45 = 3^2 \cdot 5$	$\equiv$	$-32 = -2^5$	$\longleftrightarrow$	$(0\ 0\ 0\ 1) \equiv (1\ 1\ 0\ 0)$
$50 = 2 \cdot 5^2$	$\equiv$	$-27 = -3^3$	$\longleftrightarrow$	$(0\ 1\ 0\ 0) \equiv (1\ 0\ 1\ 0)$
$75 = 3 \cdot 5^2$	$\equiv$	-2	$\longleftrightarrow$	$(0\ 0\ 1\ 0) \equiv (1\ 1\ 0\ 0)$
$320 = 2^6 \cdot 5$	$\equiv$	$243 = 3^5$	$\longleftrightarrow$	$(0\ 0\ 0\ 1) \equiv (0\ 0\ 1\ 0)$
$384 = 2^7 \cdot 3$	$\equiv$	-1	$\longleftrightarrow$	$(0\ 1\ 1\ 0) \equiv (1\ 0\ 0\ 0)$
			$\downarrow\downarrow\downarrow\downarrow$	$\downarrow\downarrow\downarrow\downarrow$
			$(0\ 0\ 0\ 0)$	$(0\ 0\ 0\ 0)$

So we have

$$3^2 \cdot 5 \cdot 2 \cdot 5^2 \cdot 3 \cdot 5^2 \cdot 2^6 \cdot 5 \cdot 2^7 \cdot 3 \equiv -2^5 \cdot -3^3 \cdot -2 \cdot 3^5 \cdot -1 \longleftrightarrow (2^7 \cdot 3^2 \cdot 5^3)^2 \equiv (2^3 \cdot 3^4)^2,$$

thus, $\gcd(2^7 \cdot 3^2 \cdot 5^3 \pm 2^3 \cdot 3^4, 77) = (7, 11)$.

Based on the above idea, the trick common to CFRAC, QS and NFS is to find the congruence (also called *relation*) of the form

$$x_k^2 \equiv (-1)^{e_{0k}} p_1^{e_{1k}} p_2^{e_{2k}} \cdots p_m^{e_{mk}} \pmod{N} \tag{4.43}$$

where each p_i is a "small" prime number (the set of all such p_i, for $1 \leq i \leq m$, forms a *factor base*, denoted by FB, as just mentioned). If we find sufficiently many such congruences, by Gaussian elimination over $\mathbb{Z}/2\mathbb{Z}$ we may hope to find a relation of the form

$$\sum_{1 \leq k \leq n} \epsilon_k (e_{0k}, e_{1k}, e_{2k}, \cdots, e_{mk}) \equiv (0, 0, 0, \cdots, 0) \pmod{2} \tag{4.44}$$

where ϵ is either 1 or 0, and then

$$x = \prod_{1 \leq k \leq n} x_k^{\epsilon_k}, \tag{4.45}$$

$$y = (-1)^{v_0} p_1^{v_1} p_2^{v_2} \cdots p_m^{v_m} \tag{4.46}$$

where

$$\sum_k \epsilon_k (e_{0k}, e_{1k}, e_{2k}, \cdots, e_{mk}) = 2(v_0, v_1, v_2, \cdots, v_m). \tag{4.47}$$

It is clear that we now have $x^2 \equiv y^2 \pmod{N}$. This splits N if in addition, $x \not\equiv \pm y \pmod{N}$.

Now we are in a position to introduce our first general purpose factoring method, CFRAC.

4.2.4   Continued FRACtion Method (CFRAC)

The continued fraction method is perhaps the first *modern, general* purpose integer factorization method, although its original idea may go back to M. Kraitchik in the 1920's or even earlier to A. M. Legendre. It was used by D. H. Lehmer and R. E. Powers to devise a new technique in the 1930's, but however the method was not very useful and applicable at the time because it was unsuitable for desk calculators. About 40 years later, it was first implemented on a computer by M. A. Morrison and J. Brillhart [101], who used it to successfully factor the seventh Fermat number

$$F_7 = 2^{2^7} + 1 = 59649589127497217 \cdot 5704689200685129054721$$

on the morning of 13 September 1970.

The CFRAC method looks for small values of $|W|$ such that $x^2 \equiv W \pmod{N}$ has a solution. Since W is small (specifically $W = \mathcal{O}(\sqrt{N})$), it has a reasonably good chance of being a product of primes in our factor base FB. Now if W is small and $x^2 \equiv W \pmod{N}$, we then can write $x^2 = W + kNd^2$ for some k and d, hence $(x/d)^2 - kN = W/d^2$ will be small. In other words, the rational number x/d is an approximation to $\sqrt{kN}$. This suggests looking at the continued fraction expansion of $\sqrt{kN}$, since continued fraction expansions of real numbers give good rational approximations. This is exactly the idea behind the CFRAC method! We first obtain a sequence of approximations (i.e., convergents) P_i/Q_i to $\sqrt{kN}$ for a number of values of k, such that

$$\left| \sqrt{kN} - \frac{P_i}{Q_i} \right| \leq \frac{1}{Q_i^2}. \tag{4.48}$$

Putting $W_i := P_i^2 - Q_i^2 kN$, then we have

$$W_i = (P_i + Q_i\sqrt{kN})(P_i - Q_i\sqrt{kN}) \sim 2Q_i\sqrt{kN}\frac{1}{Q_i} \sim 2\sqrt{kN}. \tag{4.49}$$

Hence, $P_i^2 \bmod N$ are small and more likely to be smooth, as desired. Then, we try to factor the corresponding integers $W_i := P_i^2 - Q_i^2 kN$ over our factor base FB; on each success, we obtain a new congruence of the form

$$P_i^2 \equiv W_i \Longleftrightarrow x^2 \equiv (-1)^{e_0} p_1^{e_1} p_2^{e_2} \cdots p_m^{e_m} \pmod{N}. \tag{4.50}$$

Once we have obtained at least $m + 2$ such congruences, then by Gaussian elimination over $\mathbb{Z}/2\mathbb{Z}$ we have obtained a congruence $x^2 \equiv y^2 \pmod{N}$. That is, if $(x_1, e_{01}, e_{11}, \cdots, e_{m1}), \cdots, (x_r, e_{0r}, e_{1r}, \cdots, e_{mr})$ are solutions of (4.50) such that the vector sum

$$(e_{01}, e_{11}, \cdots, e_{m1}) + \cdots + (e_{0r}, e_{1r}, \cdots, e_{mr}) = (2e'_0, 2e'_1, \cdots, 2e'_m) \quad (4.51)$$

is even in each component, then

$$x \equiv x_1 x_2 \cdots x_r \pmod{N} \quad (4.52)$$
$$y \equiv (-1)^{e'_0} p_1^{e'_1} \cdots p_m^{e'_m} \pmod{N} \quad (4.53)$$

is a solution to (4.41), except for the possibility that $x \equiv \pm y \pmod{N}$, and hence (usually) a nontrivial splitting of N.

Example 4.2.3 We now illustrate by an example the idea of CFRAC factoring. Let $N = 1037$. Then $\sqrt{1037} = [32, \overline{4, 1, 15, 3, 3, 15, 1, 4, 64}]$. The first ten continued fraction approximations to $\sqrt{1037}$ are as follows:

Convergent P/Q	$P^2 - N \cdot Q^2 := W$
32/1	$-13 = -13$
129/4	$49 = 7^2$
161/5	$-4 = -2^2$
$2544/79 \equiv 470/79$	$19 = 19$
$7793/242 \equiv 534/242$	$-19 = -19$
$25923/805 \equiv 1035/805$	$4 = 2^2$
$396638/12317 \equiv 504/910$	$-49 = -7^2$
$422561/13122 \equiv 502/678$	$13 = 13$
$2086882/64805 \equiv 438/511$	$-1 = -1$
$133983009/4160642 \equiv 535/198$	$13 = 13$

Now we search for squares on both sides either just by a single congruence or by a combination (i.e., multiplying together) of several congruences, and find that

$$129^2 \equiv 7^2 \Longleftrightarrow \gcd(1037, \ 129 \pm 7) = (17, 61)$$
$$1035^2 \equiv 2^2 \Longleftrightarrow \gcd(1037, \ 1035 \pm 2) = (1037, 1)$$
$$129^2 \cdot 1035^2 \equiv 7^2 \cdot 2^2 \Longleftrightarrow \gcd(1037, \ 129 \cdot 1035 \pm 7 \cdot 2) = (61, 17)$$
$$161^2 \cdot 504^2 \equiv (-1)^2 \cdot 2^2 \cdot 7^2 \Longleftrightarrow \gcd(1037, \ 161 \cdot 504 \pm 2 \cdot 7) = (17, 61)$$
$$502^2 \cdot 535^2 \equiv 13^2 \Longleftrightarrow \gcd(1037 \ 502 \cdot 535 \pm 13) = (1037, 1)$$

Three of them yield a factorization of $1037 = 17 \cdot 61$.

Exercise 4.2.4 Use the continued fraction expansion

$$\sqrt{1711} = [41, \overline{2, 1, 2, 1, 13, 16, 2, 8, 1, 2, 2, 2, 2, 2, 1, 8, 2, 16, 13, 1, 2, 1, 2, 82}]$$

and the factor base FB $= \{-1, 2, 3, 5\}$ to factor the integer 1711.

It is clear that the CFRAC factoring algorithm is essentially just a continued fraction algorithm for finding the continued fraction expansion $[q_0, q_1, \cdots, q_k, \cdots]$ of $\sqrt{kN}$, or the P_k and Q_k of such an expansion. In what follows, we shall briefly summarize the CFRAC method just discussed above in the following algorithmic form:

Algorithm 4.2.4 (CFRAC Algorithm)

[1] Let N be the integer to be factored and k any small integer (usually 1), and let the factor base, denoted by FB, be a set of small primes $\{p_1, p_2, \cdots, p_r\}$ chosen such that it is possible to find some integer x_i such that $x_i^2 \equiv kN \pmod{p_i}$. Usually, FB contains all such primes less than or equal to some limit. Note that the multiplier $k > 1$ is needed only when the period is short. For example, Morrison and Brillhart used $k = 257$ in factoring F_7.

[2] Compute the continued fraction expansion $[q_0, \overline{q_1, q_2, \cdots, q_r}]$ of $\sqrt{kN}$ for a number of values of k. This gives us good rational approximations P/Q. The recursion formulas to use for computing P/Q are as follows:

$$\frac{P_0}{Q_0} = \frac{q_0}{1}, \quad \frac{P_1}{Q_1} = \frac{q_0 q_1 + 1}{q_1},$$
$$\frac{P_i}{Q_i} = \frac{q_i P_{i-1} + P_{i-2}}{q_i Q_{i-1} + Q_{i-2}}, \quad i \geq 2.$$

This can be done by a continued fraction algorithm such as Algorithm 1.3.1 introduced in subsection 1.3.2 in Chapter 1.

[3] Try to factor the corresponding integer $W = P^2 - Q^2 kN$ on our factor base FB. Since $W < 2\sqrt{kN}$, each of these W is only about half of the length of kN. If we succeed, we get a new congruence. For each success, we obtain a congruence

$$x^2 \equiv (-1)^{e_0} p_1^{e_1} p_2^{e_2} \cdots p_m^{e_m} \pmod{N},$$

which is so because if P_i/Q_i is the ith continued fraction convergent to $\sqrt{kN}$ and $W_i = P_i^2 - N \cdot Q_i^2$, then

$$P_i^2 \equiv W_i \pmod{N}. \tag{4.54}$$

[4] Once we have obtained at least $m + 2$ such congruences, then by Gaussian elimination over $\mathbb{Z}/2\mathbb{Z}$ we have obtained a congruence $x^2 \equiv y^2 \pmod{N}$. That is, if $(x_1, e_{01}, e_{11}, \cdots, e_{m1}), \cdots, (x_r, e_{0r}, e_{1r}, \cdots, e_{mr})$ are solutions of (4.50) such that the vector sum defined in (4.51) is even in each component, then

$$x \equiv x_1 x_2 \cdots x_r \pmod{N}$$
$$y \equiv (-1)^{e_0'} p_1^{e_1'} \cdots p_m^{e_m'} \pmod{N}$$

is a solution to $x^2 \equiv y^2 \pmod{N}$, except for the possibility that $x \equiv \pm y \pmod{N}$, and hence we have

$$(d_1, d_2) = (\gcd(x + y, \ N), \ \gcd(x - y, \ N))$$

which are then possible nontrivial factors of N.

Conjecture 4.2.1 (The Complexity of the CFRAC Method): If N is the integer to be factored, then under certain reasonable heuristic assumptions, the CFRAC method will factor N in time

$$\mathcal{O}\left(\exp\left((\sqrt{2} + o(1))\sqrt{\log N \log\log N}\right)\right) = \mathcal{O}\left(N^{\sqrt{(2+o(1))\log\log N/\log N}}\right).$$
$$(4.55)$$

Remark 4.2.3 This is not a theorem; it is a conjecture. The conjecture is supported by some heuristic assumptions [27] which have not been proved.

4.2.5 Quadratic and Number Field Sieves (QS/NFS)

In this subsection, we shall briefly introduce two other powerful general purpose factoring method: quadratic sieve (QS) and number Field sieve (NFS).

(I) The Quadratic Sieve (QS)

The idea of the quadratic sieve (QS), was first introduced by Carl Pomerance in 1982. QS is somewhat similar to CFRAC except that instead of using continued fractions to produce the values for $W_k := P_k^2 - N \cdot Q_k^2$, it uses

$$W_k = (k + \lfloor \sqrt{N} \rfloor)^2 - N \equiv (k + \lfloor \sqrt{N} \rfloor)^2 \pmod{N}. \qquad (4.56)$$

Here if $0 < k < L$, then

$$0 < W_k < (2L + 1)\sqrt{N} + L^2. \tag{4.57}$$

If we get

$$\prod_{i=1}^{t} W_{n_i} = y^2, \tag{4.58}$$

then we have $x^2 \equiv y^2 \pmod{N}$ with

$$x \equiv \prod_{i=1}^{t}(\lfloor\sqrt{N}\rfloor + n_i) \pmod{N}. \tag{4.59}$$

Once such x and y are found, there is a good chance that $\gcd(x - y, N)$ is a non-trivial factor of N. For the purposes of implementation, we can use the same set FB as that used in CFRAC and the same idea as that described above to arrange that (4.58) holds. The most widely used variation of quadratic sieve is perhaps the multiple polynomial quadratic sieve (MPQS), proposed by Peter Montgomery in 1986. MPQS has been used to obtain many spectacular factorizations. One of such factorizations is the 103-digit number

$$\frac{2^{361} + 1}{3 \cdot 174763} = 68743016175348275093505757684543562450254030 \cdot p_{61}.$$

The most recent record is that by 1994 the quadratic sieve had factored the famous 129-digit RSA challenge number that has been estimated in Martin Gardner's 1976 *Scientific American* column to be safe for 40 quadrillion years.

Example 4.2.4 Use the quadratic sieve method (QS) to factor $N = 2041$. Let $W(x) = x^2 - N$, with $x = 43, 44, 45, 46$. Then we have:

$W(43)$	$=$	$-2^6 \cdot 3$
$W(44)$	$=$	$-3 \cdot 5 \cdot 7$
$W(45)$	$=$	-2^4
$W(46)$	$=$	$3 \cdot 5^2$

p	$W(43)$	$W(44)$	$W(45)$	$W(46)$
-1	1		1	0
2	0		0	0
3	1		0	1
5	0		0	0

which leads to the following congruence:

$$(43 \cdot 45 \cdot 46)^2 \equiv (-1)^2 \cdot 2^{10} \cdot 3^2 \cdot 5^2 = (2^5 \cdot 3 \cdot 5)^2.$$

This congruence gives the factorization of $2041 = 13 \cdot 157$, since

$$\gcd(2041, \ 43 \cdot 45 \cdot 46 + 2^5 \cdot 3 \cdot 5) = 157, \quad \gcd(2041, \ 43 \cdot 45 \cdot 46 - 2^5 \cdot 3 \cdot 5) = 13.$$

Conjecture 4.2.2 (The complexity of the QS/MPQS Method): If N is the integer to be factored, then under certain reasonable heuristic assumptions, the QS/MPQS method will factor N in time

$$\mathcal{O}\left(\exp\left((1+o(1))\sqrt{\log N \log\log N}\right)\right) = \mathcal{O}\left(N^{(1+o(1))\sqrt{\log\log N/\log N}}\right).$$
(4.60)

(II) The Number Field Sieve (NFS)

Before introducing the number field sieve NFS, it is interesting to briefly review some important milestones in the development of integer factorization methods. In 1970, it was barely possible to factor "hard" 20-digit numbers. In 1980, by using the CFRAC method, factoring of 50-digit numbers was becoming commonplace. In 1990, the QS method had doubled the length of the numbers that could be factored by CFRAC, with a record having 116 digits. In the spring of 1996, the NFS method had successfully split a 130-digit RSA challenge number in about 15% of the time the QS would have taken. So at present, the number field sieve (NFS) is the champion of all known factoring methods. NFS was first proposed by John Pollard, an English mathematician, in a letter to A. M. Odlyzko, dated 31 August 1988, with copies to R. P. Brent, J. Brillhart, H. W. Lenstra, C. P. Schnorr and H. Suyama, outlining an idea of factoring certain big numbers via *algebraic number fields*. His original idea however is not for any large composite, but for certain "pretty" composites that had the property that they were close to powers. He illustrated the idea with a factorization of the seventh Fermat number $F_7 = 2^{2^7} + 1$ that was first factored by CFRAC in 1970. He also speculated in the letter that "if F_9 is still unfactored, then it might be a candidate for this kind of method eventually?" The answer now is of course "yes", since F_9 had been factored by NFS in 1990. It would be specifically worthwhile pointing out that NFS is not only a method suitable for factoring numbers in a special form like F_9, but also a general purpose factoring method for any integer of a given size. There are, in fact, two forms of NFS (see [62] and [79]): the Special NFS (SNFS), tailored specifically for integers of the form $N = c_1 r^t + c_2 s^u$, and the General NFS (GNFS), applicable for any arbitrary numbers.

The fundamental idea of the NFS is the same as that of *quadratic sieve* (QS) introduced previously: by a sieving process we look for congruences modulo N by working over a factor base, and then we do a Gaussian elimination over

$\mathbb{Z}/2\mathbb{Z}$ to obtain a congruence of squares $x^2 \equiv y^2 \pmod{N}$, and hence hopefully a factorization of N. Given an odd positive integer N, NFS has four main steps in factoring N:

(i) **Polynomial Selection**: Select two irreducible polynomials $f(x)$ and $g(x)$ with small integer coefficients for which there exists an integer m such that

$$f(m) \equiv g(m) \equiv 0 \pmod{N} \tag{4.61}$$

The polynomials should not have a common factor over $\mathbb{Q}$.

(ii) **Sieving**: Finds pairs (a, b) such that $\gcd(a, b) = 1$ and both

$$b^{\deg(f)} f(a/b), \qquad\qquad b^{\deg(g)} g(a/b) \tag{4.62}$$

are smooth wrt a chosen factor base. The expressions in (4.62) are the norms of the algebraic numbers $a - b\alpha$ and $a - b\beta$, multiplied by the leading coefficients of f and g, respectively. (α denote a complex root of f and β a root of g). The principal ideals $a - b\alpha$ and $a - b\beta$ factor into products of prime ideals in the number field $\mathbb{Q}(\alpha)$ and $\mathbb{Q}(\beta)$, respectively.

(iii) **Linear Algebra**: Use techniques in linear algebra to find a set S of indices such that the two products

$$\prod_{i \in S}(a_i - b_i\alpha), \qquad\qquad \prod_{i \in S}(a_i - b_i\beta) \tag{4.63}$$

are both squares of products of prime ideals.

(iv) **Square root**: Using the set S in (4.63) try to find algebraic numbers $\alpha' \in \mathbb{Q}(\alpha)$ and $\beta' \in \mathbb{Q}(\beta)$ such that

$$(\alpha')^2 = \prod_{i \in S}(a_i - b_i\alpha), \quad (\beta')^2 = \prod_{i \in S}(a_i - b_i\beta)\prod_{i \in S}(a_i - b_i\beta) \tag{4.64}$$

Prior to the NFS, all modern factoring methods had an expected running time of at best

$$\mathcal{O}\left(\exp\left((c + o(1))\sqrt{\log\log N/\log N}\right)\right). \tag{4.65}$$

For example, the multiple polynomial quadratic sieve (MPQS) takes time

$$\mathcal{O}\left(\exp\left((1 + o(1))\sqrt{\log\log N/\log N}\right)\right).$$

Because of the Canfield-Erdős-Pomerance theorem [27], some people even believed that this could not be improved, except maybe for the term $(c + o(1))$. The invention of the NFS has however changed this belief, since under some reasonable heuristic assumptions, we have

Conjecture 4.2.3 (Complexity of NFS) Under some reasonable heuristic assumptions, the NFS method can factor integer N in time

$$\mathcal{O}\left(\exp\left((c + o(1))(\log N)^{1/3}(\log\log N)^{2/3}\right)\right) \tag{4.66}$$

where $c = (64/9)^{1/3} \approx 1.922999427$ if GNFS is used to factor arbitrary integer N, whereas $c = (32/9)^{1/3} \approx 1.526285657$ if SNFS is used to factor special integer N.

Example 4.2.5 The largest number ever factored by SNFS is

$$N = (12^{167} + 1)/13 = p_{75} \times p_{105}$$

It was announced by P. Montgomery, S. Cavallar and H. te Riele at CWI in Amsterdam on 3 September 1997. They used the polynomials $f(x) = x^5 - 144$ and $g(x) = 12^{33}x + 1$ with common root $m \equiv 12^{134}$ (mod N). The factor base bound was 4.8 million for f and 12 million for g. Both large prime bounds were 150 million, with two large primes allowed on each side. They sieved over $|a| \leq 8.4$ million and $0 < b \leq 2.5$ million. The sieving lasted 10.3 calendar days; 85 SGI machines at CWI contributed a combined 13027719 relations in 560 machine-days. It took 1.6 more calendar days to process the data. This processing includes 16 CPU-hours on a Cray C90 at SARA in Amsterdam to process a 1969262×1986500 matrix with 57942503 nonzero entries.

4.2.6 *"rho"* and *"p − 1"* Methods

In this and the next subsections, we shall introduce some special-purpose factoring methods. By special-purpose we mean that the methods depend, for their success, upon some special properties of the number being factored. The usual property is that the factors of the number is small. However, other properties might include the number or its factors having a special mathematics form. For example, If p is a prime number and if $2^p - 1$ is a composite number, then all of the factors of $2^p - 1$ must be congruent to 1 modulo $2p$. For example, $2^{11} - 1 = 23 \cdot 89$, then $23 \equiv 89 \equiv 1 \pmod{22}$. Certain factoring algorithms can take advantage of this special form of the factors. Special-purpose methods do not always succeed, but they are usefully tried first, before using the more powerful, general methods, such as CFRAC, MPQS or NFS.

(I) The "rho" Method

J. M. Pollard in 1975 [109] proposed a Monte Carlo method, now widely known as Pollard's "rho" or ρ-method, for finding a (small) non-trivial factor p in a large integer N. It uses an iteration of the form

$$\begin{cases} x_0 = \text{random}(0, \ N - 1), \\ x_{i+1} \equiv f(x_i) \pmod{N}, \quad i \geq 1 \end{cases} \qquad (4.67)$$

where x_0 is a random starting value, N is the number to be factored, and $f \in \mathbb{Z}[x]$ is a polynomial with integer coefficients; usually, we just simply choose $f(x) = x^2 + a$ with $a \neq -2, 0$. Suppose p is a prime factor of N, then the two sequences $\langle x_i \bmod N \rangle$ and $\langle x_j \bmod N \rangle$ are both ultimately periodic since, in either sequence, as soon as one value repeats, the following value must be also. It is from this behaviour that the algorithm gets its name (see Figure 4.1 for a graphical explanation); the pattern resembles the Greek letter ρ ("rho").

Let p be the smallest prime factor of N, and j the smallest positive index such that $x_{2j} \equiv x_j \pmod{p}$. Making some plausible assumptions, it is easy to show that the expected value of j is $\mathcal{O}(\sqrt{p})$. The argument is related to the well-known "birthday" paradox: suppose that $1 \leq k \leq n$ and that the numbers $x_1, x_2, \cdots, x_k$ are independently chosen from the set $\{1, 2, \cdots, n\}$. Then the probability that the numbers x_k are distinct is

$$\left(1 - \frac{1}{n}\right) \cdot \left(1 - \frac{2}{n}\right) \cdots \left(1 - \frac{k-1}{n}\right) \ \sim \ \exp\left(\frac{-k^2}{2n}\right). \qquad (4.68)$$

Note that the x_i's are likely to be distinct if k is small compared with $\sqrt{n}$, but unlikely to be distinct if k is large compared with $\sqrt{n}$. Of course, we cannot work out $x_i \bmod p$, since we do not know p in advance, but we can detect x_j by

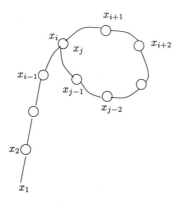

Figure 4.1: Illustration of the ρ-method

taking greatest common divisors. We simply compute $d = \gcd(x_{2i} - x_i, N)$ for $i = 1, 2, \cdots$ and stop when a $d > 1$ is found. The following is a description of the algorithm:

Algorithm 4.2.5 (Pollard's ρ) Given a composite integer $N > 1$. This algorithm tries to find a non-trivial factor of N. Suppose the polynomial to use is $f(x) = x^2 + 1$.

[1] (Initialization) Set $i \leftarrow 1$, $x_1 \leftarrow 2$, $y \leftarrow x_1$, and $k \leftarrow 2$. Set e.g., max $= 1000$ (it is wise to set the maximum number of allowed iterations to stop the infinite loop of the algorithm.

[2] (Iteration and Computation) $c \leftarrow c + 1$. $x_{i+1} \leftarrow x_i^2 + 1 \bmod N$ and $d \leftarrow \gcd(y - x_{i+1}, N)$.

[3] (Factor Found?) If $1 < d < N$, then d is a non-trivial factor of N, output d, and goto [6].

[4] (Stop Iteration?) If $c \geq$ max, then output a message saying that the algorithm fails and goto [6].

[5] (Another Iteration) If $i \neq k$, then $y \leftarrow x_{i+1}$, $k \leftarrow 2k$. Goto [2].

[6] (Exit) Terminate the algorithm.

Conjecture 4.2.4 (Complexity of the "*rho*" Method) Let p be a prime dividing N and $p = \mathcal{O}(\sqrt{p})$, then the "*rho*" algorithm has has expected running time

$$\mathcal{O}(\sqrt{p}) = \mathcal{O}(\sqrt{p}\,(\log N)^2) = \mathcal{O}(N^{1/4}(\log N)^2)$$

to find the prime factor p of N.

Remark 4.2.4 The "*rho*" method is an improvement over trial division, because in trial division, $\mathcal{O}(p) = \mathcal{O}(N^{1/4})$ divisions would be needed to find a small factor p of N. But of course, one disadvantage of the "*rho*" algorithm is that its running time is only a conjectured expected value, not a rigorous bound.

Example 4.2.6 Use the ρ-method to factor the number $N = 1387$.

[Initialization] $i = 1$, $x_1 = 2$, $y = x_1 = 2$, and $k = 2$.

[Iteration 1] $x_2 = x_1^2 + 1 = 5$, $d = \gcd(y - x_2, 1387) = \gcd(2 - 5, 1387) = 1$; since now $i = k = 2$, then $y = x_2 = 5$ and $k = 2k = 4$.

[Iteration 2] $x_3 = x_2^2 + 1 = 26$, $d = \gcd(y - x_3, 1387) = \gcd(5 - 26, 1387) = 1$; since now $i = 3$ but $k = 4$, i.e., $i \neq k$, then $y = 5$ and $k = 4$.

[Iteration 3] $x_4 = x_3^2 + 1 = 677$, $d = \gcd(y - x_4, 1387) = \gcd(5 - 677, 1387) = 1$; since now $i = 4$ but $k = 4$, then $y = x_4 = 677$ and $k = 2k = 8$.

[Iteration 4] $x_5 = x_4^2 + 1 = 620$, $d = \gcd(y - x_5, 1387) = \gcd(677 - 620, 1387) = 19$; Since now $1 < d = 19 < 1387$, then $d = 19$ is a non-trivial divisor of 1387. In fact, $1387 = 19 \cdot 73$ is the complete prime factorization of 1387.

The most successful example of the ρ-method is the complete factorization of the eighth Fermat number $F_8 = 2^{2^8} + 1$ by Brent and Pollard in 1980:

$$F_8 = 1238926361552897 \cdot p_{62}$$

where p_{62} is a 62-digit prime.

(II) The "$p-1$" Method

Pollard in 1974 invented another simple but effective factoring algorithm, now widely known as Pollard's "$p-1$" method, which can be described as follows:

Algorithm 4.2.6 (Pollard's "$p-1$") Given a composite integer $N > 1$. This algorithm attempts to find a non-trivial factor of N.

[1] (Initialization) Pick out $a \in \mathbb{Z}/N\mathbb{Z}$ at random. Select a positive integer k that is divisible by many prime powers; for example, $k = \text{lcm}(1, 2, \cdots, B)$ for a suitable bound B (the larger B is the most likely the method will succeed in producing a factor, but the longer the method will take to work).

[2] (Exponentiation) Computing $a_k = a^k \bmod N$.

[3] (Computing GCD) Computing $d = \gcd(a_k - 1, \ N)$.

[4] (Factor Found?) If $1 < d < N$, then d is a non-trivial factor of N, output d and goto [6].

[5] (Start Over?) If d is not a non-trivial factor of N and if you still want to try more experiments, then goto [2] to start all over again with a new choice of a and/or a new choice of k, else goto [6].

[6] (Exit) Terminate the algorithm.

The "$p-1$" algorithm is usually successful in the fortunate case that N has a prime divisor p for which $p-1$ has no large prime factors. Suppose that $(p-1) \mid k$ and that $p \nmid a$. Since $\#(\mathbb{Z}/p\mathbb{Z})^* = p - 1$, then $a^k \equiv 1 \pmod{p}$, thus $p \mid \gcd(a_k - 1, \ N)$. In many cases, we have $p = \gcd(a_k - 1, \ N)$, so the method finds a non-trivial factor of N.

Example 4.2.7 Use the "$p-1$" method to factor the number $N = 540143$. Choose $B = 8$ and hence $k = 840$. Choose also $a = 2$. Then we have

$$\gcd(2^{840} - 1 \bmod 540143, \ 540143) = \gcd(53046, \ 540143) = 421.$$

Thus 421 is a (prime) factor of 540143. In fact, $421 \cdot 1283$ is the complete prime factorization of 540143. It is interesting to note that by using the "$p-1$" method Baillie in 1980 found the prime factor

$$p_{25} = 1155685395246619182673033$$

of the Mersenne number $M_{257} = 2^{257} - 1$. In this case

$$p_{25} - 1 = 2^3 \cdot 3^2 \cdot 19^2 \cdot 47 \cdot 67 \cdot 257 \cdot 439 \cdot 119173 \cdot 1050151.$$

In the worst case, when $(p-1)/2$ is prime, the "$p-1$" algorithm is no better than trial division. Since the group has fixed order $p-1$ there is nothing to be done except try a different algorithm. We also mention here that there is a similar method to "$p-1$", called "$p+1$", proposed by H. C. Williams in 1982. It is suitable for the case that N has a prime factor p for which $p+1$ has no large prime factors.

4.2.7   Elliptic Curve Method (ECM)

In subsection 4.1.4, we have discussed the application of elliptic curves in primality testing. In this subsection, we shall introduce a factoring method depending on the use of elliptic curves. The method is actually obtained from Pollard's "$p-1$" algorithm: if we can choose a *random* group G with order g close to p, we may be able to perform a computation similar to that involved in Pollard's "$p-1$" algorithm, working in G rather than in F_p. If all prime factors of g are less than the bound B then we find a factor of N. Otherwise, repeat with a different group G (and hence, usually, a different g) until a factor is found. This is the motivation for H. W. Lenstra's *elliptic curve algorithm* (usually denoted "ECM"), which can be described as follows (for those who are not familiar with the basic concepts of elliptic curves, it is advised to consult Subsection 1.4.4 of Chapter 1):

Algorithm 4.2.7 (Lenstra's ECM) Given a composite integer $N > 1$, with $\gcd(N, 6) = 1$. This algorithm attempts to find a non-trivial factor of N. The method depends on the use of elliptic curves and is the analogue of Pollard's "$p-1$" method.

[1] (Choose an Elliptic Curve) Choose a random pair (E, P), where E is an elliptic curve $y^2 = x^3 + ax + b$ over $\mathbb{Z}/N\mathbb{Z}$, and $P(x, y) \in E(\mathbb{Z}/N\mathbb{Z})$ is a point on E. That is, choose $a, x, y \in \mathbb{Z}/N\mathbb{Z}$ at random, and set $b \leftarrow y^2 - x^3 - ax$. If $\gcd(4a^3 + 27b^2, N) \neq 1$, then E is not an elliptic curve, start all over to choose another pair (E, P).

[2] (Choose an Integer k) Just as in "$p-1$" method, select a positive integer k that is divisible by many prime powers; for example, $k = \text{lcm}(1, 2, \cdots, B)$ or

$k = B!$ for a suitable bound B; the larger B is the most likely the method will succeed in producing a factor, but the longer the method will take to work.

[3] (Calculate kP) Calculate the point $kP \in E(\mathbb{Z}/N\mathbb{Z})$. We use the following formula to compute $P_3(x_3, y_3) = P_1(x_1, y_1) + P_2(x_2, y_2) \bmod N$:

$$(x_3, y_3) = (\lambda^2 - x_1 - x_2 \bmod N, \quad \lambda(x_1 - x_3) - y_1 \bmod N)$$

where

$$\lambda = \begin{cases} \dfrac{m_1}{m_2} = \dfrac{3x_1^2 + a}{2y_1} \bmod N & \text{if } P_1 = P_2 \\[2mm] \dfrac{m_1}{m_2} = \dfrac{y_1 - y_2}{x_1 - x_2} \bmod N & \text{otherwise.} \end{cases}$$

The computation of $kP \bmod N$ can be done in $\mathcal{O}(\log k)$ doublings and additions; see Algorithm 3.4.3 in Chapter 3 for more information about the fast computation of $kP \bmod N$.

[4] (Computing GCD) If $kP \equiv \mathcal{O}_E \pmod{N}$, then set $m_2 = z$ and compute $d = \gcd(z, N)$, else goto [1] to start a new choice for "a" or even for a new pair (E, P).

[5] (Factor Found?) If $1 < d < N$, then d is a non-trivial factor of N, output d and goto [7].

[6] (Start Over?) If d is not a non-trivial factor of N and if you still wish to try more elliptic curves, then goto [1] to start all over again, else goto [7].

[7] (Exit) Terminate the algorithm.

As for the "$p-1$" method, one can show that a given pair (E, P) is likely to be successful in the above algorithm if N has a prime factor p for which $\#(\mathbb{Z}/p\mathbb{Z})$ is build up from small primes only. The probability for this to happen increases with the numbers of pairs (E, P) that one tries.

Example 4.2.8 Use the ECM method to factor the number $N = 187$.

(i) Choose $B = 3$, and hence $k = \mathrm{lcm}(1, 2, 3) = 6$. Let $P = (0, 5)$ be a point on the elliptic curve $E : y^2 = x^3 + x + 25$ which satisfies $\gcd(N, 4a^3 + 27b^2) = \gcd(187, 16879) = 1$ (note that here $a = 1$ and $b = 25$).

(ii) Since $k = 6 = 110_2$, then we compute $6P = 2(P + 2P)$ in the following way:

- Compute $2P = P + P = (0, 5) + (0, 5)$:

$$\begin{cases} \lambda = \frac{m_1}{m_2} = \frac{1}{10} \equiv 131 \ (\text{mod} \ 187) \\ x_3 = 144 \ (\text{mod} \ 187) \\ y_3 = 18 \ (\text{mod} \ 187) \end{cases}$$

So, $2P = (144, 18)$ with $m_2 = 10$ and $\lambda = 131$.

- Compute $3P = P + 2P = (0, 5) + (144, 18)$:

$$\begin{cases} \lambda = \frac{m_1}{m_2} = \frac{13}{144} \equiv 178 \ (\text{mod} \ 187) \\ x_3 = 124 \ (\text{mod} \ 187) \\ y_3 = 176 \ (\text{mod} \ 187) \end{cases}$$

So, $3P = (124, 176)$ with $m_2 = 144$ and $\lambda = 178$.

- Compute $6P = 2(3P) = 3P + 3P = (124, 176) + (124, 176)$:

$$\lambda = \frac{m_1}{m_2} = \frac{46129}{352} \equiv \frac{127}{165} \equiv \mathcal{O}_E \ (\text{mod} \ 187).$$

This time, $m_1 = 127$ and $m_2 = 165$, so the modular inverse for $127/165 \bmod 187$ does not exist, but this is exactly what we want! (we call this type of failure the "pretended failure"). We now set $z = m_2 = 165$.

(iii) Compute $d = \gcd(N, z) = \gcd(187, 165) = 11$. Since $1 < 11 < 187$, then 11 is a (prime) factor of 187. In fact, $187 = 11 \cdot 17$.

In 1995 Richard Brent at the Australian National University completed the factorization of the tenth Fermat number using ECM:

$$2^{2^{10}} + 1 = 2^{1024} + 1 = 45592577 \cdot 6487031809 \cdot p_{40} \cdot p_{252}$$

where the 40-digit prime p_{40} was found using ECM, and p_{252} was proved to be a 252-digit prime. Other recent ECM-records include a 38-digit prime factor in the 112-digit composite $(11^{118} + 1)/(2 \cdot 61 \cdot 193121673)$, a 40-digit prime factor of $26^{126} + 1$, a 43-digit prime factor of the partition number $p(19997)$ and a 44-digit prime factor of the partition number $p(19069)$ in the RSA Factoring Challenge List, and a 47-digit prime in c_{135} in $5^{2^8} + 1 = 2 \cdot 1655809 \cdot p_{38} \cdot c_{135}$.

Conjecture 4.2.5 Let p be the smallest prime dividing N, then the ECM method will find p of N, under some plausible assumptions, in expected running time

$$\mathcal{O}\left(\exp\left(\sqrt{(2+o(1))\log p \log\log p}\right) \cdot (\log N)^2\right) \tag{4.69}$$

In the worst case, when N is the product of two prime factors of the same order of magnitude, this is

$$\mathcal{O}\left(\exp\left(\sqrt{(2+o(1))\log N \log\log N}\right)\right) = \mathcal{O}\left(N^{\sqrt{(2+o(1))\log\log N/\log N}}\right). \tag{4.70}$$

Remark 4.2.5 What is especially interesting for us about the ECM is that its running time depends very much on p (the factor found) of N, rather than N itself. So, one advantage of the ECM is that one may use it, in a manner similar to trial division, to locate the smaller prime factors p of a number N that is much too large to factor completely.

4.3   Discrete Logarithms

The discrete logarithm problem (see equation (1.149) in Section 1.3.4 of Chapter 1 for a formal definition of discrete logarithms) can be described as follows:

$$\left\{ \begin{array}{ll} \text{Input}: & a, b, n \in \mathbb{N} \\ \text{Output}: & x \in \mathbb{N} \text{ with } a^x \equiv b \pmod{n} \\ & \text{if such a } x \text{ exists.} \end{array} \right. \tag{4.71}$$

Note that the modulus n can either be a composition or a prime. In this book we shall mainly be concerned with the case that n is a prime number p. According to Adleman [1], the Russian mathematician Bouniakowsky developed a clever algorithm to solve the congruence $a^x \equiv b \pmod{n}$, with the asymptotic complexity $\mathcal{O}(n)$ in 1870. Despite its long history, no efficient algorithm has ever emerged for the discrete logarithm problem; the best known algorithm at present, using NFS and due to Gordon [50] requires the expected running time in

$$\mathcal{O}\left(\exp\left(c(\log n)^{1/3}(\log\log n)^{2/3}\right)\right).$$

There are essentially three different categories of algorithms in use for computing discrete logarithms:

(i) Algorithms that work for arbitrary groups, that is, those that do not ex-
    ploit any specific properties of groups; the Shanks' baby-step giant-step
    method, Pollard's ρ-method (an analogue to Pollard's ρ-factoring method)
    and λ-method (also known as wild and tame Kangaroos) [110] are in this
    category.

(ii) Algorithms that work well in finite groups for which the order of the groups
    has no large prime factors; more specifically, algorithms that work for
    groups with smooth orders. A positive integer is called *smooth* if it has no
    large prime factors; it is called *y-smooth* if it has no large prime factors
    exceeding y. The well-known Silver-Pohlig-Hellman algorithm based on
    the Chinese Remainder Theorem is in this category.

(iii) Algorithms that exploit methods for representing group elements as prod-
    ucts of elements from a relatively small set (also making use of the Chi-
    nese Remainder Theorem); the typical algorithms in this category are
    Adleman's index calculus algorithm and Gordon's NFS algorithm.

In the subsections that follow, we shall introduces the basic ideas of each of
these three categories; more specifically, we shall introduce Shanks' baby-step
giant-step algorithm, Silver-Pohlig-Hellman algorithm, Adleman's index calculus algorithm as well as Gordon's NFS algorithm for computing discrete logarithms.

4.3.1   Shanks' Baby-Step Giant-Step Algorithm

Let G be a finite cyclic group of order n, a a generator of G, and $b \in G$. The
obvious algorithm of computing successive powers of a until b is found takes
$\mathcal{O}(n)$ group operations. For example, to compute $x = \log_2 15 \pmod{19}$, we
compute $2^x \bmod 19$ for $x = 0, 1, 2, \cdots, 19 - 1$ until $2^x \bmod 19 = 15$ for some x
is found, that is:

x	0	1	2	3	4	5	6	7	8	9	10	11
a^x	1	2	4	8	16	13	7	14	9	18	17	15

So $\log_2 15 \pmod{19} = 11$. It is clear that when n is large, the algorithm is
inefficient. In this section, we introduce a type of square root algorithm, called
Baby-Step Giant-Step algorithm, for taking discrete logarithms, which is better

than the above mentioned *obvious* algorithm. The algorithm works on arbitrary groups, and according to [104], its original idea is due to Shanks[5].

Let $m = \lfloor \sqrt{n} \rfloor$. The baby-step giant-step algorithm is based on the observation that if $x = \log_a b$, then we can uniquely write $x = i + jm$, where $0 \leq i, j < m$. For example, if $11 = \log_2 15 \bmod 19$, then $a = 2$, $b = 15$, $m = 5$, so we can write $11 = i + 5j$ for $0 \leq i, j < m$. Clearly here $i = 1$ and $j = 2$ that makes $11 = 1 + 5 \cdot 2$. Similarly, for $14 = \log_2 6 \bmod 19$, we can write $14 = 4 + 5 \cdot 2$, for $17 = \log_2 10 \bmod 19$. The following is a description of the algorithm:

Algorithm 4.3.1 (Baby-Step Giant-Step Algorithm): This algorithm computes the discrete logarithm x of y to the base a, modulo n, such that $y = a^x \pmod{n}$.

[1] (Initialization) Computes $s = \lfloor \sqrt{n} \rfloor$.

[2] (Computing the Baby-Step) Compute the first sequence (list), denoted by S, of pairs (ya^r, r), $r = 0, 1, 2, 3, \cdots, s - 1$:

$$S = \{(y, 0), (ya, 1), (ya^2, 2), (ya^3, 3), \cdots, (ya^{s-1}, s - 1), \bmod \ n\}. \quad (4.72)$$

and sort S by ya^r, the first element of the pairs in S.

[3] (Computing the Giant-Step) Compute the second sequence (list), denoted by T, of pairs (a^{ts}, ts), $t = 1, 2, 3, \cdots, s$:

$$T = \{(a^s, 1), (a^{2s}, 2), (a^{3s}, 3), \cdots, (a^{s^2}, s), \bmod \ n\}. \quad (4.73)$$

and sort T by a^{ts}, the first element of the pairs in T.

[4] (Searching, Comparison and Computing) Search both lists S and T for a match $ya^r = a^{ts}$ with ya^r in S and a^{ts} in T, then compute $x = ts - r$. This x is thus the required value for $\log_a y$ modulo n.

---

[5]Daniel Shanks (1917-1996) was born in the city of Chicago, where in 1937 he received his BSc in physics from the University of Chicago. He earned his PhD from the University of Maryland in 1954, and joined the Department of Mathematics of that University as an adjunct professor in 1977 and remained there until his death. Shanks was responsible for the SQUFOF method for factoring and the Baby-Step Giant-Step method for computing discrete logarithms. He served as an editor of *Mathematics of Computation* from 1959 until his death. His book, *Solved and Unsolved Problems in Number Theory* is one of the most successful books in number theory; H. C. Williams praised it in the August 1997's *Notices of the AMS* as "a charming, unconventional, provocative, and fascinating book on elementary number theory". The book was first published in 1962 by Spartan Books, Washington, DC, and the 4th edition was published in 1993 by Chelsea, New York.

The above algorithm requires a table with $\mathcal{O}(m)$ entries (assume $m = \lfloor \sqrt{n} \rfloor$, where n is the modulus). Using the well-known sorting algorithm, we can sort both the lists S and T in $\mathcal{O}(m \log m)$ operations. Thus this gives an algorithm for computing discrete logarithms that uses $\mathcal{O}(\sqrt{n} \log n)$ time and space for $\mathcal{O}(\sqrt{n})$ group elements. Note that Shanks' idea is originally for computing the order of a group element g in the group G, but here we use his idea to compute discrete logarithms. Note also that although this algorithm works on arbitrary groups, if the order of a group is larger than 10^{40}, it will be infeasible.

Example 4.3.1 Suppose we wish to compute the discrete logarithm $x = \log_2 6 \bmod 19$ such that $6 = 2^x \bmod 19$. According to Algorithm 4.3.1, we perform the following computations:

(i) $y = 6$, $a = 2$ and $n = 19$, $s = \lfloor \sqrt{19} \rfloor = 4$.

(ii) Computing the Baby-Step:

$$
\begin{aligned}
S &= \{(y, 0), (ya, 1), (ya^2, 2), (ya^3, 3), \bmod\ 19\} \\
&= \{(6, 0), (6 \cdot 2, 1), (6 \cdot 2^2, 2), (6 \cdot 2^3, 3), \bmod\ 19\} \\
&= \{(6, 0), (12, 1), (5, 2), (10, 3)\} \\
&= \{(5, 2), (6, 0), (10, 3), (12, 1)\}.
\end{aligned}
$$

(iii) Computing the Giant-Step:

$$
\begin{aligned}
T &= \{(a^s, s), (a^{2s}, 2s), (a^{3s}, 3s), (a^{4s}, 4s), \bmod\ 19\} \\
&= \{(2^4, 4), (2^8, 8), (2^{12}, 12), (2^{16}, 16), \bmod\ 19\} \\
&= \{(16, 4), (9, 8), (11, 12), (5, 16)\} \\
&= \{(5, 16), (9, 8), (11, 12), (16, 4)\}
\end{aligned}
$$

(iv) Matching and Computing: The number 5 is the common value of the first element in pairs of both lists S and T with $r = 2$ and $st = 16$, so $x = st - r = 16 - 2 = 14$. That is, $\log_2 6 \pmod{19} = 14$, or equivalently, $2^{14} \pmod{19} = 6$.

Example 4.3.2 Suppose now we wish to find the discrete logarithm $x = \log_{59} 67 \bmod 113$ such that $67 = 59^x \bmod 113$. Again by Algorithm 4.3.1, we have:

(i) $y = 67$, $a = 59$ and $n = 113$, $s = \lfloor \sqrt{113} \rfloor = 10$.

(ii) Computing the Baby-Step:

$$
\begin{aligned}
S &= \{(y,0), (ya,1), (ya^2,2), (ya^3,3), \cdots, (ya^9,9), \ \text{mod} \ 113\} \\
&= \{(67,0), (67 \cdot 59,1), (67 \cdot 59^2,2), (67 \cdot 59^3,3), (67 \cdot 59^4,4), (67 \cdot 59^5,5), \\
&\quad (67 \cdot 59^6,6), (67 \cdot 59^7,7), (67 \cdot 59^8,8), (67 \cdot 59^9,9), \ \text{mod} \ 113\} \\
&= \{(67,0), (111,1), (108,2), (44,3), (110,4), (49,5), (66,6), \\
&\quad (52,7), (17,8), (99,9)\} \\
&= \{(17,8), (44,3), (49,5), (52,7), (66,6), (67,0), (99,9), \\
&\quad (108,2), (110,4), (111,1)\}
\end{aligned}
$$

(iii) Computing the Giant-Step:

$$
\begin{aligned}
T &= \{(a^s,s), (a^{2s},2s), (a^{3s},3s), \cdots (a^{10s},10s), \ \text{mod} \ 113\} \\
&\quad \{(59^{10},10), (59^{2 \cdot 10},2 \cdot 10), (59^{3 \cdot 10},3 \cdot 10), (59^{4 \cdot 10},4 \cdot 10), (59^{5 \cdot 10},5 \cdot 10), \\
&\quad (59^{6 \cdot 10},6 \cdot 10), (59^{7 \cdot 10},7 \cdot 10), (59^{8 \cdot 10},8 \cdot 10), (59^{9 \cdot 10},9 \cdot 10), \ \text{mod} \ 113\} \\
&= \{(72,10), (99,20), (9,30), (83,40), (100,50), (81,60), (69,70), (109,80), \\
&\quad (51,90), (56,100)\} \\
&= \{(9,30), (51,90), (56,100), (69,70), (72,10), (81,60), (83,40), \\
&\quad (99,20), (100,50), (109,80)\}
\end{aligned}
$$

(iv) [Matching and Computing] The number 99 is the common value of the first element in pairs of both lists S and T with $r = 9$ and $st = 20$, so $x = st - r = 20 - 9 = 11$. That is, $\log_{59} 67 \ (\text{mod} \ 113) = 11$, or equivalently, $59^{11} \ (\text{mod} \ 113) = 67$.

Exercise 4.3.1 Use the baby-step giant-step algorithm to compute the following discrete logarithms x (if such a x exists):

(i) $x = \log_3 5 \ \text{mod} \ 29$.

(ii) $x = \log_5 96 \ \text{mod} \ 317$.

(iii) $x = \log_{37} 15 \ \text{mod} \ 123$.

The Shanks' baby-step giant-step algorithm is a type of *square root method* for computing discrete logarithms. In 1978 Pollard [110] gave also two other types of square root methods; namely the ρ-method and the λ-method for taking discrete logarithms. Pollard's methods are probabilistic but remove the necessity of precomputing the lists S and T, compared with Shanks' baby-step giant-step method. Again, Pollard's algorithm requires $\mathcal{O}(n)$ group operations and hence is infeasible if the order of the group G is larger than 10^{40}.

4.3.2   Silver-Pohlig-Hellman Algorithm

In 1978, Pohlig and Hellman [108] proposed an important special algorithm, known as Silver-Pohlig-Hellman algorithm for computing discrete logarithms over $\mathrm{GF}(q)$ using on the order of $\sqrt{p}$ operations and a comparable amount of storage, where p is the largest prime factor of $q-1$. Pohlig and Hellman showed that if

$$q - 1 = \prod_{i=1}^{k} p_i^{\alpha_i}, \tag{4.74}$$

where p_i are distinct primes and α_i are natural numbers, and if $r_1, \cdots, r_k$ are any real numbers with $0 \leq r_i \leq 1$, then logarithms over $\mathrm{GF}(q)$ can be computed in

$$\mathcal{O} \left(\sum_{i=1}^{k} \left(\log q + p_i^{1-r_i} \left(1 + \log p_i^{r_i}\right)\right) \right) \tag{4.75}$$

field operations, using

$$\mathcal{O} \left(\log q \sum_{i=1}^{k} \left(1 + p_i^{r_i}\right) \right) \tag{4.76}$$

bit of memory, provided that a precomputation requiring

$$\mathcal{O} \left(\sum_{i=1}^{k} p_i^{r_i} \log p_i^{r_i} + \log q \right) \tag{4.77}$$

field operations is performed first. This algorithm is very efficient if q is "smooth", i.e., all the prime factors of $q - 1$ are small. We shall give a brief description of the algorithm as follows:

Algorithm 4.3.2 This algorithm computes the discrete logarithm $x = \log_a b \bmod q$.

[1] Factor $q - 1$ into its prime decomposition form:

$$q - 1 = \prod_{i=1}^{k} p_1^{\alpha_1} p_2^{\alpha_2} \cdots p_k^{\alpha_k}.$$

[2] Precompute the table $r_{p_i,j}$ for a given field:

$$r_{p_i,j} = a^{j(q-1)/p_i} \bmod q, \quad 0 \le j < p_i. \tag{4.78}$$

This only needs to be done once for any given field.

[3] Compute the discrete logarithm of b to the base a modulo q, i.e., compute $x = \log_a b \bmod q$:

[3-1] Using the similar idea in the Baby-Step Giant-Step algorithm to find the individual discrete logarithms $x \bmod p_i^{\alpha_i}$: To compute $x \bmod p_i^{\alpha_i}$, we consider the presentation of this number to the base p_i:

$$x \bmod p_i^{\alpha_i} = x_0 + x_1 p_i + \cdots + x_{\alpha_i - 1} p_i^{\alpha_i - 1}, \tag{4.79}$$

where $0 \le x_n < p_i - 1$.

- To find x_0, we compute $b^{(q-1)/p_i}$ which equals $r_{p_i,j}$ for some j, and set $x_0 = j$ for which

$$b^{(q-1)/p_i} \bmod q = r_{p_i,j}.$$

This is so because

$$b^{(q-1)/p_i} \equiv a^{x(q-1)/p} \equiv a^{x_0(q-1)/p} \bmod q = r_{p_i,x_0}$$

- To find x_1, compute $b_1 = ba^{-x_0}$. If

$$b_1^{(q-1)/p_i^2} \bmod q = r_{p_i,j}$$

then set $x_1 = j$. This is so because
$$b_1^{(q-1)/p_i^2} \equiv a^{(x-x_0)(q-1)/p_i^2} \equiv a^{x_1 + x_2 p_i + \cdots)(q-1)/p_i}$$
$$\equiv a^{x_1(q-1)/p} \bmod q = r_{p_i,x_1}$$

- To obtain x_2, consider the number $b_2 = ba^{-x_0-x_1 p_i}$ and compute

$$b_2^{(q-1)/p_i^3} \bmod q.$$

The procedure is carried on inductively to find all $x_0, x_1, \cdots, x_{\alpha-1}$.

[3-2] Using the Chinese Remainder Theorem to find the unique value of x from the congruences $x \bmod p_i^{\alpha_i}$.

We now give an example how the above algorithm works:

Example 4.3.3 Suppose we wish to compute the discrete logarithm $x = \log_2 62 \bmod 181$. Now we have $a = 2$, $b = 62$ and $q = 181$ (2 is the generator of $\mathbb{F}_{181}^*$). We follow the computation steps described in the above algorithm:

(i) Factor $q - 1$ into its prime decomposition form:

$$180 = 2^2 \cdot 3^2 \cdot 5.$$

(ii) Using the following formula to precompute the table $r_{p_i,j}$ for the given field $\mathbb{F}_{181}^*$:

$$r_{p_i,j} = a^{j(q-1)/p_i} \bmod q, \quad 0 \le j < p_i.$$

This only needs to be done once for the field.

(a) Compute $r_{p_1,j} = a^{j(q-1)/p_1} \bmod q = 2^{90j} \bmod 181$ for $0 \le j < p_1 = 2$:

$$r_{2,0} = 2^{90 \cdot 0} \bmod 181 = 1,$$

$$r_{2,1} = 2^{90 \cdot 1} \bmod 181 = 180.$$

(b) Compute $r_{p_2,j} = a^{j(q-1)/p_2} \bmod q = 2^{60j} \bmod 181$ for $0 \le j < p_2 = 3$:

$$r_{3,0} = 2^{60 \cdot 0} \bmod 181 = 1,$$

$$r_{3,1} = 2^{60 \cdot 1} \bmod 181 = 48,$$

$$r_{3,2} = 2^{60 \cdot 2} \bmod 181 = 132.$$

(c) Compute $r_{p_3,j} = a^{j(q-1)/p_3} \bmod q = 2^{36j} \bmod 181$ for $0 \le j < p_3 = 5$:

$$r_{5,0} = 2^{36 \cdot 0} \bmod 181 = 1,$$

$$r_{5,1} = 2^{36 \cdot 1} \bmod 181 = 59,$$

$$r_{5,2} = 2^{36 \cdot 2} \bmod 181 = 42,$$

$$r_{5,3} = 2^{36 \cdot 3} \bmod 181 = 125,$$

$$r_{5,4} = 2^{36 \cdot 4} \bmod 181 = 135.$$

Construct the $r_{p_i,j}$ table as follows:

p_i	j				
	0	1	2	3	4
2	1	180			
3	1	48	132		
5	1	59	42	125	135

This table is manageable if all p_i are small. The table only needs to be constructed once for the field $\mathbb{F}_{181}^*$.

(iii) Compute for example the discrete logarithm of 62 to base 2 modulo 181, i.e., compute $x = \log_2 62 \bmod 181$. Here $a = 2$ and $b = 62$:

(a) Find the individual discrete logarithms $x \bmod p_i^{\alpha_i}$ using

$$x \bmod p_i^{\alpha_i} = x_0 + x_1 p_i + \cdots + x_{\alpha_i-1} p_i^{\alpha_i-1}, \quad 0 \le x_n < p_i - 1.$$

(a-1) Find the discrete logarithms $x \bmod p_1^{\alpha_1}$, i.e., $x \bmod 2^2$:

$$x \bmod 181 \Longleftrightarrow x \bmod 2^2 = x_0 + 2x_1.$$

- To find x_0, we compute

$$b^{(q-1)/p_1} \bmod q = 62^{180/2} \bmod 181 = 1 = r_{p_1,j} = r_{2,0}$$

hence, $x_0 = 0$.

- To find x_1, compute first $b_1 = ba^{-x_0} = b = 62$, then compute

$$b_1^{(q-1)/p_1^2} \bmod q = 62^{180/4} \bmod 181 = 1 = r_{p_1,j} = r_{2,0}$$

hence, $x_1 = 0$. So,

$$x \bmod 2^2 = x_0 + 2x_1 \Longrightarrow x \bmod 4 = 0.$$

(a-2) Find the discrete logarithms $x \bmod p_2^{\alpha_2}$, i.e., $x \bmod 3^2$:

$$x \bmod 181 \iff x \bmod 3^2 = x_0 + 2x_1.$$

- To find x_0, we compute
$$b^{(q-1)/p_2} \bmod q = 62^{180/3} \bmod 181 = 48 = r_{p_2,j} = r_{3,1}$$
hence, $x_0 = 1$.
- To find x_1, compute first $b_1 = ba^{-x_0} = 62 \cdot 2^{-1} = 31$, then compute
$$b_1^{(q-1)/p_2^2} \bmod q = 31^{180/3^2} \bmod 181 = 1 = r_{p_2,j} = r_{3,0}$$
hence, $x_1 = 0$. So,
$$x \bmod 3^2 = x_0 + 2x_1 \implies x \bmod 9 = 1.$$

(a-3) Find the discrete logarithms $x \bmod p_3^{\alpha_3}$, i.e., $x \bmod 5^1$:

$$x \bmod 181 \iff x \bmod 5^1 = x_0.$$

- To find x_0, we compute
$$b^{(q-1)/p_3} \bmod q = 62^{180/5} \bmod 181 = 1 = r_{p_3,j} = r_{5,0}$$
hence, $x_0 = 0$. So we conclude that
$$x \bmod 5 = x_0 \implies x \bmod 5 = 0.$$

(b) To find the x in

$$x \bmod 181$$

such that

$$\begin{cases} x \bmod 4 = 0, \\ x \bmod 9 = 1, \\ x \bmod 5 = 0. \end{cases}$$

To do this, we just use the Chinese Remainder Theorem to solve the following system of congruences:

$$\begin{cases} x \equiv 0 \pmod{4} \\ x \equiv 1 \pmod{9} \\ x \equiv 0 \pmod{5} \end{cases}$$

The unique value of x for this system of congruences is thus $x = 100$. (This can be easily done by using the Maple function chrem([0, 1, 0], [4,9, 5])). So the value of x in the congruence $x \bmod 181$ is 100. Hence $x = \log_2 62 = 100$.

4.3.3 Subexponential Algorithms

In this subsection, we shall briefly introduce two subexponential algorithms for taking discrete logarithms $x = \log_a b \bmod q$, namely Adleman's index calculus algorithm and Gordon's number field sieve method.

In 1979, Adleman [1] proposed a general purpose, subexponential algorithm for taking discrete logarithms with the following expected running time:

$$\mathcal{O}\left(\exp\left(c\sqrt{\log q \log\log q}\right)\right).$$

His algorithm can be briefly described as follows:

Algorithm 4.3.3 Adleman's algorithm for computing discrete logarithms $x = \log_a b \bmod q$ on input a, b, q, where a and b are generators and q a prime.

[1] Factor $q - 1$ into its prime decomposition form:

$$q - 1 = p_1^{\alpha_1} p_2^{\alpha_2} \cdots p_k^{\alpha_k}$$

[2] For each $p_k^{\alpha_k} \mid n$ proceed the following steps until m_l is obtained:

- (by guessing and checking) find r_i, s_i such that $a^{r_i} \bmod q$ and $ba^{s_i} \bmod q$ are smooth with respect the bound $2^{(\log q \log\log q)^{1/2}}$.

- (using Gaussian elimination) check if over the finite field $\mathbb{Z}_{p_l^{\alpha_l}}$, $ba^{s_i} \bmod q$ is dependent on

$$\{a^{r_1} \bmod q, \cdots, a^{r_i} \bmod q\}.$$

If yes, calculate β_j's such that

$$ba^{s_i} \bmod q \equiv \left(\sum_{j=1}^{i} \beta_j a^{r_i} \bmod q\right) \bmod p_l^{\alpha_l}$$

then

$$m_l = \left(\sum_{j=1}^{i} \beta_j r_j\right) \bmod p_l^{\alpha_l} - s_i$$

[3] (using the Chinese Remainder Theorem) calculate and output x such that

$$x \equiv m_l \bmod p_l^{\alpha_l}, \quad l = 1, 2, \cdots k.$$

Note that the above algorithm can also be easily generalized to the case that q is not a prime, or a or b are not generators [1].

For more than ten years since its invention of the above algorithm and its variants, they were the fastest algorithms for computing discrete logarithms. But the situation changed when Gordon in 1993 [50] developed an algorithm for computing discrete logarithms in GF(p). Gordon's algorithm is based on Number Field Sieve (NSF) for integer factorization, with the heuristic expected running time

$$\mathcal{O}\left(\exp\left(c(\log p)^{1/3}(\log\log p)^{2/3}\right)\right),$$

the same as that used in factoring. The algorithm can be briefly described as follows:

Algorithm 4.3.4 Gordon's algorithm for computing discrete logarithms $a^x \equiv b \pmod{p}$ on input a, b, n, where a and b are generators and p is prime.

[1] (Precomputation): finding the discrete logarithms of a factor base of small rational primes, which only must be done once for a given n.

[2] (Computing Individual Logarithms): finding the logarithm for an individual $b \in \mathbb{F}_p$ by finding the logarithms of a number of "medium-sized" primes.

[3] (Computing the Final Logarithm): Combining all the individual logarithms to find the logarithm of b.

Interested readers are referred to Gordon's paper [50] for more detailed information. Note also that Gordon, co-authored by McCurley in [49], discussed some implementation issues of massively parallel computations of discrete logarithms over GF(2^n).

4.3.4 Factoring versus Discrete Logarithms

The computation of discrete logarithms has much in common with the factorization of integers. First of all, they are both intractable problems, no efficient algorithms have been found for both of the problems. Interestingly enough, they seem to have the same degree of intractability; as a rule of thumb, the complexity of factoring a number n with β digits is almost the same as that of taking discrete logarithm modulo a prime p with $\beta - 10$ digits. Secondly, the methods used for integer factorization are similar to (or sometimes the same as) the methods used for computing discrete logarithms. In short, many methods for integer factorization are applicable for computing discrete logarithms. To illustrate the point, let us briefly compare various methods used for factoring integers and computing discrete logarithms:

Factoring Integers	$\Longleftrightarrow$	**Computing Discrete Logarithms**
Trial Divisions	$\Longleftrightarrow$	Baby-Step Giant-Step
Pollard's ρ-method	$\Longleftrightarrow$	Pollard's λ-method
CFRAC/MPQS	$\Longleftrightarrow$	Index Calculus
NFS	$\Longleftrightarrow$	NFS
Quantum Algorithms	$\Longleftrightarrow$	Quantum Algorithms

Note that quantum computation is a new model of computations, which is suitable for (or applicable to) both integer factorization and discrete logarithms. We shall introduce some quantum algorithms for integer factorization and discrete logarithms in Chapter 5 of this book. Finally, both integer factorization and discrete logarithms have important applications in public-key cryptography and computer security; the security of such cryptographic schemes are based on the intractability of factoring large integers and computing discrete logarithms. We shall discuss this topic in the next section.

4.4   Cryptology and Systems Security

In this section, we discuss some applications of number-theoretic computations in cryptology and systems security.

4.4.1   Private/Public-Key Cryptosystems

Cryptology is about secure communications over insecure channels such as computer networks and telephone lines. The history of cryptography is as old as writing. Modern cryptography is the study of "mathematical" systems for solving two main types of security problems: *privacy* and *authentication*. A privacy system prevents the extraction of information by unauthorised parties from messages transmitted over a public channel, thus assuring the sender of a message that it is only being read by the intended receiver. An authentication system prevents the unauthorised injection of messages into a public channel, assuring the receiver of a message of the legitimacy of its sender. It is interesting to note that the computational engine, designed and built by a British group led by Alan Turing, to crack the German "Enigma" code are considered to be the first real "computers"; thus one could argue that modern cryptography is the mother (or at least the midwife) of computer science. There are essentially two different types of cryptographic systems: conventional *private key* cryptographic systems and modern *public key* cryptographic systems. In a conventional cryptosystem (see Figure 4.2), both the sender and the receiver use the same key K to cipher

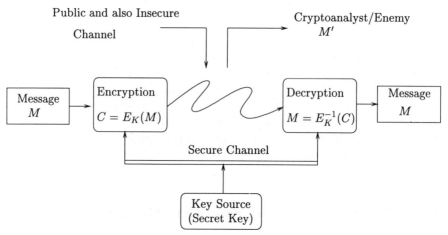

Figure 4.2: Conventional Private-key Cryptosystems

or decipher the message M. The sender uses an invertible transformation

$$S_K : M \to C \tag{4.80}$$

to produce the cipher text $C = S_K(M)$ and transmits it on the public *insecure* channel to the receiver. The Key K should also be transmitted to the legitimate receiver but via a *secure* channel. Since the legitimate receiver knows the key K, he can decipher C by a transformation

$$S_K^{-1} : C \to M \tag{4.81}$$

and obtain

$$SK^{-1}(C) = SK^{-1}(S_K(M)) = M \qquad (4.82)$$

the original plain-text message. The secure channel cannot be used to transmit M itself for reasons of capacity and delay. The secure channel might be a weekly courier, whereas the insecure channel is normally a computer network or a telephone line. Now suppose that there are n users registered in a computer network. Potentially, there are $n(n + 1)/2$ pairs of users who may wish to communicate privately. It is certainly unrealistic to assume either that a pair of users with no prior acquaintance will be able to wait for a key to be sent by some secure physical means, or that keys for all $n(n + 1)/2$ pairs of users can be arranged in advance. Whitfield Diffie[6] and Martin Hellman[7] at Stanford University in 1976 proposed an excellent idea to solve this problem [35], leading to the birth of public key cryptosystems. In a public key cryptosystem (see Figure 4.3), there are two sets of keys $K = \{K_{pub}, K_{sec}\}$ with $K_{pub} \subset K_{sec}$, where K_{pub} represents the encipher keys and K_{sec} the deciphering keys. Since K_{pub} is only used for encryption, then it can be made public; only K_{sec} must be kept a secret for decryption. The implementation of public key cryptosystems is based on the so-called *one-way functions* (also known as *trapdoor functions*).

---

[6]Whitfield Diffie (1944–), a Distinguished Engineer at Sun Microsystems in Palo Alto, California, is perhaps best known for his 1975 discovery of the concept of public key cryptography, for which he was awarded a Doctorate in Technical Sciences (Honoris Causa) by the Swiss Federal Institute of Technology in 1992. He received a BSc degree in mathematics from the Massachusetts Institute of Technology in 1965. Prior to becoming interested in cryptography, he worked on the development of the Mathlab symbolic manipulation system — sponsored jointly at Mitre and the MIT Artificial Intelligence Laboratory — and later on proof of correctness of computer programs at Stanford University. Dr. Diffie is the recipient of the IEEE Information Theory Society Best Paper Award 1979 for the paper "New Directions in Cryptography", the IEEE Donald E. Fink award 1981 for expository writing for the paper "Privacy and Authentication" (both papers co-authored by Martin Hellman), and the National Computer Systems Security Award for 1996.

[7]Martin E. Hellman (1945–) received his BEng from New York University in 1966, and his MSc and PhD from Stanford University in 1967 and 1969, respectively, all in Electrical Engineering. Dr. Hellman was on the research staff at IBM's Watson Research Center from 1968-69 and on the faculty of Electrical Engineering at MIT from 1969-71. He returned to Stanford as a faculty member in 1971, where he served on the regular faculty until becoming Professor Emeritus in 1996. He has authored over 60 technical papers, 5 US and a number of foreign patents. His work, particularly the invention of public key cryptography, has been covered in the popular media including Scientific American and Time magazine. He is the recipient of an IEEE Centennial Medal (1984).

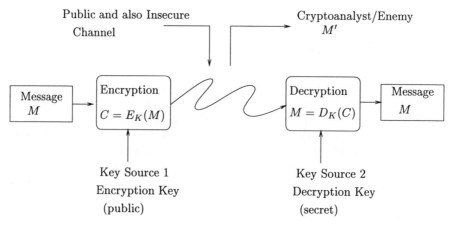

Figure 4.3: Modern Public-key Cryptosystems

Definition 4.4.1 Let S be a finite set. A one-way function is an invertible function

$$f: \ S \to S \tag{4.83}$$

such that $f(x)$ is easy to compute but the inverse function $f^{-1}(y)$ is hard to compute.

Example 4.4.1 Let P be the plain-text, and C the cipher-text. Then f is a one-way function if

(i) $f: \ P \to C$ is feasible/tractable,

(ii) $f^{-1}: \ C \to P$ is infeasible/intractable.

Thus f is a one-way function (at least at present) if it is based on one of the following functions:

(i) Integer factorization:

- $f: \ p \cdot q \to n,$ feasible (multiplying two integers is easy),
- $f^{-1}: \ n \to p \cdot q,$ infeasible (factor a composite into its prime decomposition form is hard).

(ii) Discrete logarithm:

- $f : g^a \bmod p \to b$,   feasible (modular exponentiation is easy),
- $f^{-1} : \log_g b \to a$,   infeasible (taking discrete logarithm is hard).

Remark 4.4.1 Up to date, the discrete logarithms and integer factorization are essentially the only hard problems on which to build one-way functions in practice.

Remark 4.4.2 Public-key cryptosystems generally require significantly more computing time than traditional private-key cryptosystems such as Data Encryption Standard (known as DES) — DES was designed at IBM and approved as a standard by the US National Bureau of Standards in 1977. It would be nice if we can make a combined use of the ideas of both private-key and public-key into a single cryptosystem for secure communications.

In the subsections that follow, we shall introduce several important and widely used public-key cryptosystems and cryptoprotocols.

4.4.2   Discrete Logarithm Based Cryptosystems

(I) The Diffie-Hellman Key-Exchange Protocol

Recall that the discrete logarithms problem is the problem that

$$\left\{ \begin{array}{ll} \text{Input :} & a, b, n \in \mathbb{N} \\ \text{Output :} & x \in \mathbb{N} \text{ with } a^x \equiv b \pmod{n} \\ & \text{if such a } x \text{ exists} \end{array} \right.$$

The idea of public-key cryptography was first proposed in 1976 by Whitfield Diffie and Martin Hellman. In their seminal paper "New directions in Cryptography" [35], they proposed the following public-key cryptographic scheme based on the difficult discrete logarithm problem. Their scheme was not to send secret messages, but for two parties to find a common private key over public networks to be used later in exchanging messages through conventional cryptography. Thus, the Diffie-Hellman scheme has the nice property that a very fast scheme such as DES can be used for actual encryption, yet it still enjoys one of the main advantages of public-key cryptography.

The Diffie-Hellman key-exchange protocol works in the following way:

(i) A prime q and a generator g are made public (assume all users have agreed upon a finite group over a fixed finite field $\mathbb{F}_q$),

(ii) Alice chooses a random number $a \in \{1, 2, \cdots, q - 1\}$ and sends $g^a \bmod q$ to Bob,

(iii) Bob chooses a random number $b \in \{1, 2, \cdots, q - 1\}$ and sends $g^b \bmod q$ to Alice,

(iv) Alice and Bob both compute $g^{ab} \bmod q$ and use this as a key for future communications.

Clearly, an eavesdropper has g, q, $g^a \bmod q$ and $g^b \bmod q$, so if he can take discrete logarithms, he can calculate $g^{ab} \bmod q$ and comprise communications. That is, if the eavesdropper can use his knowledge of g, q, $g^a \bmod q$ and $g^b \bmod q$ to recover the integer a, then he can easily break the Diffie-Hellman codes. So the security of Diffie-Hellman system is based on the following assumption:

Diffie-Hellman Assumption: It is computationally infeasible to compute g^{ab} from g^a and g^b.

In theory, there could be a way to use knowledge of g^a and g^b to find g^{ab}. But at present we simply cannot imagine a way to go from g^a and g^b to g^{ab} without essentially solving the discrete logarithm problem.

Example 4.4.2 The following example, taking from [89], shows how Diffie-Hellman's scheme works in a real situation:

(i) Let $q = (7^{149} - 1)/6$ and $p = 2 \cdot 739 \cdot q + 1$. (It can be shown that both p and q are primes [89]).

(ii) A chooses a random number residue x modulo p, computing $7^x \pmod{p}$, and sends the result to B, keeping x secret.

(iii) B receives

$$7^x = 12740218011997394682426924433432284974938204258693162165$$
$$4557735290322914679095998681860978813046595166455458144280588076766033781$$

(iv) B chooses a random number residue y modulo p, computing $7^y \pmod{p}$, and sends the result to A, keeping y secret.

(v) A receives

$$7^y = 18016228528745310244478283483679989501596704669534669731$$
$$30251217340599537720584759581769106253806921016518486623$$
$$62137934026803049$$

(vi) Now both A and B can compute the security key 7^{xy} (mod p).

McCurley offers a prize of \$100 in 1990 to the first person who finds the secret-key constructed from the above communication. The prize is still unclaimed to this day.

As we have already mentioned earlier that Diffie-Hellman scheme is not intended to be used for actual secure communications, but only for key-exchanges. There are, however, other cryptosystems also based on discrete logarithms that can be used for secure message transmissions.

(II) The ElGamma Cryptosystem for Secure Communications

In 1985, ElGamma proposed a public-key cryptosystem based on discrete logarithms:

(i) A prime q and a generator $g \in \mathbb{F}_q^*$ are made public.

(ii) A chooses a secret integer $a = a_A \in \{1, 2, \cdots, q-1\}$. This a is the secret deciphering key. The public enciphering key is $g^a \in \mathbb{F}_q$.

(iii) Suppose now B wishes to send a message to A, he chooses a random number $b \in \{1, 2, \cdots, q-1\}$ and sends A the following pair of elements of $\mathbb{F}_q$:

$$(g^b, Mg^{ab})$$

where M is the message.

(iv) Since A knows the secret deciphering key a, then he can recover M from this pair by computing g^{ab} (mod q) and dividing this result into the second element, i.e., Mg^{ab}.

Remark 4.4.3 Someone who can solve the discrete logarithm problem in $\mathbb{F}_q$ breaks the cryptosystem by finding the secret deciphering key a from the public enciphering key g^a. In theory, there could be a way to use knowledge of g^a

and g^b to find g^{ab} and hence break the cipher without solving the discrete logarithm problem. But as we have already seen in Diffie-Hellman's scheme, there is no known way to go from g^a and g^b to g^{ab} without essentially solving the discrete logarithm problem. So, the ElGamma cryptosystem is equivalent to Diffie-Hellman key-exchange system.

(III) The Massey-Omura Cryptosystem for Message Transmissions

Yet, there is another popular cryptosystem, called Massey-Omura system, based on discrete logarithms. The system works in the following way:

(i) All the users have agreed upon a finite group over a fixed finite field $\mathbb{F}_q$ with q a prime power.

(ii) Each user secretly selects a random integer e between 0 and $q - 1$ such that $\gcd(e, q - 1) = 1$, and compute $d = e^{-1} \bmod (q - 1)$ by using the extended Euclid's algorithm.

(iii) Now suppose user Alice wishes to send a secure message M to user Bob, then they follow the following procedure:

- Alice first sends M^{e_A} to Bob,

- On receiving Alice message, Bob sends $M^{e_A e_B}$ back to Alice, (note that at this point, Bob cannot read Alice message M)

- Alice sends $M^{e_A e_B d_A} = M^{e_B}$ to Bob,

- Bob then computes $P^{d_B e_B} = M$, and hence recovers Alice's original message M.

4.4.3 RSA Public-Key Cryptosystems

In 1978, just shortly after Diffie and Hellman proposed the first public-key exchange protocol, Rivest[8], Shamir[9] and Adleman[10], all at MIT at the time, proposed the first public-key cryptosystem, now widely known as RSA public-key cryptosystem. The RSA system is based on the following assumption:

RSA Assumption: It is not so difficult to find two large prime numbers, but it is very difficult to factor a large composite into its prime decomposition form.

An example of a one-way function $f : S \rightarrow S$ of the form used in RSA cryptosystem can be defined as follows [18]:

$$f(x) \equiv x^k \pmod{N} \tag{4.84}$$

where

$$\begin{cases} N = pq \ (p \text{ and } q \text{ are two large distinct primes}) \\ S = \{0, 1, 2, \cdots, N - 1\} \\ \lambda = \text{lcm}(p - 1, \ q - 1) = \dfrac{(p-1)(q-1)}{\gcd(p-1, \ q-1)} \\ k > 1, \quad \gcd(k, \lambda) = 1. \end{cases}$$

We assume that k and N are publicly known but p, q and λ are not. Note that the λ defined above is the Carmichael's function, i.e., $\lambda = \lambda(N)$. In RSA cryptosystem, we are essentially only interested in the case when the modulus $N = pq$. In general, instead of considering the modulus $N = pq$, we can consider an arbitrary integer N, which is larger than any number representing a message block. The inverse function of $f(x)$ is defined by

$$f^{-1}(y) \equiv y^{k'} \pmod{N}, \quad \text{with} \ kk' \equiv 1 \pmod{\lambda}. \tag{4.85}$$

---

[8]Ronald L. Rivest obtained his PhD in computer science from Stanford University in 1973 and is now Professor in the Department of Electrical Engineering and Computer Science at the Massachusetts Institute of Technology and Associate Director of the MIT's Laboratory for Computer Science.

[9]Adi Shamir is currently Professor in the Department of Applied Mathematics and Computer Science at the Weizmann Institute of Science, Israel.

[10]Leonard Adleman received his BSc in mathematics and PhD in computer science all from the University of California at Berkeley in 1972 and 1976, respectively. He is currently Professor in the Department of Computer Science at the University of Southern California. His main research activities are in theoretical computer science with particular emphasis on the complexity of number theoretic problems.

It should be easy to compute $f^{-1}(y) \equiv y^{k'} \pmod{N}$ if k' is known, provided that $f^{-1}(y)$ exists. The assumption underlying the RSA cryptosystem is that it is hard to compute $f^{-1}(y)$ without knowing k'. Note that the knowledge of p, q or λ makes it easy to compute k'. In what follows, we shall present a briefly discussion of the existence of the inverse function $f^{-1}(y)$ defined in (4.85) for all y. We first introduce a useful theorem due to Riesel [118]:

Theorem 4.4.1 If N is a product of distinct primes, then for all a,

$$a^{\lambda(N)+1} \equiv a \pmod{N}. \tag{4.86}$$

Note that if N contains multiple prime factors, then (4.86) need no longer be true; say, for example, let $N = 12 = 2^2 \cdot 3$, then $9^{\lambda(12)+1} = 9^3 \equiv 9 \pmod{12}$, but $10^{\lambda(12)+1} = 10^3 \equiv 4 \not\equiv 10 \pmod{12}$.

Now, let k and N have been chosen suitably as follows:

$$N = pq, \quad \text{with } p, q \text{ distinct primes} \tag{4.87}$$

$$a^{kk'} \equiv a \pmod{N}, \quad \text{for all } a. \tag{4.88}$$

Then, by Theorem 4.4.1, the inverse function $f^{-1}(y)$ defined in (4.85) exists for all y. It follows also immediately from (4.86) that

$$a^{m\lambda(N)+1} \equiv a \pmod{N}, \tag{4.89}$$

which is the form needed in a RSA cryptosystem.

Remark 4.4.4 The following results, although of no practical consequence for RSA systems, are interesting. Lidl and Pilz [81] pointed out that for arbitrary integer N,

$$a^{k+u} \equiv a^k \pmod{N}, \quad \text{for all } a \tag{4.90}$$

if and only if N is free of $(k+1)$-th powers and $\lambda(N) \mid u$. More recently Freeman and Jackson [41] conjectured and proved that for arbitrary integer N and $m \geq 1$, a necessary and sufficient condition for (4.89) to have a solution a is that

$$\gcd(a^2, N) \mid a, \tag{4.91}$$

or equivalently,

$$\gcd(a, N/d) = 1, \quad \text{where } d = \gcd(a, N). \tag{4.92}$$

Remark 4.4.5 Both the Lidl-Pilz and Freeman-Jackson results can be generalized to (communicated by Pleasants [107]): a necessary and sufficient condition for

$$a^{m\lambda(N)+k} \equiv a^k \pmod{N} \tag{4.93}$$

is

$$\gcd(a^{k+1}, N) \mid a^k, \tag{4.94}$$

or equivalently,

$$\gcd(a, N/d) = 1, \quad \text{where } d = \gcd(a^k, N). \tag{4.95}$$

The proof of the above generalized result is as follows. Let p be prime and $p^\alpha \parallel N$. Let β be such that $p^\beta \parallel a$. We assume that $p \mid N$, that is $\alpha > 0$. There are three cases:

(i) $\beta = 0$: we have $a^{m\lambda(N)+k} \equiv a^k \pmod{p^\alpha}$, by Euler's theorem,

(ii) $0 < k\beta < \alpha$: we have $a^t \not\equiv a^k \pmod{p^\alpha}$ for all $t > k$, obviously,

(iii) $k\beta \geq \alpha$: we have $a^t \equiv a^k \pmod{p^\alpha}$ for all $t > k$, obviously.

We conclude that $a^{m\lambda(N)+k} \equiv a^k \pmod{N}$ if and only if we are never in the second case for all primes $p \mid N$. Never being in the second case is equivalent to the condition $\gcd(a^{k+1}, N) \mid a^k$.

It is clear that one of the most important tasks in the construction of RSA cryptosystems is to find two large primes, say each with at least 100 digits. This can be done by the following algorithm:

Algorithm 4.4.1 (Large prime generation)

[1] (Initialization) Randomly generate an odd integer n with e.g., 100 digits;

[2] (Primality Testing – Probabilistic Method) Use a combination of Rabin-Miller's test and a Lucas test to determine if n is a probable prime. If it is, goto [3], else goto [1] to get another 100-digit odd integer.

[3] (Primality Proving – Elliptic Curve Method) Use the elliptic curve method to verify whether or not n is indeed a prime. If it is, then report that n is prime, and save it for later use; or otherwise, goto [1] to get another 100-digit odd integer.

[4] (done?) If you need more primes, goto [1], else terminate the algorithm.

How many primes with 100 digits do we have? Chebyshev[11] showed 100 years ago, even before the *Prime Number Theorem* was proved, that if N is large, then

$$0.9\frac{N}{\ln N} < \pi(N) < 1.1\frac{N}{\ln N}. \tag{4.96}$$

Since

$$0.9\frac{10^{99}}{\ln 10^{99}} < \pi(10^{99}) < 1.1\frac{10^{99}}{\ln 10^{99}}$$

$$0.9\frac{10^{100}}{\ln 10^{100}} < \pi(10^{100}) < 1.1\frac{10^{100}}{\ln 10^{100}}$$

then the difference $\pi(10^{100}) - \pi(10^{99})$ will give the numbers of primes with exactly 100 digits:

$$3.42 \cdot 10^{97} < \pi(10^{100}) - \pi(10^{99}) < 4.38 \cdot 10^{97}.$$

The above procedure for prime generation depends on primality testing and proving, but we do have some other methods which do not rely on primality testing and proving. One such method is based on Pocklington's theorem (i.e., Theorem 4.1.10), that can automatically lead to primes, say with 100 digits [116]. We re-state the theorem in a slightly different style to Theorem 4.1.10 as follows:

Theorem 4.4.2 Let p be an odd prime, k a natural number such that p does not divide k and $1 < k < 2(p + 1)$; and let $N = 2kp + 1$. Then the following conditions are equivalent:

(i) N is prime.

(ii) There exists a natural number $a, 2 \leq a < N$, such that

$$a^{kp} \equiv -1 \pmod{N} \tag{4.97}$$

and

$$\gcd(a^k + 1, N) = 1. \tag{4.98}$$

---

[11]Pafnuty Lvovich Chebyshev (1821–1894), a Russian mathematician who contributed to several branches of mathematics, and perhaps best known for his *Chebyshev polynomials*. He was also largely responsible for founding the influential Russian school of mathematical probability. In number theory, he proved, among other things that, for $N > 3$, there is at least one prime between N and $2N - 2$.

Algorithm 4.4.2 (Large prime generation based on Pocklington's theorem)

[1] Choose, e.g., a prime p_1 with $d_1 = 5$ digits. Find $k_1 < 2(p_1 + 1)$ such that $p_2 = 2k_1p_1 + 1$ has $d_2 = 2d_1 = 10$ digits or $d_2 = 2d_1 - 1 = 9$ digits and there exists $a_1 < p_2$ satisfying the conditions $a_1^{k_1p_1} \equiv -1 \pmod{p_2}$ and $\gcd(a_1^{k_1} + 1, p_2) = 1$. By Pocklington's Theorem, p_2 is prime.

[2] Repeat the same procedure starting from p_2 to obtain the primes $p_3, p_4, \cdots$. In order to produce a prime with 100 digits, the process must be iterated five times. In the last step, k_5 should be chosen so that $2k_5p_5 + 1$ has 100 digits.

As pointed out in [116], for all practical purposes, the above algorithm for producing primes of a given size will run in polynomial time, even though this has not yet been supported by a proof.

Now let us return to the construction of a good trapdoor function [18] based on the above two algorithms:

Algorithm 4.4.3 (Construction of a Trapdoor Function)

[1] Using Algorithm 4.4.2 or Algorithm 4.4.1 to find two large primes p and q, each with at least 100 digits such that:

    [1.1] $|p - q|$ is large;

    [1.2] $p \equiv -1 \pmod{12}, q \equiv -1 \pmod{12}$; and

    [1.3] $p' = (p-1)/2$, $p'' = (p+1)/12$, $q' = (q-1)/2$, $q'' = (q+1)/12$ are all primes.

[2] Compute $N = pq$ and $\lambda = 2p'q'$.

[3] Choose a random integer k relatively prime to λ such that $k - 1$ is not a multiple of p' or q'.

[4] Apply the extended Euclid's algorithm to k and λ to find k' and λ' such that $0 < k' < \lambda$ and
$$kk' + \lambda\lambda' = 1.$$

[5] Destroy all evidence of p, q, λ, λ'.

[6] Make (k, N) public but keep k' secret.

According to the Prime Number Theorem, the probability that a randomly chosen integer in $[1, N]$ is prime is $\sim 1/\ln N$. Thus, the expected number of random trails required to find p (or p', or p''; assume that p, p', and p'' were independent) is conjectured to be $\mathcal{O}\left((\log N)^3\right)$. Based on this assumption, the expected time required to construct the above one-way trapdoor function is $\mathcal{O}\left((\log N)^6\right)$.

Suppose now the sender, say, e.g., Alice wants to send a message M to the receiver, say, e.g., Bob. Bob will have already chosen a trapdoor function f described above, and published his *public-key* (k, N), so we can assume that both Alice and any potential adversary know (k, N). Alice splits the message M into blocks of $\lfloor \log N \rfloor$ bits or less (padded on the right with zeroes for the last block), and treats each block as integer $x \in \{0, 1, 2, \cdots, N - 1\}$. Alice computes

$$y \equiv x^k \pmod{N} \tag{4.99}$$

and transmits y to Bob. Bob, who knows the secret-key k', computes

$$x \equiv y^{k'} \pmod{N}. \tag{4.100}$$

An adversary who intercepts the encrypted message should be unable to decrypt it without knowledge of k'. There is no known way of cracking the RSA system without essentially factoring N, so it is clear that the security of the RSA system depends on the difficulty of factoring N. Some authors, say, e.g., Woll [145] observed that finding the RSA deciphering key k' is random polynomial-time equivalent to factorization. More recently, Pinch [106] showed that an algorithm $A(N, k)$ for obtaining k' given N and k can be turned into an algorithm which obtains p and q with positive probability.

Example 4.4.3 Suppose the plain-text message is "NATURAL NUMBERS ARE MADE BY GOD" and suppose the encoding alphabet is defined as follows:

⊔	A	B	C	D	E	F	G	H	I	J	K	L	M
00	01	02	03	04	05	06	07	08	09	10	11	12	13
N	O	P	Q	R	S	T	U	V	W	X	Y	Z	
14	15	16	17	18	19	20	21	22	23	24	25	26	

Then we transform the message into e.g., four message blocks, each with 15 digits as follows:

$x = \langle 140020211801120, 014211302051800, 011805001301040, 500022500071504 \rangle$.

and perform the following computation steps:

(i) Select two primes p and q, compute $N = pq$ and $\lambda(N)$:

$p = 440334654777631, \quad q = 145295143558111$
$N = pq = 63978486879527143858831415041$
$\lambda(N) = 71087207643918398032258977 0.$

(ii) Determine the keys k and k': we try to factorize $m\lambda(N) + 1$ for $m = 1, 2, 3, \cdots$ until we find a "good" factorization that can be used to obtain suitable k and k':

$\lambda(N) + 1 = 1193 \cdot 2990957 \cdot 209791523 \cdot 17107 \cdot 55511$
$2\lambda(N) + 1 = 47 \cdot 131 \cdot 199 \cdot 3322357 \cdot 1716499 \cdot 203474209$
$3\lambda(N) + 1 = 674683 \cdot 1696366781 \cdot 297801601 \cdot 6257$
$4\lambda(N) + 1 = 17 \cdot 53 \cdot 5605331 \cdot 563022035211575351$
$5\lambda(N) + 1 = 17450633 \cdot 13017248387079301 \cdot 15647$
$6\lambda(N) + 1 = 1261058128567 \cdot 49864411 \cdot 2293 \cdot 29581$
$7\lambda(N) + 1 = 19 \cdot 2619002386881204138030593 89$
$8\lambda(N) + 1 = 15037114930441 \cdot 378195992902921$
$9\lambda(N) + 1 = 11 \cdot 13200581 \cdot 8097845885549501 \cdot 5441$
$10\lambda(N) + 1 = 71087207643918398032258977 01$
$11\lambda(N) + 1 = 2131418173 \cdot 7417510211 \cdot 494603657$
$12\lambda(N) + 1 = 4425033337657 \cdot 1927774158146113$
$13\lambda(N) + 1 = 23 \cdot 6796296973884340591 \cdot 59120027$
$14\lambda(N) + 1 = 14785772846857861 \cdot 673093599721$
$15\lambda(N) + 1 = 500807 \cdot 647357777401277 \cdot 17579 \cdot 1871.$

Suppose now we wish to use the 15th factorization $15\lambda(N) + 1$ to obtain $\langle k, k' \rangle = \langle 17579, 60658064432491948943846 9 \rangle$ such that $kk' = 1 + 15\lambda(N)$.

(iii) Encrypt the message $x \mapsto x^k \bmod N = y$ (using the fast modular exponentiation method, e.g., Algorithm 3.4.1):

$140020211801120^{17579} \bmod N = 60379537366647508826042726177$
$014211302051800^{17579} \bmod N = 47215464067987497433568498485$
$011805001301040^{17579} \bmod N = 20999327573397550148935085516$
$500022500071504^{17579} \bmod N = 37746963038639759803119392704.$

(iv) Decrypt the message $y \mapsto y^{k'} \bmod N = x^{kk'} \bmod N = x$ (again using e.g., Algorithm 3.4.1):

$$6037953736664750882604272617^{6065806443249194894384 69} \bmod N = 140020211801120$$
$$4721546406798749743356849848^{5606580644324919489438469} \bmod N = 014211302051800$$
$$2099932757339755014893508551^{6606580644324919489438469} \bmod N = 011805001301040$$
$$3774696303863975980311939270^{4606580644324919489438469} \bmod N = 500022500071504.$$

Exercise 4.4.1 Let the public-key $\langle N, k \rangle$, obtained from the above 6th factorization $6\lambda(N) + 1 = 1261058128567 \cdot 49864411 \cdot 2293 \cdot 29581$, as follows:

$$\langle N, k \rangle = \langle 6397848687952714385883141 5041, 2293 \rangle.$$

(i) Find the secret decryption key k' such that $kk' \equiv 1 \pmod{\lambda(N)}$;

(ii) Use the public-key $\langle N, k \rangle$ to encrypt the message "NUMBER THEORY IS THE QUEEN OF MATHEMATICS", using the same encoding alphabet as defined in Example 4.4.3;

(iii) Use the secret decryption key k' to decrypt the encrypted message from step (ii).

Remark 4.4.6 *The RSA method is slow.* Compared with the most famous cryptosystem Data Encryption Standard (DES), the RSA system is slow, this is because the main RSA operations – modular multiplication and exponentiation are relatively slow, even using specialized hardware. RSA method is said to be efficient only because it is compared with the exponential complexity $\mathcal{EXP}$, since $\mathcal{P} \subset \mathcal{EXP}$.

Remark 4.4.7 *Possible attacks on RSA.* (i) Wiener's attack on short RSA secret-key [143]: it is important that the secret-key k' should be large (nearly as many bits as the modulus N), or otherwise, there is an attack due to Wiener and based on properties of continued fractions, that can find the secret-key k' in time polynomial in the length of the modulus N, and hence decrypt the message. (ii) iterated encryption or fixed-point attack ([92] and [106]): Suppose k has order r in the multiplicative group modulo $\lambda(N)$. Then $k^r \equiv 1 \pmod{\lambda(N)}$, so $x^{k^r} \equiv x \pmod{N}$. This is just the r-th iterate of the encryption of x. So we must ensure that r is large.

Example 4.4.4 We now give a large RSA example. In one of his series papers of Mathematical Games in *Scientific American*, 8(1977), pp 120–124, Martin Gardner reported a RSA challenge with \$100 to decode the following message (it was estimated in 1977 by RSA that it would need 23000 years to break the code):

$C = 96869613754622061477140922254355882905759991124574319874695120930_{\text{-}}$
$08162982251457083569314766228839896280133919905518299451578151\allowbreak54$

The public-key N

$N = 11438162575788886766923577997614661201021829672124236256256184293$
$5706935245733897830597123563958705058989075147599290026879543541$

is a "random" 129-digit number (RSA-129). It was factored by Derek Atkins, Michael Graff, Arjen K. Lenstra, Paul Leyland, et al. in 2 April 1994 to win the $100 prize offered by RSA in 1977. Its two prime factors are as follows:

$3490529510847650949147849619903898133417764638493387843990820577,$

$32769132993266709549961988190834461413177642967992942539798288533.$

They used the double large prime variation of the Multiple Polynomial Quadratic Sieve (MPQS) factoring method. The sieving step took approximately 5000 mips years, and was carried out in 8 months by about 600 volunteers from more than 20 countries, on all continents except Antarctica. As we have explained in the previous example, to encipher an RSA-coded message, we only need to use the public-key (N, k) to compute

$$y \equiv x^k \pmod{N}.$$

But decoding an RSA-message requires factorization of N if one does not know the secret deciphering key. This means that if we can factor N, then we can compute the secret-key k', and get back the original message by calculating

$$x \equiv y^{k'} \pmod{N}.$$

Since now $N = $ RSA-129 has been factored, then it is trivial to compute the secret-key k', which in fact is

$k' = 10669861436857802444286877132892015478070990663393786280122622440_{\text{-}}$
$966310631259117744708733401685974623065539685445132771090536060\allowbreak95$

So we shall be able to compute

$$M = C^{k'} \pmod{N}.$$

To use the fast exponential method for $C^{k'} \pmod{N}$, we first let $d = k'$ and write d in its binary form $d_1 d_2 \cdots d_{\text{size}}$ (where size is the number of the bits of d) as follows:

$d = d_1 d_2 \cdots d_{426} =$
1001110110011111100101001100100010000010000011101001111001001100 1001_
1110100111000000000000011111111010000110101011000101110111101010 00011_
1110110000001000001110110101010101111010101001111110110110100001 111110_
1000000111101001100010110010110011010010100011001001110101100000 10111_
0100101011010000011100000001110001110101010011011101000111101001 11100_
0110101101010101001001110101000100111100000010011101001100011011 1110_
101100100011001111

and perform the following computation:

    $M \leftarrow 1$
    for i from 1 to 426 do
        $M \leftarrow M^2 \bmod N$
        if $d_i = 1$ then $M \leftarrow M \cdot C \bmod N$
    print M

which gives

$M =$2008050013010709030023151804190001180500191721050113091908_
    00151919090618010705

and hence the original message:

THE MAGIC WORDS ARE SQUEAMISH OSSIFRAGE

via the encoding alphabet $\sqcup = 00, A = 01, B = 02, \cdots, Z = 26$.

Of course, by the public enciphering key $k = 9007$, we can compute $C \equiv M^k \pmod{N}$; first let $k = e$ and write e in the binary form $e = e_1 e_2 \cdots e_{14} = 10001100101111$, then perform the following procedure:

    $C \leftarrow 1$
    for i from 1 to 14 do
        $C \leftarrow C^2 \bmod r$
        if $e_i = 0$ then $C \leftarrow C \cdot M \bmod r$
    print C

which shall give the enciphered text in the beginning of this example:

$C =$ 9686961375462206147714092225435588290575999112457431987469512093_
    0816298225145708356931476622883989628013391990551829945157815154

Remark 4.4.8 In fact, anyone else who can factor the integer RSA-129 can decipher the message, too. Thus, deciphering the message is essentially factoring the 129-digit integer. The factorization of RSA-129 implies that it is possible to factor a random 129-digit integer. It should be also noted that on 10 April 1996, Arjen K. Lenstra, et al. also factored the following RSA-130

1807082086874048059516561644059055662781025167694013491701270214_
5005666254024440483873411275908123033717818879665631820132148480557

which has the following two prime factors:

39685999459597454290161126162883786067576449112810064832555157243,
45534498646735972188403686897274408864356301263205069600999044599

This factorization was found using the Number Field Sieve (NFS) factoring algorithm, and beats the above mentioned 129-digit record by the Quadratic Sieve (QS) factoring algorithm. The amount of computer time spent on this 130-digit NFS-record is only a fraction of what was spent on the old 129-digit QS-record.

Thus, it follows from that

Corollary 4.4.1 The composite number N used in the RSA cryptosystem should have more than 130 decimal digits.

Now we consider a more general and more realistic case of secure communications. Consider a computer network with n nodes, it is apparent that there are

$$\binom{n}{2} = n(n-1)/2$$

ways of communications between two nodes in the network. Suppose one of the nodes (users), say A, wants to send a secure message M to another node, say B, or vice versa. Then A uses B's encryption key e_B to encrypt his message M

$$C = M^{e_B} \bmod N_B \qquad (4.101)$$

and sends the encrypted message C to B; on receiving A's message M, B uses his own decryption key d_B to decrypt A's message C:

$$M = C^{d_B} \bmod N_B. \qquad (4.102)$$

Since only B has the decryption key d_B, then only B (at least from a theoretical point of view) can recover the original message. B can of course send a secure message M to A in a similar way. Figure 4.4 shows diagrammatically the idea of secure communications between any two parties, say, e.g., A and B.

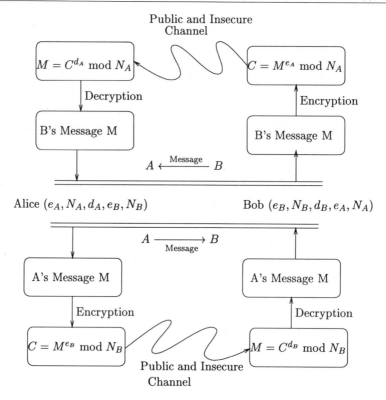

Figure 4.4: Sending Encrypted Messages Using RSA Public-Key Cryptography

4.4.4   Elliptic Curve Public-Key Cryptosystems

We have discussed some novel applications of elliptic curves in primality testing
and integer factorization in the previous sections. In this subsection, we shall
discuss a novel application of elliptic curves in public-key cryptography ([70],
[91], and [93]).

(I) Basic Computations of Elliptic Curves for Cryptology

We shall first present some preliminaries for elliptic curve computations for
cryptology.

(i) Embedding Messages on Elliptic Curves: Our aim here is to do cryptography with elliptic curve groups in place of $\mathbb{F}_q$. More specifically, we wish to embed plaintext messages as points on an elliptic curve defined over a finite field $\mathbb{F}_q$, with $q = p^r$ and $p \in$ Primes. Let our message units m be integers $0 \le m \le M$, let also κ be a large enough integer so that we are satisfied with an error probability of $2^{-\kappa}$ when we attempt to embed a plaintext message m. In practice, $30 \le \kappa \le 50$. Now let us take $\kappa = 30$ and an elliptic curve $E : y^2 = x^3 + ax + b$ over $\mathbb{F}_q$. Given a message number m, we compute a set of values for x

$$x = \{m\kappa + j, \; j = 0, 1, 2, \cdots\} = \{30m, \; 30m + 1, \; 30m + 2, \; \cdots\} \quad (4.103)$$

until we find $x^3 + ax + b$ is a square modulo p, giving us a point $(x, \sqrt{x^3 + ax + b})$ on E. To convert a point (x, y) on E back to a message number m, we just simply compute $m = \lfloor x/30 \rfloor$. Since $x^3 + ax + b$ is a square for approximately 50% of all x, there is only about a $2^{-\kappa}$ probability that this method will fail to produce a point on E over $\mathbb{F}_q$. In what follows, we shall give a simple example of how to embed a message number by a point on an elliptic curve. Let E be $y^2 = x^3 + 3x$, $m = 2174$ and $p = 4177$ (in practice, we select $p > 30m$). Then we calculate $x = \{30 \cdot 2174 + j, \; j = 0, 1, 2, \cdots\}$ until $x^3 + 3x$ is a square modulo 4177. We find that when $j = 15$

$$
\begin{aligned}
x &= 30 \cdot 2174 + 15 \\
&= 65235 \\
x^3 + 3x &= (30 \cdot 2174 + 15)^3 + 3(30 \cdot 2174 + 15) \\
&= 277614407048580 \\
&\equiv 1444 \bmod 4177 \\
&\equiv 38^2
\end{aligned}
$$

So we get the message point for $m = 2174$ as follows

$$(x, \sqrt{x^3 + ax + b}) = (65235, 38).$$

To convert the message point $(65235, 38)$ on E back to its original message number m, we just simply compute

$$m = \lfloor 65235/30 \rfloor = \lfloor 2174.5 \rfloor = 2174.$$

(ii) Multiples of Points on Elliptic Curves over $\mathbb{F}_q$: We have discussed the multiplication of $kP \in E$ over $\mathbb{Z}/N\mathbb{Z}$. In elliptic curve public-key cryptography, we shall be mainly interested in the multiplication of $kP \in E$ over $\mathbb{F}_q$, which can be done in $\mathcal{O}(\log k \log^3 q)$ bit operations by the *repeated doubling method*. If we happen to know the number N of points on our elliptic curve E and if $k > N$, then the coordinate of kP on E can be computed in $\mathcal{O}(\log^4 q)$ bit operations [70]; recall that the number N of points on E satisfies $N \leq q + 1 + 2\sqrt{q} = \mathcal{O}(q)$ and can be computed by René Schoof's algorithm in $\mathcal{O}(\log^8 q)$ bit operations.

(iii) Discrete Logarithms on Elliptic Curves: Given an elliptic curve E over $\mathbb{F}_q$, and B a point on E. Then the *discrete logarithm* on E is the problem, given a point $P \in E$, of finding an integer $x \in \mathbb{Z}$ such that $xB = P$ if such an integer x exists.

It is likely that the discrete logarithm problem on elliptic curves over $\mathbb{F}_q$ is likely more intractable than the discrete logarithm problem in $\mathbb{F}_q$. It is this feature that makes the cryptographic systems based on elliptic curves even more secure than that based on the discrete logarithm problem. In the rest of this subsection, we shall discuss elliptic curve analogues of some important public-key cryptosystems.

(II) Elliptic Curve Analogues of Some Public-Key Cryptosystems

In what follows, we shall present the corresponding elliptic curve analogues of some widely used public-key cryptosystems, namely, Diffie-Hellman key exchange system, Massey-Omura, ElGamal and RSA public-key cryptosystems.

(i) Analogue of Diffie-Hellman Key Exchange System:

(a) Alice and Bob publicly choose a finite field $\mathbb{F}_q$ with $q = p^r$ and $p \in$ Primes, an elliptic curve E over $\mathbb{F}_q$, and a random *base* point $P \in E$ such that P generates a large subgroup of E, preferably of the same size as E itself. All of this is public information.

(b) To agree on a secret key, Alice and Bob choose two secret random integers a and b. Alice computes $aP \in E$ and sends aP to Bob; Bob computes $bP \in E$ and send bP to Alice. Both aP and bP are, of course, public, but a and b are not.

(c) Now both Alice and Bob compute the secret-key $abP \in E$, and use it for further secure communications.

There is no known fast way to compute abP if one only knows P, aP and bP – this is the discrete logarithm problem on E.

(ii) Analogue of Massey-Omura Cryptosystem:

(a) Alice and Bob publicly choose an elliptic curve E over $\mathbb{F}_q$ with q large, and we suppose also that the number of points (denoted by N) is publicly known.

(b) Alice chooses a secret pair of numbers (e_A, d_A) such that $d_A e_A \equiv 1 \pmod{N}$. Similarly, Bob chooses (e_B, d_B).

(c) If Alice wants to send a secret message-point $P \in E$ to Bob, the procedure is as follows:

  - Alice sends $e_A P$ to Bob,
  - Bob sends $e_B e_A P$ to Alice,
  - Alice sends $d_A e_B e_A P = e_B P$ to Bob,
  - Bob computes $d_B e_B P = P$.

Note that an eavesdropper would know $e_A P$, $e_B e_A P$, and $e_B P$. So if he could solve the discrete logarithm problem on E, he could determine e_B from the first two points and then compute $d_B = e_B^{-1} \bmod N$ and hence get $P = d_B(e_B P)$.

(iii) Analogue of ElGamal Cryptosystem:

(a) Alice and Bob publicly choose an elliptic curve E over $\mathbb{F}_q$ with $q = p^r$ and $p \in \text{Primes}$, and a random *base* point $P \in E$.

(b) Alice chooses an random integer r_a and computes $r_a P$; Bob also chooses an random integer r_b and computes $r_b P$.

(c) To send a message-point M to Bob, Alice chooses a random integer k and sends the pair of points $(kP, \ M + k(r_b P))$.

(d) To read M, Bob computes

$$M + k(r_b P) - r_b(kP) = M. \qquad (4.104)$$

An eavesdropper who can solve the discrete logarithm problem on E can, of course, determine r_b from the publicly known information P and r_bP. But as everybody knows, there is no efficient way to compute discrete logarithms, the systems is thus secure.

(iv) Analogue of RSA Cryptosystem:

RSA, the most popular cryptosystem in use, also has the following elliptic curve analogue:

(a) $N = pq$ is a public-key which is the product of the two large distinct secret primes p and q.

(b) Choose two random integers a and b such that $E : y^2 = x^3 + ax + b$ defines an elliptic curve both mod p and mod q.

(c) To encrypt a message-point P, just perform $eP \bmod N$, where e is the public (encryption) key. To decrypt, one needs to know the number of points on E modulo both p and q.

4.4.5   Digital Signatures

The idea of public-key cryptography (suppose we are using the RSA public-key scheme) can also be used to obtain digital signatures. Suppose Alice wishes to send Bob a secure message as well as a secure signature. Alice uses Bob's encryption key to encrypt her message, and then, she uses her own decryption key to encrypt her signature, and finally send out her message and signature to Bob. At the other end, Bob first uses Alice's encryption key to decrypt Alice's signature, and then, uses his own decryption key to decrypt Alice's message. Figure 4.5 shows how A (Alice) and B (Bob) can send secure message/signature each other over the insecure channel.

Example 4.4.5 (Digital Signature) To verify the \$100 offer in Example 4.4.4 actually came RSA, the following signature had been added:

$S =$ 1671786115038084424601527138916839824543690103235831121783503844469_
29062655448792237114490509578608655662496577974840004057020373

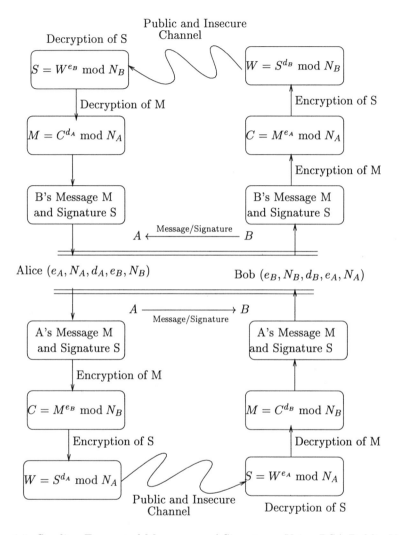

Figure 4.5: Sending Encrypted Messages and Signatures Using RSA Public-Key Cryptography

It was enciphered by $S = M^d$ (mod N), where d is the secret key, as in Example 4.4.4. To deciphering the signature, we use $M = S^e$ (mod N) by performing the following procedure (also same as in Example 4.4.4):

$C \leftarrow 1$
$e = (10001100101111)_2$
for i from 1 to 14 do
    $C \leftarrow C^2$ mod N
    if $e[i] = 0$ then $C \leftarrow C * M$ mod N
    print C

which shall give the following deciphered text:

609181920001915122205180023091419001514050008211404180504000415121201819

It translates to

FIRST SOLVER WINS ONE HUNDRED DOLLARS

Since this signature was encoded by RSA's secret key, it cannot be forged by an eavesdropper or even by RSA themselves.

In Example 4.4.5, the message and the signature are actually two different texts. We can, of course directly sign the signature on the message. This can be done in the following way: Suppose A (Alice) wants to send B (Bob) a signed message. Suppose also that

(i) A has her own public and secret keys $(e_A, N_A; d_A)$ as well as B's public-key e_B and N_B from a public domain;

(ii) B has his own public and secret keys $(e_B, N_B; d_B)$ as well as A's public-key e_A and N_A from a public domain.

To send a signed message from A to B:

(i) Alice uses B's public-key e_B and N_B to encrypt her message M:

$$C = M^{e_B} \bmod N_B; \tag{4.105}$$

(ii) Alice signs the message using her own secret-key d_A:

$$S = C^{d_A} \bmod N_A \tag{4.106}$$

and sends this signed message to B over the network.

Upon receiving A's signed message,

(i) B uses A's public-key e_A to decrypt A's signature:

$$C = S^{e_A} \bmod N_A; \qquad (4.107)$$

(ii) B uses his own secret-key d_B to decrypt A's encrypted message:

$$M = C^{d_B} \bmod N_B. \qquad (4.108)$$

B can make sure that the message he just received indeed comes from A, since the signature on the message is enciphered by A's own secret-key, which is only known to A. Once the message is sent out, A cannot deny the message. Similarly, Bob can send a signed message to Alice in the symmetric way. The above process can be best described in Figure 4.6.

Example 4.4.6 (Digital Signature) Suppose now Alice wants to send Bob the following signed message "Number Theory is the Queen of Mathematics". The process can be as follows:

(i) Suppose Alice has the following information at her hand:

$M_A=$ 14211302051800200805151825000919002008050017210505140015060013012008051301200090319

$N_A=$ 18070820886874048059516561644059055662781025167694013491701270214500566625402440483873411275908123033717818879665631820132148805570 (130 digits)

$=$ 396859994595974542901611261628837860675764491128100648325551572430 * 4553449864673597218840368689727440886435630126320506960099044599

$e_A=$ 2617

$d_A=$ 96465176839751796481255776143486819873538754907407477447102309852757971788488016357111391440322426247791075740923050236448593109

and suppose Bob has the following information at his hand:

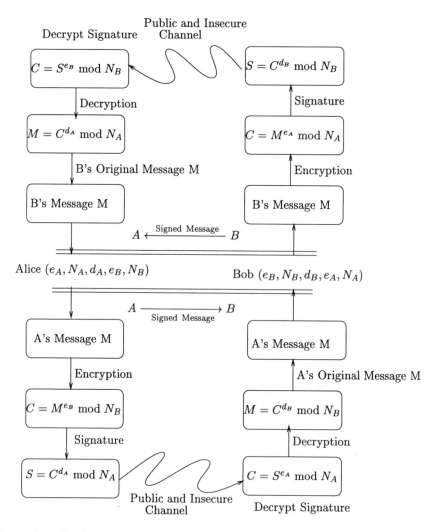

Figure 4.6: Sending Encrypted and Signed Messages Using RSA Public-Key Cryptography

$N_B=$ 11438162575788886766923577997614661201021829672124236256256_
61842935706935245733897830597123563958705058989075147599296_
0026879543541 (129 digits)

$=$ 34905295108476509491478496199038981334177646384933878439906_
820577 * 32769132993266709549961988190834461413177642967996_
2942539798288533

$e_B=$ 9007

$d_B=$ 10669861436857802444286877132892015478070990663393786280126_
26224496631063125911774470873340168597462306553968544513276_
7109053606095

(ii) Alice first encrypts the message M_A using e_B and N_B to get $C_A = M_A^{e_B} \bmod N_B$ via the following process:

$C_A \leftarrow 1$
$e_B \leftarrow (10001100101111)_2$
for i from 1 to 14 do
    $C_A \leftarrow C_A^2 \bmod N_B$
    if $e_B[i] = 1$ then $C_A \leftarrow C_A \cdot M_A \bmod N_B$
Save C_A

(iii) Alice then signs the message C_A using d_A and N_A to get $S_A = C_A^{d_A} \bmod N_A$ via the following process:

$S_A \leftarrow 1$
$d_A \leftarrow (10110010 \cdots 11010101)_2$
for i from 1 to 429 do
    $S_A \leftarrow S_A^2 \bmod N_A$
    if $d_A[1, i] = 1$ then $S_A \leftarrow S_A \cdot C_A \bmod N_A$
Send S_A

(iv) Upon receiving Alice's message, Bob first decrypts Alice's signature using e_A and N_A to get $C_A = S_A^{e_A} \bmod N_A$ via the following process:

$C_A \leftarrow 1$
$e_A \leftarrow (101000111001)_2$
for i from 1 to 12 do

$C_A \leftarrow C_A^2 \bmod N_A$
if $e_A[i] = 1$ then $C_A \leftarrow C_A \cdot S_A \bmod N_A$
Save C_A

(v) Bob then decrypts Alice's message using d_B and N_B to get $M_A = C_A^{d_B} \bmod N_B$ via the following process:

$M_A \leftarrow 1$
$d_B \leftarrow (1001110110 \cdots 1001111)_2$
for i from 1 to 426 do
$\quad M_A \leftarrow M_A^2 \bmod N_B$
$\quad$ if $d_B[i] = 0$ then $M_A \leftarrow M_A \cdot C_A \bmod N_B$
print M_A

In 1991, the US government's National Institute of Standards and Technology (NIST) proposed a Digital Signature Standard (DSS). The role of DSS is expected to be analogous to that of the Data Encryption Standard (DES). The DSS is similar to a signature scheme proposed by Schnorr; it is also similar to a signature scheme of ElGamal. For more information about the DSS, readers are suggested to consult [70].

We have already noted that almost every public-key cryptosystem has an elliptic curve analogue. It should be also noted that digital signature schemes can also be analogued by elliptic curves over $\mathbb{F}_q$ with q a prime power or over $\mathbb{Z}/n\mathbb{Z}$ with $n = pq$ and $p, q \in$ Primes in exactly the same way as that for public-key cryptography; several elliptic curve analogues of digital signature schemes have already been proposed, say, e.g., [94]. As an exercise, you are asked to develop an elliptic curve analogue of an existing signature scheme for obtaining and checking digital signatures.

4.4.6 Database Security

Databases pose a special challenge to the designer of secure information systems. Databases are meant to be shared. The sharing is often complex. In many organizations, there are many "rules" concerning the access to different *fields* (or parts) of a database. For example, the payroll department may have access to the name, address and salary fields, while the insurance office may have access to the health field of that individual. In this subsection, we shall introduce a method for database protection; it encrypts the entire database but the individual fields may be decrypted and read without effecting the security of other fields in the database.

Let

$$D = \langle F_1, F_2, \cdots F_n \rangle \tag{4.109}$$

where D is the database and each F_i is an individual file (or record). As in RSA encryption, each file in D can be regarded as an integer. To encrypt D, we first select n distinct primes $m_1, m_2, \cdots, m_n$, where $m_i > F_i$, for $i = 1, 2, \cdots, n$. Then by solving the following system of congruences:

$$\begin{cases} C \equiv F_1 \pmod{m_1} \\ C \equiv F_2 \pmod{m_2} \\ \quad \cdots \cdots \\ \quad \cdots \cdots \\ C \equiv F_n \pmod{m_n} \end{cases} \tag{4.110}$$

we get C, the encrypted text of D. According to the Chinese Remainder Theorem (i.e., Theorem 1.3.25 in Chapter 1), such a C always exists and can be found. Let

$$M = m_1 m_2 \cdots m_n \tag{4.111}$$

$$M_i = M/m_i, \tag{4.112}$$

$$e_i = M_i \left[M_i^{-1} \bmod m_i \right] \tag{4.113}$$

for $i = 1, 2, \cdots, n$. Then, C can be obtained as follows:

$$C = \sum_{i=1}^{n} e_j F_j \pmod{M}, \quad 0 \le C < M. \tag{4.114}$$

The integers $e_1, e_2, \cdots e_n$ is used as the *write-keys*. To retrieve the i-th file F_i from the encrypted text C of D, we simply perform the following operation:

$$F_i \equiv C \pmod{m_i}, \quad 0 \le F_i < m_i. \tag{4.115}$$

The moduli $m_1, m_2, \cdots m_n$ are called the *read-keys*. Only the people knowing the read-key m_i can read file F_i, but not other files. To read other files, say, e.g., F_{i+2}, it is necessary to know the read-key other than m_i. We present in the following an algorithm for database encryption and decryption.

Algorithm 4.4.4 Given $D = \langle F_1, F_2, \cdots F_n \rangle$, this algorithm will first encrypt the database D into its encrypted text C. To retrieve information from the encrypted database C, the user uses the appropriate read-key m_i to read file F_i.

[1] Database Encryption. The database administrators (DBA) perform the following operations to encrypt the database D:

    [1-1] Select n distinct primes $m_1, m_2, \cdots, m_n$ with $m_i > F_i$, for $i = 1, 2, \cdots, n$.

    [1-2] Use the Chinese Remainder Theorem to solve the following system of congruences:

$$\begin{cases} C \equiv F_1 \ (\text{mod } m_1) \\ C \equiv F_2 \ (\text{mod } m_2) \\ \quad \cdots \cdots \\ \quad \cdots \cdots \\ C \equiv F_n \ (\text{mod } m_n) \end{cases} \qquad (4.116)$$

and to get

$$C = \sum_{i=1}^{n} e_j F_j \ (\text{mod } M), \quad 0 \le C < M \qquad (4.117)$$

where

$$\begin{cases} M = m_1 m_2 \cdots m_n \\ e_i = M_i \left[M_i^{-1} \bmod m_i \right] & \text{for } i = 1, 2, \cdots n \\ M_i = M/m_i & \text{for } i = 1, 2, \cdots n \end{cases}$$

    [1-3] Distribute the read-key m_i to the appropriate database user U_i.

[2] (Database Decryption) At this stage, the database user U_i is supposed to have access to the encrypted database C as well as to have the read-key m_i, so he performs the following operation:

$$F_i \equiv C \ (\text{mod } m_i), \quad 0 \le F_i < m_i. \qquad (4.118)$$

The required file F_i should be now readable to user U_i.

Example 4.4.7 (Database Encryption and Decryption) Let

$$\begin{aligned} D &= \langle F_1, F_2, F_3, F_4, F_5 \rangle \\ &= \langle 198753, 217926, 357918, 377761, 391028 \rangle. \end{aligned}$$

Choose five primes m_1, m_2, m_3, m_4 and m_5 as follows:

$$\begin{cases} m_1 = 350377 > F_1 = 198753, \\ m_2 = 364423 > F_2 = 217926, \\ m_3 = 376127 > F_2 = 357918, \\ m_4 = 389219 > F_4 = 377761, \\ m_5 = 391939 > F_5 = 391028. \end{cases}$$

According to (4.116), we have:

$$\begin{cases} C \equiv F_1 \ (\text{mod } m_1) \Longrightarrow C \equiv 198753 \ (\text{mod } 350377) \\ C \equiv F_2 \ (\text{mod } m_2) \Longrightarrow C \equiv 217926 \ (\text{mod } 364423) \\ C \equiv F_3 \ (\text{mod } m_3) \Longrightarrow C \equiv 357918 \ (\text{mod } 376127) \\ C \equiv F_4 \ (\text{mod } m_4) \Longrightarrow C \equiv 377761 \ (\text{mod } 389219) \\ C \equiv F_5 \ (\text{mod } m_5) \Longrightarrow C \equiv 391028 \ (\text{mod } 391939) \end{cases}$$

Using the Chinese Remainder Theorem to solve the above system of congruences, we get

$$C = 58262627076918016013522277219.$$

Since $0 \le C < M$ with

$$M = 350377 \cdot 364423 \cdot 376127 \cdot 389219 \cdot 391939 = 73263623028327268830024522697$$

So, C is the required encrypted text of D. Now suppose user U_2 has the read-key $m_2 = 364423$. Then he can simply perform the following computation and get F_2:

$$F_2 \equiv C \ (\text{mod } m_i)$$

Since

$$\begin{aligned} C \ (\text{mod } m_2) &= 58262627076918016013522277219 \text{ mod } 364423 \\ &= 217926 \\ &= F_2. \end{aligned}$$

which is exactly what the user U_2 wanted. Similarly, a user can read F_5 if he knows m_5, since

$$\begin{aligned} C \ (\text{mod } m_5) &= 58262627076918016013522277219 \text{ mod } 391939 \\ &= 391028 \\ &= F_5. \end{aligned}$$

Remark 4.4.9 In Example 4.4.7, we have not explicitly given the computing processes for the write keys e_i and the encrypted text C, because we assume that readers are familiar with the Chinese Remainder Theorem (CRT) discussed in Chapter 1. For those who are not very familiar with CRT, we give the detailed computing processes as follows:

$$
\begin{aligned}
e_1 &= M_1 \cdot \left(M_1^{-1} \bmod m_1 \right) \\
&= 20909940729079611056161 \cdot \left(20909940729079611056161^{-1} \bmod 350377 \right) \\
&= 30405772112376534825095539493 \\
e_2 &= M_2 \cdot \left(M_2^{-1} \bmod m_2 \right) \\
&= 20104006341072673467439 \cdot \left(20104006341072673467439^{-1} \bmod 364423 \right) \\
&= 28303827407405984794600334493 \\
e_3 &= M_3 \cdot \left(M_3^{-1} \bmod m_3 \right) \\
&= 19478426975018349873911 \cdot \left(19478426975018349873911^{-1} \bmod 376127 \right) \\
&= 19918834208923514764560127 71 \\[2mm]
e_4 &= M_4 \cdot \left(M_4^{-1} \bmod m_4 \right) \\
&= 18823239109171769320163 \cdot \left(18823239109171769320163^{-1} \bmod 389219 \right) \\
&= 60680287683845941039716261 47 \\
e_1 &= M_5 \cdot \left(M_5^{-1} \bmod m_5 \right) \\
&= 18692608550903908217923 \cdot \left(18692608550903908217923^{-1} \bmod 391939 \right) \\
&= 72185246441025622365153249 1.
\end{aligned}
$$

So

$$
\begin{aligned}
C &= \left(e_1 F_1 + e_2 F_2 + e_3 F_3 + e_4 F_4 + e_5 F_5 \right) \bmod M \\
&= (30405772112376534825095539493 \cdot 198753 \\
&\quad + 28303827407405984794600334493 \cdot 217926 \\
&\quad + 19918834208923514764560127 71 \cdot 357918 \\
&\quad + 60680287683845941039716261 47 \cdot 377761 \\
&\quad + 72185246441025622365153249 1 \cdot 391028) \\
&\quad \bmod \ 732636230283272688302 4522697 \\
&= 582626270769180160135227721 9.
\end{aligned}
$$

Exercise 4.4.2 Let the database D be

$$\begin{aligned} D &= \langle F_1, F_2, F_3, F_4 \rangle \\ &= \langle 9853, 6792, 3761, 5102 \rangle. \end{aligned}$$

and the four read keys be

$$\begin{cases} m_1 = 9901 > F_1 = 9853, \\ m_2 = 7937 > F_2 = 6792, \\ m_3 = 5279 > F_3 = 3761, \\ m_4 = 6997 > F_4 = 5102. \end{cases}$$

(i) What are the four write keys e_1, e_2, e_3 and e_4 used in the encryption process?

(ii) What is the encrypted text C corresponding to D?

(iii) If F_1 is changed from $F_1 = 9853$ to $F_1 = 9123$, what is the new value of the encrypted text C?

4.4.7 Random Number Generation

A sequence of numbers is random if each number in the sequence is independent of the preceding numbers; there are no patterns to help us to predict any number of the sequence. Of course, truly *"random"* numbers are hard to come by, or even impossible to get. So, the random numbers we are talking about are actually *pseudorandom numbers*. Random numbers are useful in many different kinds of applications, such as sampling, simulation, numerical analysis, computer programming, decision making. They are valuable source; in some cases, they can speed up the computations, they can improve the rate of communication of partial information between two users (communication complexity), and they can also be used to solve problems in asynchronous distributed computation that can be proved impossible of solution by deterministic means. We have, in fact, already seen some applications of random numbers in the previous sections of this chapter. For example, the Pollard's ρ-method is essentially a random number generator. In this section, we shall briefly introduce some methods for generating random numbers.

(I) Middle-Square Method

Since the invention of the first electronic computer, researchers have been trying to find *efficient* ways to generate random numbers on computers; among them are John von Neumann who first suggested in around 1946 the arithmetic method, called *middle-square method* to obtain random numbers. (John von Neumann once said: "anyone who considers arithmetic methods of producing random digits is, of course, in a state of sin"). The algorithmic description of the middle-square method is as follows:

Algorithm 4.4.5 (von Neumann's middle-square method for generating random numbers)

[1] Let m be the number of random numbers we wish to generate (all with e.g., 10-digits), and set $i \leftarrow 0$.

[2] Randomly choose a starting 10-digit number n_0.

[3] Square n_i to get an intermediate number M, with 20 or less digits.

[4] Set $i = i + 1$ and take the middle ten digits of M as the new random number n_i.

[5] If $i < m$ then goto step [3] to generate a new random number, else stop the generating process.

Example 4.4.8 Let $n_0 = 9524101765$, and $m = 10$. Then by Algorithm 4.4.5 we have

$$
\begin{array}{lll}
9524101765^2 = 90708514430076115225 & \Longrightarrow & n_1 = 5144300761 \\
5144300761^2 = 26463830319625179121 & \Longrightarrow & n_2 = 8303196251 \\
8303196251^2 = 68943067982620455001 & \Longrightarrow & n_3 = 0679826204 \\
0679826204^2 = 462163667645049616 & \Longrightarrow & n_4 = 6366764504 \\
6366764504^2 = 40535690249394366016 & \Longrightarrow & n_5 = 6902493943 \\
6902493943^2 = 47644422633151687249 & \Longrightarrow & n_6 = 4226331516 \\
4226331516^2 = 17861878083134858256 & \Longrightarrow & n_7 = 8780831348 \\
8780831348^2 = 77102999162019497104 & \Longrightarrow & n_8 = 9991620194 \\
9991620194^2 = 99832474101148597636 & \Longrightarrow & n_9 = 4741011485 \\
4741011485^2 = 22477189900901905225 & \Longrightarrow & n_{10} = 1899009019.
\end{array}
$$

A serious problem with the middle-square method is that for many choices of the initial integer, the method produces the same small set of numbers over and over. For example, working with numbers that have four digits, staring from 4100, we obtain the sequence $8100, 6100, 2100, 4100, 8100, 6100, 2100, \cdots$. The most commonly used methods for generating random numbers are the methods based on the residue arithmetic.

(II) Linear Congruential Method

Residue arithmetic is helpful for generating a list of random numbers. At present, the most popular random number generators is use are special cases of the so-called *linear congruential method* (LCG for short), introduced by D. H. Lehmer in 1949.

In this method, we first choose four "magic" numbers as follows:

m:	the modulus;	$m > 0$
x_0:	the seed;	$0 \leq x_0 \leq m$
a:	the multiplier;	$0 \leq b \leq m$
b:	the increment;	$0 \leq b \leq m$

then the sequence of random numbers is defined recursively by:

$$x_{n+1} \equiv ax_n + b \pmod{m}, \quad n > 0. \tag{4.119}$$

So the algorithm for this method can be described as follows:

Algorithm 4.4.6 (Linear Congruential Generator) This algorithm will generate a sequence of random numbers $\{x_1, x_2, \cdots, x_n\}$.

[1] (Initialization) Input x_0, a, b, m and n. Set $j \leftarrow 1$.

[2] (Generation) Compute $x_j \leftarrow (ax_{j-1} + b) \pmod{m}$, and print the x_j.

[3] (Increase j) $j \leftarrow j + 1$. If $j \geq n$, then goto step [4], else goto step [2].

[4] (Exit) Terminate the algorithm.

Example 4.4.9 Let $x_0 = 5$, $a = 11$, $b = 73$, $m = 1399$ and $n = 10$. Then by Algorithm 4.4.6 we have:

$$
\begin{aligned}
x_0 &= 5 \\
x_1 &\equiv ax_0 + b \pmod{m} &&\Longrightarrow &&x_1 = 128 \\
x_2 &\equiv ax_1 + b \pmod{m} &&\Longrightarrow &&x_2 = 82 \\
x_3 &\equiv ax_2 + b \pmod{m} &&\Longrightarrow &&x_3 = 975 \\
x_4 &\equiv ax_3 + b \pmod{m} &&\Longrightarrow &&x_4 = 1005 \\
x_5 &\equiv ax_4 + b \pmod{m} &&\Longrightarrow &&x_5 = 1335 \\
x_6 &\equiv ax_5 + b \pmod{m} &&\Longrightarrow &&x_6 = 768 \\
x_7 &\equiv ax_6 + b \pmod{m} &&\Longrightarrow &&x_7 = 127 \\
x_8 &\equiv ax_7 + b \pmod{m} &&\Longrightarrow &&x_8 = 71 \\
x_9 &\equiv ax_8 + b \pmod{m} &&\Longrightarrow &&x_9 = 854 \\
x_{10} &\equiv ax_9 + b \pmod{m} &&\Longrightarrow &&x_{10} = 1073.
\end{aligned}
$$

To make LCG a good random number generator, it is necessary to find good values for the four magic numbers (i.e., parameters) that define the linear congruential sequence. Interested readers are suggested to consult [69] for a thorough discussion about the choice of the parameters. There are many congruential generators based on LCG, say, for example, the *power generator* defined by

$$x_{n+1} \equiv (x_n)^d \pmod{m} \tag{4.120}$$

where (d, m) are parameters describing the generator and x_0 the seed, and the *discrete exponential generator* defined by

$$x_{n+1} \equiv g^{x_n} \pmod{m} \tag{4.121}$$

where (g, m) are parameters describing the generator, and x_0 the seed; for a survey of various congruential generators, see [75] for more information.

(III) Stream-Cipher and Random Bit Generators

Unlike the RSA cryptosystem introduced earlier, a stream-cipher is a private-key cryptosystem [148], based on random bit generation (see Figure 4.7). The plaintext is encrypted on a bit-by-bit basis:

plaintext M	0	1	1	0	0	0	1	1	1	1	1	1	1	0	1	0	1	0	$\cdots$
Secret Key K	1	0	0	1	1	0	0	1	0	0	0	1	0	1	1	1	0	1	$\cdots$
Ciphertext C	1	1	1	1	1	0	1	0	1	1	1	0	1	1	0	1	1	1	$\cdots$

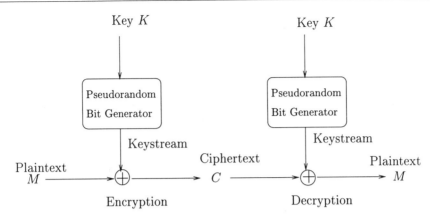

Figure 4.7: A Stream Cipher

The key is fed into the random bit generator to create a long sequence of binary signals. This "key-stream" is then mixed with the plaintext, usually by an Exclusive-OR (XOR bit-wise modulo-2 addition) gate to produce the ciphertext. The decryption is done by XORing with the same secret-key.

Ciphertext C	1	1	1	1	1	0	1	0	1	1	1	0	1	1	0	1	1	1	$\cdots$
Secret Key K	1	0	0	1	1	0	0	1	0	0	0	1	0	1	1	1	0	1	$\cdots$
plaintext M	0	1	1	0	0	0	1	1	1	1	1	1	1	0	1	0	1	0	$\cdots$

As a good random bit generator, it must be secure enough so that the output bit is hard to predict. Fortunately, there exit number-theoretic random bit generators whose security under next-bit tests can be proved to be equivalent to special cases of three apparently difficult number-theoretic problems: inverting RSA, factoring large integers, and taking discrete logarithms. We present in the following just three of the number-theoretic random bit generators (more examples can be found e.g., in [75]):

(i) RSA bit generator. Given $k \geq 2$ and $m \geq 1$, select odd primes p and q uniformly from the range $2^k \leq p, q < 2^{k+1}$ and form $m = pq$. Select e uniformly from $[1, m]$ subject to $\gcd(e, \phi(m)) = 1$. Set

$$x_{n+1} \equiv (x_n)^e \pmod{m} \qquad (4.122)$$

and let the bit z_{n+1} be given by

$$z_{n+1} \equiv x_{n+1} \pmod{2}. \qquad (4.123)$$

Then $\{z_n : 1 \leq n \leq k^m + m\}$ are the random bits generated by the seed x_0 of the length $2k$ bits.

(ii) Rabin bit generators. Given $k \geq 2$, and select odd primes p and q uniformly from primes in the range $2^k \leq p,\ q < 2^{k+1}$ and form $m = pq$, such that $p \equiv q \equiv 3 \pmod{4}$ (this assumption is used to guarantee that -1 is a quadratic nonresidue for both p and q). Let

$$x_{n+1} = \begin{cases} (x_n)^2 \pmod{m} & \text{if it lies in } [0, m/2), \\ m - (x_n)^2 \pmod{m} & \text{otherwise}, \end{cases} \tag{4.124}$$

so that $0 \leq x_{n+1} < m/2$, and the bit z_{n+1} is given by

$$z_{n+1} \equiv x_{n+1} \pmod{2}. \tag{4.125}$$

Then $\{z_n : 1 \leq n \leq k^m + m\}$ are the random bits generated by the seed x_0 of the length $2k$ bits.

(iii) Discrete exponential bit generator. Let $k \geq 2$ and $m \geq 1$, and select an odd primes p uniformly from primes in the range $[2^k,\ 2^{k+1}]$, provided with a complete factorization of $p - 1$ and a primitive root g. Set

$$x_{n+1} \equiv g^{x_n} \pmod{p} \tag{4.126}$$

and let the bit z_{n+1} be the most significant bit

$$z_{n+1} \equiv \left\lceil \frac{x_{n+1}}{2^k} \right\rceil \pmod{2}. \tag{4.127}$$

Then $\{z_n : 1 \leq n \leq k^m + m\}$ are the random bits generated by the seed x_0.

(iv) Elliptic curve bit generator. Elliptic curves, as we have already seen, have important applications in primality testing, integer factorization and cryptography. It is interesting to note that elliptic curves can also be used to generate random bits; the interested reader is referred to [65] for a more detailed discussion.

4.5 High-Speed Computation

Efficiency in machine computation involving large numbers is often limited by machine's memory (the amount of space available for storing data) and by the length of time required to perform computation. Even simple operations such as addition, subtraction, and multiplication of integers can overwhelm the memory or take an unreasonable amount of time if the integers are large and if millions of these operations must be performed in a given computation. Residue arithmetic can be used to increase the efficiency. The idea is to have several "small" pairwise relatively prime "moduli"

$$\langle m_1, m_2 \cdots m_k \rangle$$

and to work indirectly with "residues"

$$\langle x \bmod m_1, \ x \bmod m_2, \ \cdots, \ x \bmod m_k \rangle$$

rather than directly with the integer x. It is clearly easy to compute $\langle x_1, x_2, \cdots, x_k \rangle$ from the integer x just by

$$\langle x_1, \ x_2, \ \cdots, \ x_k \rangle = \langle x \bmod m_1, \ x \bmod m_2, \ \cdots, \ x \bmod m_k \rangle$$

It is also important to note that no information will be lost in this process, since we can recompute x from $\langle x_1, x_2, \cdots, x_k \rangle$. That is;

$$x \longleftrightarrow \langle x_1, x_2, \cdots, x_k \rangle.$$

This is the consequence of the Chinese Remainder Theorem; any positive integer x not exceeding $m_1 m_2 \cdots m_k$ can be represented uniquely by its least positive residues modulo m_i. In the next two subsections, we shall first introduce some basic concepts of residue arithmetic, then we shall discuss the fast computing processes in residue arithmetic. We will also indicate that it is possible to design fast parallel arithmetic units specifically suitable for residue arithmetic operations.

4.5.1 Residue Number Systems (RNS)

The way we do arithmetic on numbers is intimately related to the way we represent the numbers. There are essentially two different types of methods to represent numbers: *nonpositional* and *positional*. The roman numerals i, ii, iii, iv, v, vi, vii, viii, ix, x, xi, xii, xiii, $\cdots$, are a classical example of nonpositional number system; whereas the familiar *decimal* or *binary* number system is a good example of positional number system. The positional number system using base b (or *radix* b) is defined by the rule

$$(\cdots a_3 a_2 a_1 a_0 a_{-1} a_{-2} a_{-3} \cdots)_b = \cdots + a_3 b^3 + a_2 b^2 + a_1 b^1 + a_0 b^0 +$$
$$a_{-1} b^{-1} a_{-2} b^{-2} a_{-3} b^{-3} + \cdots \tag{4.128}$$

It is clear that when $b = 10$, it is the decimal system, whereas $b = 2$ represents the binary system. This type of positional number system is said to have a *fixed-base* or *fixed-radix*. A positional number system which is not fixed-base is said to be *mixed-base*. The number systems – residue number system we shall study in this section is a type of mixed-base systems.

First let us recall the Fundamental Theorem of Arithmetic that any positive integer $n \in \mathbb{N}_{>1}$ can be uniquely written as

$$n = p_1^{\alpha_1} p_2^{\alpha_2} \cdots p_k^{\alpha_k} = n_1 n_2 \cdots n_k \tag{4.129}$$

where $p_1, p_2, \cdots, p_k$ are distinct primes, $\alpha_1, \alpha_2, \cdots, \alpha_k$ are natural numbers, $n_i = p_i^{\alpha_i}$, $i = 1, 2, \cdots, k$, and $\gcd(n_i, n_j) = 1$ for $i \neq j$. The prime decomposition of n can be used to represent any number in $\mathbb{Z}/n\mathbb{Z}$ in terms of the numbers in $\mathbb{Z}/n_i\mathbb{Z}$, for $i = 1, 2, \cdots, k$.

Definition 4.5.1 Let x be any number in $\mathbb{Z}/n\mathbb{Z}$ and

$$\begin{cases} x \equiv a_1 \pmod{n_1} \\ x \equiv a_2 \pmod{n_2} \\ \cdots\cdots \\ \cdots\cdots \\ x \equiv a_k \pmod{n_k} \end{cases} \tag{4.130}$$

then the k-tuple

$$\langle a_1, a_2, \cdots, a_k \rangle = \langle x \bmod n_1, \ x \bmod n_2, \ \cdots, \ x \bmod n_k \rangle \tag{4.131}$$

is called the residue (or congruence, modular) representation of x. For simplicity, we often write the residue representation of x as follows:

$$x \longleftrightarrow \langle x \bmod n_1, \ x \bmod n_2, \ \cdots, \ x \bmod n_k \rangle \tag{4.132}$$

Example 4.5.1 Let $n_1 = 3$, $n_2 = 5$, $n_3 = 7$, then the residue representation of the integer 103 will be

$$\begin{cases} 103 \equiv 1 \ (\text{mod } 3) \\ 103 \equiv 3 \ (\text{mod } 5) \\ 103 \equiv 5 \ (\text{mod } 7) \end{cases}$$

That is;

$$103 \longleftrightarrow (1, 3, 5).$$

Note that the residue representation of an integer x *wrt* moduli $n_1, n_2, \cdots, n_k$ is unique. However, the inverse is not true.

Example 4.5.2 Let again $n_1 = 3$, $n_2 = 5$, $n_3 = 7$, then all the numbers in the form

$$105t + 103, \quad \text{for } t \in \mathbb{N}$$

will have the same residue representation $(1, 3, 5)$. That is,

$$105t + 103 \longleftrightarrow (1, 3, 5).$$

Definition 4.5.2 Let $(\mathbb{Z}/n\mathbb{Z})^*$ be the "direct-product" decomposition of $\mathbb{Z}/n\mathbb{Z}$. That is,

$$(\mathbb{Z}/n\mathbb{Z})^* = \mathbb{Z}/n_1\mathbb{Z} \times \mathbb{Z}/n_2\mathbb{Z} \times \cdots \times \mathbb{Z}/n_k\mathbb{Z} \tag{4.133}$$

where $n_i = p_i^{\alpha_i}$ for $i = 1, 2, \cdots, k$ is the prime decomposition of n.

Theorem 4.5.1 Let $m_1 > 0, m_2 > 0, \cdots, m_k > 0$, and $\gcd(m_i, m_j) = 1$ with $0 < i < j \leq k$. Then two integers x and x' have the same residue representation if and only if

$$x \equiv x' \ (\text{mod } M) \tag{4.134}$$

where $M = m_1 m_2 \cdots m_k$.

So if we restrict $0 \leq x < M = m_1 m_2 \cdots m_k$, then different integers x will have different residue representation moduli $m_1, m_2, \cdots, m_k$.

Theorem 4.5.2 Let $f : \mathbb{Z}/n\mathbb{Z} \to (\mathbb{Z}/n\mathbb{Z})^*$ be such that for any $x \in \mathbb{Z}/n\mathbb{Z}$, we have

$$\begin{aligned} f(x) &= \langle a_1, a_2, \cdots, a_k \rangle \\ &= \langle x \bmod n_1, \ x \bmod n_2, \ \cdots, \ x \bmod n_k \rangle \end{aligned} \tag{4.135}$$

then f is a bijection (one-to-one onto).

Remark 4.5.1 Theorem 4.5.1 is just another form of the Chinese Remainder Theorem.

Example 4.5.3 Let $m = 30$, so that $m_1 = 2$, $m_2 = 3$, $m_3 = 5$ with $(\mathbb{Z}/30\mathbb{Z})^* = \mathbb{Z}/2\mathbb{Z} \times \mathbb{Z}/3\mathbb{Z} \times \mathbb{Z}/5\mathbb{Z}$. Then the residue representations for integers in $\mathbb{Z}/30\mathbb{Z}$ will be:

$$
\begin{array}{ll}
0 \longleftrightarrow (0,0,0) & 1 \longleftrightarrow (1,1,1) \\
2 \longleftrightarrow (0,2,2) & 3 \longleftrightarrow (1,0,3) \\
4 \longleftrightarrow (0,1,4) & 5 \longleftrightarrow (1.2,0) \\
6 \longleftrightarrow (0,0,1) & 7 \longleftrightarrow (1,1,2) \\
8 \longleftrightarrow (0,2,3) & 9 \longleftrightarrow (1,0,4) \\
10 \longleftrightarrow (0,1,0) & 11 \longleftrightarrow (1,2,1) \\
12 \longleftrightarrow (0,0,2) & 13 \longleftrightarrow (1,1,3) \\
14 \longleftrightarrow (0,2,4) & 15 \longleftrightarrow (1,0,0) \\
16 \longleftrightarrow (0,1,1) & 17 \longleftrightarrow (1,2,2) \\
18 \longleftrightarrow (0,0,3) & 19 \longleftrightarrow (1,1,4) \\
20 \longleftrightarrow (0,2,0) & 21 \longleftrightarrow (1,0,1) \\
22 \longleftrightarrow (0,1,2) & 23 \longleftrightarrow (1,2,3) \\
24 \longleftrightarrow (0,0,4) & 25 \longleftrightarrow (1,1,0) \\
26 \longleftrightarrow (0,2,1) & 27 \longleftrightarrow (1,0,2) \\
28 \longleftrightarrow (0,1,3) & 29 \longleftrightarrow (1,2,4).
\end{array}
$$

Once the residue representation

$$\langle a_1, a_2, \cdots, a_k \rangle = \langle x \bmod n_1, \ x \bmod n_2, \ \cdots, \ x \bmod n_k \rangle$$

of an integer x is given, then we can uniquely solve x by using the Chinese Remainder Theorem (see the following example).

Example 4.5.4 Suppose we have the residue representations of x as follows:

$$\langle x \bmod 3, \ x \bmod 5, \ x \bmod 7 \rangle = (1,3,5),$$

That is, we have the following systems of congruences:

$$
\left\{
\begin{array}{l}
x \equiv 1 \ (\bmod \ 3), \\
x \equiv 3 \ (\bmod \ 5), \\
x \equiv 5 \ (\bmod \ 7).
\end{array}
\right.
$$

So, by using the Chinese Remainder Theorem, the above system of congruences can be solved as follows:

$$x = 103.$$

On most computers the word size is a large power of 2, with 2^{35} a common value. So to use residue arithmetic and the Chinese Remainder Theorem to do arithmetic, we will need the moduli less than e.g., 2^{35} that are pairwise relatively prime which multiply together to give a large integer.

4.5.2   Fast Computation in RNS

In this subsection, we shall discuss the fast arithmetic operations in residue number systems.

Let $n = n_1 n_2 \cdots n_k$ and

$$(\mathbb{Z}/n\mathbb{Z})^* = \mathbb{Z}/n_1\mathbb{Z} \times \mathbb{Z}/n_2\mathbb{Z} \times \cdots \times \mathbb{Z}/n_k\mathbb{Z}$$

We first provide a definition for arithmetic operations of addition $\oplus$, subtraction $\ominus$, and multiplication $\odot$ on $(\mathbb{Z}/n\mathbb{Z})^*$ in terms of the corresponding operations in $\mathbb{Z}/n_i\mathbb{Z}$, for $i = 1, 2, \cdots, k$.

Definition 4.5.3 For each $x = \langle a_1, a_2, \cdots, a_k \rangle$ and $y = \langle b_1, b_2, \cdots, b_k \rangle$ in $\mathbb{Z}/n\mathbb{Z}$, let $\oplus$, $\ominus$ and $\odot$ be binary operations defined on $(\mathbb{Z}/n\mathbb{Z})^*$. Then

$$
\begin{aligned}
x + y &= \langle a_1, a_2, \cdots, a_k \rangle \oplus \langle b_1, b_2, \cdots, b_k \rangle \\
&= f(x) \oplus f(y) \\
&= \langle (x \bmod n_1) +_{n_1} (y \bmod n_1), \\
&\qquad (x \bmod n_2) +_{n_2} (y \bmod n_2), \\
&\qquad \cdots\cdots \\
&\qquad \cdots\cdots \\
&\qquad (x \bmod n_k) +_{n_k} (y \bmod n_k) \rangle
\end{aligned}
\tag{4.136}
$$

$$
\begin{aligned}
x - y &= \langle a_1, a_2, \cdots, a_k \rangle \ominus \langle b_1, b_2, \cdots, b_k \rangle \\
&= f(x) \ominus f(y) \\
&= \langle (x \bmod n_1) -_{n_1} (y \bmod n_1), \\
&\qquad (x \bmod n_2) -_{n_2} (y \bmod n_2), \\
&\qquad \cdots\cdots \\
&\qquad \cdots\cdots \\
&\qquad (x \bmod n_k) -_{n_k} (y \bmod n_k) \rangle
\end{aligned}
\tag{4.137}
$$

$$\begin{aligned}
x \cdot y &= \langle a_1, a_2, \cdots, a_k \rangle \odot \langle b_1, b_2, \cdots, b_k \rangle \\
&= f(x) \odot f(y) \\
&= \langle (x \bmod n_1) \cdot_{n_1} (y \bmod n_1), \\
&\quad\; (x \bmod n_2) \cdot_{n_2} (y \bmod n_2), \\
&\quad\;\; \cdots\cdots \\
&\quad\;\; \cdots\cdots \\
&\quad\; (x \bmod n_k) \cdot_{n_k} (y \bmod n_k) \rangle
\end{aligned}$$
(4.138)

Theorem 4.5.3 The function $f : \mathbb{Z}/n\mathbb{Z} \to (\mathbb{Z}/n\mathbb{Z})^*$ that maps each x in $\mathbb{Z}/n\mathbb{Z}$ into its residue representation is an isomorphism of $\langle \mathbb{Z}/n\mathbb{Z}, +_n \rangle$ onto $\langle (\mathbb{Z}/n\mathbb{Z})^*, \oplus \rangle$, or of $\langle \mathbb{Z}/n\mathbb{Z}, -_n \rangle$ onto $\langle (\mathbb{Z}/n\mathbb{Z})^*, \ominus \rangle$, or of $\langle \mathbb{Z}/n\mathbb{Z}, \cdot_n \rangle$ onto $\langle (\mathbb{Z}/n\mathbb{Z})^*, \odot \rangle$.

We now give two examples of adding and multiplying two large integers in residue number systems, both are involved in the application of the Chinese Remainder Theorem.

Example 4.5.5 Compute $z = x + y = 123684 + 413456$ on a computer of word size 100.

Firstly we have

$$\begin{cases}
x \equiv 33 \;(\bmod\; 99), \\
x \equiv \;8 \;(\bmod\; 98), \\
x \equiv \;9 \;(\bmod\; 97), \\
x \equiv 89 \;(\bmod\; 95),
\end{cases} \qquad
\begin{cases}
y \equiv 32 \;(\bmod\; 99), \\
y \equiv 92 \;(\bmod\; 98), \\
y \equiv 42 \;(\bmod\; 97), \\
y \equiv 16 \;(\bmod\; 95),
\end{cases}$$
(4.139)

so that

$$\begin{cases}
z = x + y \equiv 65 \;(\bmod\; 99), \\
z = x + y \equiv \;2 \;(\bmod\; 98), \\
z = x + y \equiv 51 \;(\bmod\; 97), \\
z = x + y \equiv 10 \;(\bmod\; 95).
\end{cases}$$
(4.140)

Now we use the Chinese Remainder Theorem to find

$$x + y \bmod 99 \times 98 \times 97 \times 95.$$

Note that the solution to (4.140) is

$$z \equiv \sum_{i=1}^{4} M_i M_i' z_i \pmod{m}. \tag{4.141}$$

where

$$\begin{cases} m = m_1 m_2 \cdots m_k \\ M_i = m/m_i, \quad i = 1, 2, 3, 4 \\ M_i' M_i \equiv 1 \pmod{m_i}. \end{cases}$$

Now we have

$$M = 99 \times 98 \times 97 \times 95 = 89403930,$$

and

$$\begin{cases} M_1 = M/99 = 903070, \\ M_2 = M/98 = 912285, \\ M_3 = M/97 = 921690, \\ M_4 = M/95 = 941094. \end{cases}$$

We need to find the inverse M_i' for $i = 1, 2, 3, 4$. To do this, we solve the following four congruences

$$\begin{cases} 903070 M_1' \equiv 91 M_1' \equiv 1 \pmod{99}, \\ 912285 M_2' \equiv \ 3 M_2' \equiv 1 \pmod{98}, \\ 921690 M_3' \equiv 93 M_3' \equiv 1 \pmod{97}, \\ 941094 M_4' \equiv 24 M_4' \equiv 1 \pmod{95}. \end{cases}$$

We find that

$$\begin{cases} M_1' \equiv 37 \pmod{99}, \\ M_2' \equiv 38 \pmod{98}, \\ M_3' \equiv 24 \pmod{97}, \\ M_4' \equiv 4 \pmod{95}. \end{cases}$$

Hence, we get:

$$x + y \equiv \sum_{i=1}^{4} z_i M_i M_i' \pmod{m}$$

$$\equiv 65 \times 903070 \times 37 + 2 \times 912285 \times 33 + 51 \times 921690 \times 24$$
$$+10 \times 941094 \times 4 \pmod{89403930}$$
$$\equiv 3397886480 \pmod{89403930}$$
$$\equiv 537140 \pmod{89403930}$$

Since $0 < x + y = 537140 < 89403930$, we conclude that $x + y = 537140$ is the correct answer.

Example 4.5.6 Suppose now we want to multiply $x = 123684$ and $y = 413456$ on a computer of word size 100. We then have

$$\begin{cases} x \equiv 60 \ (\text{mod } 101), \\ x \equiv 33 \ (\text{mod } 99), \\ x \equiv \ 8 \ (\text{mod } 98), \\ x \equiv \ 9 \ (\text{mod } 97), \\ x \equiv 89 \ (\text{mod } 95), \\ x \equiv 63 \ (\text{mod } 89), \end{cases} \qquad \begin{cases} y \equiv 63 \ (\text{mod } 101), \\ y \equiv 32 \ (\text{mod } 99), \\ y \equiv 92 \ (\text{mod } 98), \\ y \equiv 42 \ (\text{mod } 97), \\ y \equiv 16 \ (\text{mod } 95), \\ y \equiv 51 \ (\text{mod } 89), \end{cases}$$

so that

$$\begin{cases} x \cdot y \equiv 43 \ (\text{mod } 101), \\ x \cdot y \equiv 66 \ (\text{mod } 99), \\ x \cdot y \equiv 50 \ (\text{mod } 98), \\ x \cdot y \equiv 87 \ (\text{mod } 97), \\ x \cdot y \equiv 94 \ (\text{mod } 95), \\ x \cdot y \equiv \ 9 \ (\text{mod } 89). \end{cases}$$

Now using the Chinese Remainder Theorem to solve the above system of congruences, we get

$$x \cdot y = 51137891904.$$

Since

$$0 < x \cdot y = 51137891904 < 803651926770 = n_1 n_2 n_3 n_4 n_5 n_6$$

we conclude that $x \cdot y = 51137891904$ is the correct answer.

To summarize the processes of residue arithmetic in $\mathbb{Z}/n\mathbb{Z}$, we have the following general rule to follow:

(i) **Convert integers to their residue representation**: Represent each integer, e.g., x and y as an element of multiplicative group $(\mathbb{Z}/n\mathbb{Z})^*$ where

$$(\mathbb{Z}/n\mathbb{Z})^* = \mathbb{Z}/n_1\mathbb{Z} \times \mathbb{Z}/n_2\mathbb{Z} \times \cdots \times \mathbb{Z}/n_k\mathbb{Z}$$

by taking the congruence class of x or y modulo each n_i; for example, the following is the residue representation of x and y modulo each n_i:

$$\langle x \equiv x_1 \ (\text{mod } n_1), \ x \equiv x_2 \ (\text{mod } n_2), \ \cdots, \ x \equiv x_k \ (\text{mod } n_k) \rangle.$$

$$\langle y \equiv y_1 \ (\text{mod } n_1), \ y \equiv y_2 \ (\text{mod } n_2), \ \cdots, \ y \equiv y_k \ (\text{mod } n_k) \rangle.$$

(ii) Perform the residue arithmetic for each $\mathbb{Z}/n_i\mathbb{Z}$: For example, let $\star$ denote one of the three binary operations $+$, $-$ and $\cdot$, then we need to perform the following operations in $\mathbb{Z}/n_i\mathbb{Z}$:

$$\langle x_1 \star y_1 \ (\text{mod } n_1), \ x_2 \star y_2 \ (\text{mod } n_2), \ \cdots, x_k \star y_k \ (\text{mod } n_k)\rangle.$$

(iii) Convert the residue representations back to integers: Use the Chinese Remainder Theorem to covert the computation results for each $\mathbb{Z}/n_i\mathbb{Z}$ into their integer form in $\mathbb{Z}/n\mathbb{Z}$

$$x \star y \equiv \sum_{i=1}^{k} M_i M_i' z_i \ (\text{mod } M). \tag{4.142}$$

where

$$\begin{cases} M = n_1 n_2 \cdots n_k \\ M_i = M/n_i, \quad i = 1, 2, \cdots, k \\ M_i' \equiv M_i^{-1} \ (\text{mod } n_i) \\ z_i \equiv x_i \star y_i \ (\text{mod } n_i). \end{cases}$$

It should be noted that the above computing processes can be implemented entirely in special computer hardware, which is the subject matter of the next subsection.

4.5.3   Residue Computers

The conventional "binary computers" have a serious problem that restricts the speed of performing arithmetic operations, caused by e.g., carry propagation and time delay. Fortunately, the residue number system (RNS) is not a fixed-base numbers systems, and all arithmetic operations (except division) in RNS are inherently carry-free; that is, each digit in the computing result is a function of only the corresponding digits of the operands. Consequently, addition, subtraction and multiplication can be performed in "residue computers" in less time than that would be in equivalent binary computers.

The construction of residue computers are much easier than that of binary computers; for example, to construct fast adders of a residue computer for

$$(\mathbb{Z}/n\mathbb{Z})^* = \mathbb{Z}/n_1\mathbb{Z} \times \mathbb{Z}/n_2\mathbb{Z} \times \cdots \times \mathbb{Z}/n_k\mathbb{Z}$$

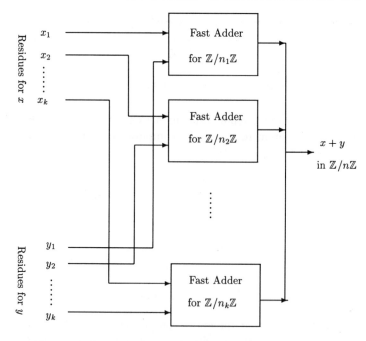

Figure 4.8: Fast Adders for Residue Arithmetic

it is sufficient to just construct some smaller adders for each $\mathbb{Z}/n_i\mathbb{Z}$, $(i = 1, 2, \cdots, k)$ (see Figure 4.8). More generally, we can construct residue computers performing fast additions, subtractions and multiplications as in Figure 4.9. Since n_i is substantially less than n, computations in each $\mathbb{Z}/n_i\mathbb{Z}$ will certainly be much easier than that in $\mathbb{Z}/n\mathbb{Z}$. More importantly, additions, subtractions and multiplications in each $\mathbb{Z}/n_i\mathbb{Z}$ are carry-free, so residue computers will be substantially faster than conventional binary computers inherently with carry propagation. The idea of decompositing a large computation in $\mathbb{Z}/n\mathbb{Z}$ to several smaller computations in $\mathbb{Z}/n_i\mathbb{Z}$ is exactly the idea of "divide-and-conquer" as used in algorithm design. Of course, the central idea of residue arithmetic and residue computers is the Chinese Remainder Theorem which enables us to combine separate results in each $\mathbb{Z}/n_i\mathbb{Z}$ to a final result in $\mathbb{Z}/n\mathbb{Z}$. So, if Euclid's algorithm is regarded as the first non-trivial algorithm, then the Chinese Remainder Theorem should be regarded as the first non-trivial divide-

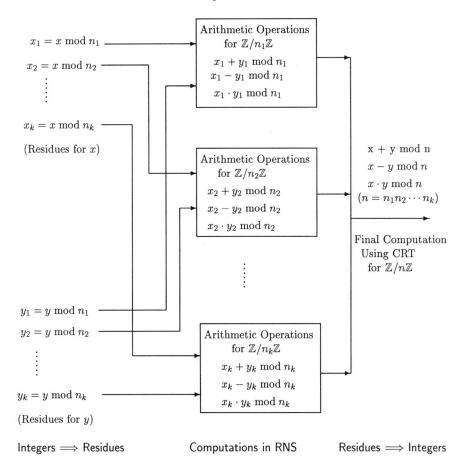

Figure 4.9: A Model of Residue Computers

and-conquer algorithm. It should be also noted that the Chinese Remainder Theorem is also an algorithm, not just a mathematics theorem.

Residue computers are a special type of high-speed computer, that has found many important applications in several central areas of computer science and electrical engineering, particularly in image and digital signal processing [74].

4.6   Three More Applications in Computing

Number theory has long been considered as the *purest* branch of mathematics, with very few applications to other areas. However, recent years has seen a great deal increase of interest in several central topics in number theory, precisely because of their importance in other areas. Today, number theory has been applied to such diverse areas as physics, biology, chemistry, computing, cryptography, digital communications, acoustics, graphics design, and even music. For example, modular arithmetic in number theory has been used in constructing perpetual calendars, scheduling round-robin tournaments, splicing telephone cables, devising systematic methods for storing computer files, constructing magic squares, and generating random numbers. More importantly, number theory is also an important source for the development of modern computer science; noting the title of Turing's seminal paper "On Computable Numbers" and Church's seminal paper "An Unsolved Problem of Elementary Number Theory". In this final section of the chapter, we shall briefly introduce three more applications of number-theoretic computations in computer science.

(I) Program Testing

The Galileo spacecraft is somewhere near Jupiter, but its main radio antenna is not working, so communication with it is slow [19]. Suppose we want to check that a critical program in Galileo's memory is correct. How can we do this without transmitting the whole program from/to Galileo? Here is a method for checking out the Galileo's program, based on some number-theoretic ideas:

Let P_g be the program in Galileo and P_e on Earth, each represented as an integer. Assuming P_e is correct, this algorithm will try to determine whether or not P_g is correct.

[1] Choose a prime number $10^9 < p < 2 \cdot 10^9$ and transmit p (p has no more than 32 bits) to Galileo to ask it to compute $r_g \leftarrow P_g \bmod p$ and send the remainder r_g back to Earth (r_g has no more than 32 bits).

[2] On Earth, we compute $r_e \leftarrow P_e \bmod p$, and check if $r_g = r_e$.

[3] If $r_g \neq r_e$, we conclude that $P_g \neq P_e$. That is, Galileo's program has been corrupted!

[4] If $r_g = r_e$, we conclude that P_g is *probably* correct. That is, if P_g is not correct, there is only a small probability of $< 10^{-9}$ that $r_g = r_e$. If this error probability is too large to accept for the quality-assurance team, just goto step [1] to start the process all over again, else terminate the algorithm by saying that P_g is "almost surely" correct! It is clear that if we repeat the process e.g., 10 times on 10 different random primes, then the error probability will be less than 10^{-90}, an extremely small number.

As we can see that the idea of the above method for program testing is exactly the same as that of the probabilistic method for primality testing discussed in Section 4.1.

(II) Minimal Perfect Hashing

The computation $r_e \leftarrow P_e \bmod p$ discussed in the previous section can also be considered as a hash function (see Equation (1.89) in Chapter 1), a common technique used to map words into a small set of integers which can be used as indices to address a table.

First let us recall that a hash function is a function

$$h: W \to I \qquad (4.143)$$

where $W = \{w_0, w_1, \cdots, w_{m-1}\}$ is a finite set of words w_j, each is a finite string of symbols over a finite alphabet Σ, and $I = \{0, 1, \cdots, k-1\}$ and k is a fixed integer (the table size). A hash function is *perfect* if there are no collisions; a collision occurs if two words w_1 and w_2 map to the same address, i.e, $h(w_1) = h(w_2)$. For a perfect hash function, we must have $k \geq m$. If $k = m$, then the hash function is called a *minimal perfect hash function*. A minimal perfect hash function is also called a minimal *collision-free* hash function. It can be shown that

Theorem 4.6.1 For a given finite set W (without loss the generality, we assume that W is a finite set of positive integers), there exist three constants C, D and E, such that the function h defined by

$$h(w) \equiv \lfloor C/(Dw + E) \rfloor \pmod{k - 1}, \quad |W| = k - 1 \qquad (4.144)$$

is a minimal perfect hash function.

The function is clearly a bijection from W onto the set $I = \{0, 1, \cdots, k-1\}$. The proof of this theorem can be done by using a generalization of the Chinese Remainder Theorem (CRT) for non-pairwise relatively prime moduli (see [63]); first note that for a given set $W = \{w_0, w_1, \cdots, w_{k-1}\}$ of positive integers there exist two integer constants D and E such that

$$Dw_0 + E, \ Dw_1 + E, \ \cdots, \ Dw_{k-1} + E$$

are pairwise relatively prime, then note that by CRT there exist an integer C such that

$$\begin{cases} C \equiv a_0 \ (\text{mod} \ (k-1)(Dw_0 + E)) \\ C \equiv a_1 \ (\text{mod} \ (k-1)(Dw_1 + E)) \\ \qquad \cdots \cdots \\ \qquad \cdots \cdots \\ C \equiv a_{k-1} \ (\text{mod} \ (k-1)(Dw_{k-1} + E)). \end{cases} \qquad (4.145)$$

(III) Checking Bits/Digits for Errors

Finally we discuss one more application of congruences in checking for errors in strings of bits or digits.

It is evident that manipulating and transmitting bit strings can introduce errors. A simple error detection method, called *parity check* works in the following way (suppose the bit string to be sent is $x_1 x_2 \cdots x_n$):

[1] [Preparation] Append the bit string with a *parity check bit* x_{n+1} by

$$x_{n+1} \equiv x_1 + x_2 + \cdots + x_n \ (\text{mod} \ 2), \qquad (4.146)$$

so that

$$\begin{cases} x_{n+1} = 0, & \text{if an even number of the first n bits in the string are 1} \\ x_{n+1} = 1, & \text{if an odd number of these bits are 1.} \end{cases}$$
$$\qquad (4.147)$$

The appended string $x_1 x_2 \cdots x_n x_{n+1}$ should satisfy the following congruence

$$x_1 + x_2 + \cdots + x_n + x_{n+1} \equiv 0 \ (\text{mod} \ 2). \qquad (4.148)$$

[2] [Error Detection] Suppose now we send the string $x = x_1 x_2 \cdots x_n x_{n+1}$ and the string $y = y_1 y_2 \cdots y_n y_{n+1}$ is received. If $x = y$, then there are no errors, but if $x \neq y$, there will be errors. We check whether or not

$$y_1 + y_2 + \cdots + y_n + y_{n+1} \equiv 0 \ (\text{mod} \ 2). \tag{4.149}$$

holds. If this congruence fails, at least one error is presented; but if it holds, errors may still exist. It can be shown that in general, we can detect an odd number of errors, but not an even number of errors.

Remark 4.6.1 The above method can be easily extended to checking for errors in strings of *digits*, rather than just bits. For example, suppose the identification number of a student card in a university is six digits, say, e.g., 972211. A checking digit may be chosen so that

$$\begin{aligned}
x_7 &\equiv 7x_1 + 3x_2 + x_3 + 7x_4 + 3x_5 + x_6 \ (\text{mod} \ 10) \\
&\equiv 7 \cdot 9 + 3 \cdot 7 + 1 \cdot 2 + 7 \cdot 2 + 3 \cdot 1 + 1 \cdot 1 \ (\text{mod} \ 10) \\
&\equiv 4 \ (\text{mod} \ 10).
\end{aligned}$$

So, the check digit is 5, and the seven-digit number 9722114 is printed on the student card. Note that we can always detect a single error in a student identification number appended with a check digit computed in this way.

The use of check digits with identification numbers for error detection is now a standard practice, notable examples include social security numbers, telephone numbers, serial numbers on currency predate computers, Universal Product Codes (UPC) on grocery items, and International Standard Book Numbers (ISBN) on published books. Interested readers are suggested to consult [43] for more information about applications (particularly in areas of business applications) of checking digits.

Further Reading

In this chapter, we have discussed some basic computational problems in number theory and their applications in computer science, particularly in public-key cryptography, information security and computer systems design. For those who desire a more detailed exposition in this interesting area, we recommend the following references for further reading; [10], [15], [27], [46], [69], [70], [74], [102], [117], [118], [122], [127], [142], and [147]. Richard Pinch [106] has written a very readable survey on mathematics for cryptography.

Chapter 5

New Models of Computation

The best is yet to come. We've only scratched the surface, computers can only do what we tell them now, but it will be different in the future.

– CLAUDE ELWOOD SHANNON (1916–)

We have studied in the previous chapters the computability and complexity, particularly the feasibility and infeasibility within the theoretical framework of Turing machines. It is widely believed, although not yet proved, that problems in $\mathcal{NP}$ are infeasible (or intractable), whereas problems in $\mathcal{P}$ are feasible (or tractable). For example, the integer factorization problem in number theory is believed to be intractable and the traveling salesman problem (TSP) in operations research are proved to be intractable on classical Turing machines. In this chapter, we shall study some new models of computation, namely quantum computation, and DNA molecular biological computation. In particular, we shall show that, although quantum and biological computations do not offer new computability-theoretic power over Turing machines, they have some more complexity-theoretic power than classical Turing machines. In fact, the well-known infeasible problems of integer factorization and discrete logarithms can be solved efficiently by quantum computers in polynomial time, and TSP can be solved in polynomial time by DNA molecular biological computers.

5.1 Quantum Computation

5.1.1 Physics and the New Computation

Computers can be viewed as physical objects and computations as physical processes. Quantum computers are machines that rely on characteristically quantum phenomena, such as quantum interference and quantum entanglement in order to perform computation. The classical theory of computation usually does not refer to physics but purely mathematical subjects. Until very recently, the fundamental connection between the laws of physics and computation was not properly studied (see [33] and [129]).

A *quantum computer* is a quantum analog of a digital computer, which can be described by a quantum Turing machine, the quantum counterpart of classical Turing machines. Just as the classical Turing machine, at each step of the computation of the quantum Turing machine, the computation is in the superposition of all configurations $|c_1\rangle, \cdots, |c_k\rangle$ corresponding to the nodes of the computation tree, each $|c_j\rangle$ having amplitude α_j. This superposition is denoted by

$$|\Psi\rangle = \sum_i \alpha_i |c_i\rangle, \tag{5.1}$$

where the amplitudes $\alpha_i \in \mathbb{C}$ such that $\sum_i ||\alpha_i||^2 = 1$, and each $|c_i\rangle$ is a *basis vector* of the *Hilbert space*. Suppose the computation is performed on a one-bit quantum computer, then the superposition will be

$$|\Psi\rangle = \alpha |0\rangle + \beta |1\rangle, \tag{5.2}$$

where $\alpha, \beta \in \mathbb{C}$ subject to that $||\alpha||^2 + ||\beta||^2 = 1$. The different possible states are $|0\rangle = \begin{pmatrix} 1 \\ 0 \end{pmatrix}$ and $|1\rangle = \begin{pmatrix} 0 \\ 1 \end{pmatrix}$. Let the unitary matrix M be

$$M = \frac{1}{\sqrt{2}} \begin{pmatrix} 1 & 1 \\ -1 & 1 \end{pmatrix} \tag{5.3}$$

Then the quantum operations on a bit can be defined by

$$M|0\rangle = \frac{1}{\sqrt{2}} \begin{pmatrix} 1 & 1 \\ -1 & 1 \end{pmatrix} \begin{pmatrix} 1 \\ 0 \end{pmatrix} = \frac{1}{\sqrt{2}}|0\rangle - \frac{1}{\sqrt{2}}|1\rangle$$

$$M|1\rangle = \frac{1}{\sqrt{2}} \begin{pmatrix} 1 & 1 \\ -1 & 1 \end{pmatrix} \begin{pmatrix} 0 \\ 1 \end{pmatrix} = \frac{1}{\sqrt{2}}|0\rangle + \frac{1}{\sqrt{2}}|1\rangle$$

$$M^2 \, | \, 0 \rangle = \begin{pmatrix} 0 & 1 \\ -1 & 0 \end{pmatrix} \begin{pmatrix} 1 \\ 0 \end{pmatrix} = 0 \, | \, 0 \rangle - 1 \, | \, 1 \rangle = -1 \, | \, 1 \rangle$$

$$M^2 \, | \, 0 \rangle = \begin{pmatrix} 0 & 1 \\ -1 & 0 \end{pmatrix} \begin{pmatrix} 0 \\ 1 \end{pmatrix} = 1 \, | \, 0 \rangle + 0 \, | \, 1 \rangle = | \, 0 \rangle$$

which is actually the quantum gate, similar to the classical logic gate:

$$| \, 0 \rangle \to \tfrac{1}{\sqrt{2}} \, | \, 0 \rangle - \tfrac{1}{\sqrt{2}} \, | \, 1 \rangle$$

$$| \, 1 \rangle \to \tfrac{1}{\sqrt{2}} \, | \, 0 \rangle + \tfrac{1}{\sqrt{2}} \, | \, 1 \rangle$$

Similarly, we can define the quantum gate for the two bits as follows:

$$| \, 00 \rangle \to | \, 00 \rangle$$

$$| \, 01 \rangle \to | \, 01 \rangle$$

$$| \, 10 \rangle \to \tfrac{1}{\sqrt{2}} \, | \, 10 \rangle + \tfrac{1}{\sqrt{2}} \, | \, 11 \rangle$$

$$| \, 11 \rangle \to \tfrac{1}{\sqrt{2}} \, | \, 10 \rangle - \tfrac{1}{\sqrt{2}} \, | \, 11 \rangle$$

or equivalently by giving the unitary matrix of the quantum operation:

$$M = \begin{pmatrix} 1 & 0 & 0 & 0 \\ 0 & 1 & 0 & 0 \\ 0 & 0 & \frac{1}{\sqrt{2}} & \frac{1}{\sqrt{2}} \\ 0 & 0 & \frac{1}{\sqrt{2}} & -\frac{1}{\sqrt{2}} \end{pmatrix} \tag{5.4}$$

This matrix is actually the counterpart of the truth table of Boolean logic used for digital computers. Suppose now the computation is in the superposition of the states

$$\frac{1}{\sqrt{2}} \, | \, 10 \rangle - \frac{1}{\sqrt{2}} \, | \, 11 \rangle \tag{5.5}$$

Using the unitary transformations defined above, we have

$$\tfrac{1}{\sqrt{2}} \, | \, 10 \rangle - \tfrac{1}{\sqrt{2}} \, | \, 11 \rangle = \tfrac{1}{\sqrt{2}} \left(\tfrac{1}{\sqrt{2}} \, | \, 10 \rangle + \tfrac{1}{\sqrt{2}} \, | \, 11 \rangle \right) - \tfrac{1}{\sqrt{2}} \left(\tfrac{1}{\sqrt{2}} \, | \, 10 \rangle - \tfrac{1}{\sqrt{2}} \, | \, 11 \rangle \right)$$

$$= \tfrac{1}{\sqrt{2}} \left(| \, 10 \rangle + | \, 11 \rangle \right) - \tfrac{1}{\sqrt{2}} \left(| \, 10 \rangle + | \, 11 \rangle \right)$$

$$= | \, 11 \rangle$$

5.1.2   Algorithms for Quantum Computation

We are now in a position to discuss quantum algorithms for solving some hard number-theoretic problems, namely integer factorization and discrete logarithms.

Suppose first we wish to factor an odd positive composite $n > 1$. The inner loop of many fast methods is the search for finding

$$y = f(x) \equiv x^2 \ (\text{mod} \ n) \tag{5.6}$$

so that n factors completely over a given factor base. To do this, we need only to find the order r of an element x in the multiplicative group $G = (\mathbb{Z}/n\mathbb{Z})^*$, denoted by order$(x, n)$. That is, the order r of x in group G modulo n is the smallest positive integer r such that $x^r \equiv 1 \ (\text{mod} \ n)$. Finding the order of an element $x \in G$ is therefore, in theory, not a problem: just keep multiplying until you get to "1", the identity element of the multiplicative group G. For example, let $n = 179359$ and $x = 3 \in G = (\mathbb{Z}/179359\mathbb{Z})^*$ such that $\gcd(3, 179359) = 1$. To find the order r of $x \in G$ such that $x^r \equiv 1 \ (\text{mod} \ n)$, we just keep multiplying until we get to "1":

$$
\begin{array}{lllll}
3^1 & \text{mod} & 179359 & = & 3 \\
3^2 & \text{mod} & 179359 & = & 9 \\
3^3 & \text{mod} & 179359 & = & 27 \\
& \cdots\cdots & & & \\
& \cdots\cdots & & & \\
3^{1000} & \text{mod} & 179359 & = & 31981 \\
3^{1001} & \text{mod} & 179359 & = & 95943 \\
3^{1002} & \text{mod} & 179359 & = & 108470 \\
& \cdots\cdots & & & \\
& \cdots\cdots & & & \\
3^{14716} & \text{mod} & 179359 & = & 99644 \\
3^{14717} & \text{mod} & 179359 & = & 119573 \\
3^{14718} & \text{mod} & 179359 & = & 1.
\end{array}
$$

Thus, the order r of 3 in the multiplicative group $(\mathbb{Z}/179359\mathbb{Z})^*$ is 14718, that is, order$(3, 179359) = 14718$. Once the order of an element in the multiplicative group modulo n is found, it is then trivial to factor n by just calculating

$$\{(\gcd(x^{r/2} + 1, n), \ (\gcd(x^{r/2} - 1, n)\}$$

which, as we have showed, can always be performed in polynomial time. For instance, for $x = 3$, $r = 14718$ and $n = 179359$, we have

$$\left\{ \gcd(3^{14718/2} + 1, 179359) = 67, \quad \gcd(3^{14718/2} - 1, 179359) = 2677 \right\},$$

and hence, the factorization of n

$$179359 = 67 \cdot 2677.$$

If one of the factors is not prime, then we can invoke the above process recursively until a complete prime factorization of n is obtained. Of course, we can choose other elements x in $(\mathbb{Z}/179359\mathbb{Z})^*$, rather than 3. For example, we can choose $x = 5$. In this case, we have $\operatorname{order}(5, 179359) = 29436$. Then we have

$$\left\{ \gcd(5^{29436/2} + 1, 179359) = 2677, \quad \gcd(5^{29436/2} - 1, 179359) = 67 \right\},$$

which also leads to the factorization of n: $179359 = 67 \cdot 2677$.

Exercise 5.1.1 First compute the order r of the element 13 in the multiplicative group modulo 35179 such that $13^r \equiv 1 \pmod{35179}$, and then find the prime factorization of 35179.

But in practice, however, the above computation for finding the order of $x \in (\mathbb{Z}/n\mathbb{Z})^*$ may not work, since for element x in a large group with n having more than 200 digits, the computation of r may require more than 10^{150} multiplications. Even if these multiplications could be carried out at the rate of 1000 billion per second, it would take approximately $3 \cdot 10^{80}$ years to arrive at the answer[1]. This is why integer factorization is difficult. Fortunately, Peter Shor[2] has discovered an efficient quantum mechanical way to find the order of an element $x \in (\mathbb{Z}/n\mathbb{Z})^*$ and hence the complete factorization of n. In what follows, we shall introduce Shor's algorithm for factoring large integers [129]):

Algorithm 5.1.1 (Quantum Algorithm for Integer Factorization) Given integers x and n, the algorithm will find the order of x, i.e., the smallest positive integer r such that $x^r \equiv 1 \pmod{n}$. It uses two quantum registers which hold integers in binary form.

---

[1]There is however a "quick" way to find the order of an element x in the multiplicative group G modulo n if the order $|G|$ (where $|G| = \#(\mathbb{Z}/n\mathbb{Z})^* = \phi(n)$) of G as well as the prime factorization of $|G|$ are known, since, by Lagrange's theorem, $r = \operatorname{order}(x, n)$ is a divisor of $|G|$. But finding $|G|$ (i.e., $\phi(n)$) is as hard as finding $r = \operatorname{order}(x, n)$ where $x \in G$. So this quick way is essentially useless in practice.

[2]Peter Shor is a distinguished researcher in theoretical computer science at AT&T Research Laboratories in Florham Park, New Jersey.

[1] Find a q, a power of 2, with $n^2 < q < 2n^2$.

[2] Put the first register in the uniform superposition of states representing numbers $a \pmod{q}$. This leaves the machine in the state $|\Psi_1\rangle$:

$$|\Psi_1\rangle = \frac{1}{\sqrt{q}} \sum_{a=0}^{q-1} |a\rangle |0\rangle. \tag{5.7}$$

What this step does is put each bit in the first register into the superposition

$$\frac{1}{\sqrt{2}} (|0\rangle + |1\rangle).$$

[3] Compute $x^a \pmod{n}$ in the second register. This leaves the machine in state $|\Psi_2\rangle$:

$$|\Psi_2\rangle = \frac{1}{\sqrt{q}} \sum_{a=0}^{q-1} |a\rangle |x^a \pmod{n}\rangle. \tag{5.8}$$

This step can be done reversibly since a is kept in the first register.

[4] Perform a Fourier transform A_q on the first register, mapping $|a\rangle$ to

$$\frac{1}{\sqrt{q}} \sum_{c=0}^{q-1} \exp(2\pi i a c/q) |c\rangle. \tag{5.9}$$

That is, we apply the unitary matrix with the (a, c) entry equal to $\frac{1}{\sqrt{q}} \exp(2\pi i a c/q)$. This leaves the machine in the state $|\Psi_3\rangle$:

$$|\Psi_3\rangle = \frac{1}{q} \sum_{a=0}^{q-1} \sum_{c=0}^{q-1} \exp(2\pi i a c/q) |c\rangle |x^a \pmod{n}\rangle. \tag{5.10}$$

[5] Finally observe the machine and measure both arguments of this superposition, obtaining the values of $|c\rangle$ in the first register and $|x^a \pmod{n}\rangle$ in the second register. Given the pure state $|\Psi_3\rangle$, the probability of different results for this measurement will be given by the probability distribution: $|c, x^k \pmod{n}\rangle$ (where we may assume $0 \le k < r$) is

$$\text{Prob}(c, x^k) = \left| \frac{1}{q} \sum_a \exp(2\pi i a c/q) \right|^2 \tag{5.11}$$

where a satisfies the condition that

$$x^a \equiv x^k \ (\text{mod} \ n).$$

Independent of k, $\text{Prob}(c, x^k)$ is periodic in c with period q/r, but since q is known, then we can deduce r with just a few trial executions, and hence, the factors of n.

Compared with the best known factoring algorithm NFS with asymptotic running time, as we already know, $\mathcal{O}\left(\exp\left(c(\log n)^{1/3}(\log\log n)^{2/3}\right)\right)$ for some constant c depending on detailed implementation, the quantum factoring algorithm takes asymptotically $\mathcal{O}\left((\log n)^2(\log\log n)(\log\log\log n)\right)$ steps on a quantum computer and $\mathcal{O}(\log n)$ amount of post-processing time on a classical computer that converts the output of the quantum computers. That is, the quantum method can factor integers in polynomial time $\mathcal{O}\left((\log n)^{2+\epsilon}\right)$.

As the finding of the order of x mod n is close to the computation of discrete logarithms: given a prime p, a generator g of the multiplicative group $\mathbb{Z}/p\mathbb{Z}$, and an x modulo p, find an r such that $g^r \equiv x \ (\text{mod} \ p)$, the quantum algorithm for factoring integers can be used, of course with some modifications, to compute discrete logarithms. The following is the algorithm:

Algorithm 5.1.2 (Quantum Algorithm for Computing Discrete Logarithms) Given $g, x \in \mathbb{N}$ and p prime. This algorithm will find the integer r such that $g^r \equiv x \ (\text{mod} \ p)$ if r exists. It will use three quantum registers.

[1] Find a q, a power of 2, with $p < q < 2p$.

[2] Put the first two registers of the quantum computer in the uniform superposition of all $|a\rangle$ and $|\ (\text{mod} \ p-1)\rangle$. Compute $g^a x^{-b} \ (\text{mod} \ p)$ in the third register. This leaves the machine in the state $|\Psi_1\rangle$:

$$|\Psi_1\rangle = \frac{1}{p-1}\sum_{a=0}^{p-2}\sum_{b=0}^{p-2}|a, b, g^a x^{-b} \ (\text{mod} \ p)\rangle. \tag{5.12}$$

[3] Use the Fourier transform A_q to map $|a\rangle \to |c\rangle$ and $|b\rangle \to |d\rangle$ with probability amplitude

$$\frac{1}{q}\exp\left(\frac{2\pi i}{q}(ac+bd)\right).$$

That is, we take the state $|a, b\rangle$ to the state $|\Psi_2\rangle$:

$$|\Psi_2\rangle = \frac{1}{q} \sum_{c=0}^{q-1} \sum_{d=0}^{q-1} \exp\left(\frac{2\pi i}{q}(ac + bd)\right) |c, d\rangle. \tag{5.13}$$

This leaves the machine in the state $|\Psi_3\rangle$:

$$|\Psi_3\rangle = \frac{1}{(p-1)q} \sum_{a,b=0}^{p-2} \sum_{c,d=0}^{q-1} \exp\left(\frac{2\pi i}{q}(ac + bd)\right) |c, d, g^a x^{-b} \pmod{p}\rangle. \tag{5.14}$$

[4] Finally observe the state of the quantum computer and extract the required information. It has been shown that the probability of observing a state $|c, d, y\rangle$ with $y \equiv g^k \pmod{p}$ is

$$\text{Prob}(c, d, g^k) = \left| \frac{1}{(p-1)q} \sum_{a,\ b} \exp\left(\frac{2\pi i}{q}(ac + bd)\right) \right|^2 \tag{5.15}$$

where a and b satisfy the condition that

$$a - rb \equiv k \pmod{p-1}.$$

Compared with Gordon's NFS algorithm for taking discrete logarithms with expected asymptotic time complexity $\mathcal{O}\left(\exp\left(c(\log p)^{1/3}(\log\log p)^{2/3}\right)\right)$, the above quantum discrete logarithm algorithm used only *two* modular exponentiations and *two* quantum Fourier transformations.

As many important computational problems have been proven to be $\mathcal{NP}$-complete. Thus, quantum computers will likely not become widely useful unless they can solve $\mathcal{NP}$-complete problems. In this next section, we shall study yet another non-classical model of computation, DNA biological computation, which claimed to be able to solve $\mathcal{NP}$-complete problems, such as directed Hamiltonian path problem (HPP).

5.2 Biological Computation

In November 1994, Leonard Adleman at the University of Southern California published a ground-breaking paper [3] in *Science*, Volume 266, entitled "Molecular Computation of Solutions to Combinatorial Problems". In this paper, he showed that molecules can be used to create a computer with a very high degree of parallelism. More specifically he showed that instances of the famous Hamiltonian Path Problem (HPP) can be solved by using biological experiments. Since HPP is known to be $\mathcal{NP}$-complete, it then follows that biology can be used to solve any problem in $\mathcal{NP}$. But, of course, it does not mean that all the instances of $\mathcal{NP}$ problem can be solved in a *feasible* way. Note also that this does not constitute a solution to the so-called "$\mathcal{P} \neq \mathcal{NP}$?" problem. Later on in December 1994, Richard Lipton at Princeton University [84] extended Adleman's result to that biology can be used to *directly* solve any problem in $\mathcal{NP}$. In this section, we shall first introduce some ideas and algorithms for using biology to solve some $\mathcal{NP}$-complete problems, then we shall show some methods for constructing molecular computers.

5.2.1 Biological Algorithms for HPP/SAT

DNA (deoxyribonucleic acid), found in every living creatures, is the storage medium for genetic information. It is composed of units called nucleotides, distinguished by the four chemical groups or bases attached to them:

(i) adenine (A)

(ii) cytosine (C)

(iii) guanine (G)

(iv) thymine (T)

These bases form an alphabet $\Sigma = \{A, C, G, T\}$ for biological information encoding, just the same as the alphabet $\Sigma = \{0, 1\}$ for the classical binary encoding. Single nucleotides are linked together end-to-end to form DNA strands which are just sequences $\alpha_1, \alpha_2, \cdots, \alpha_k$ over the alphabet $\Sigma = \{A, C, G, T\}$. Double DNA strands consist of two DNA sequences $\alpha_1, \alpha_2, \cdots, \alpha_k$ and $\beta_1, \beta_2, \cdots, \beta_k$ that satisfy the Watson-Crick complementary condition: for each $i = 1, 2, \cdots, k$, α_i and β_i must be complements, that is, $A \leftrightarrow T$ and $C \leftrightarrow G$. Complementary sequences anneal in an antiparallel fashion, where $5'$ and $3'$ refer to the chemically distinct ends of the DNA strands:

$$
\begin{array}{cccccccc}
5' & - & \alpha_1 & - & \alpha_2 & - & \alpha_3 & \cdots & - & 3' \\
 & - & \updownarrow & & \updownarrow & & \updownarrow & & \\
3' & - & \beta_1 & - & \beta_2 & - & \beta_3 & \cdots & - & 5'
\end{array}
\tag{5.16}
$$

Now suppose we wish to solve a small instance of HPP shown in Figure 5.1. The graph is small enough that we can easily find the unique Hamiltonian path

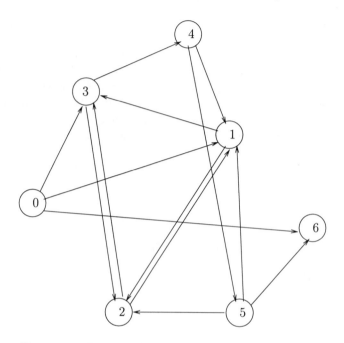

Figure 5.1: Directed Graph with $v_{in} = 0$ and $v_{out} = 6$

$0 \rightarrow 1$, $1 \rightarrow 2$, $2 \rightarrow 3$, $3 \rightarrow 4$, $4 \rightarrow 5$ and $5 \rightarrow 6$. For each vertex i in the graph, a random 20-mer oligonucleotide O_i is generated (shown are O_2, O_3, O_4 for vertices $2, 3, 4$). For each edge $i \rightarrow j$ in the graph, an oligonucleotide $O_{i \rightarrow j}$ is derived from the 3' 10-mer of O_i and from the 5' 10-mer of O_j (see Figure 5.2; shown are $O_{2 \rightarrow 3}$ for edge $2 \rightarrow 3$ and $O_{3 \rightarrow 4}$ for edge $3 \rightarrow 4$). For each vertex i in the graph, $\overline{O_i}$ is the Watson-Crick complement of O_i (see Figure 5.2; shown is $\overline{O_3}$ for complement O_3, all oligonucleotide are written 5' to 3', except $\overline{O_3}$). $\overline{O_3}$ serves a splint to bind $O_{2 \rightarrow 3}$ and $O_{3 \rightarrow 4}$ in preparation for ligation, and the result of ligation is the formation of DNA molecules encoding random paths through the graph. If there is much more DNA than possible paths, then at least one of the Hamiltonian paths, satisfying the following conditions, should be formed. So, we need to:

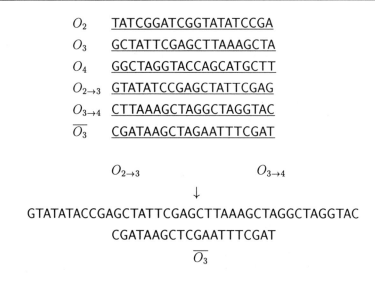

O_2 TATCGGATCGGTATATCCGA

O_3 GCTATTCGAGCTTAAAGCTA

O_4 GGCTAGGTACCAGCATGCTT

$O_{2\to3}$ GTATATCCGAGCTATTCGAG

$O_{3\to4}$ CTTAAAGCTAGGCTAGGTAC

$\overline{O_3}$ CGATAAGCTAGAATTTCGAT

$O_{2\to3}$ $O_{3\to4}$

$\downarrow$

GTATATACCGAGCTATTCGAGCTTAAAGCTAGGCTAGGTAC

CGATAAGCTCGAATTTCGAT

$\overline{O_3}$

Figure 5.2: Encoding a Graph in DNA

(i) keep only those paths that begin with vertex v_{in} and end with vertex v_{out},

(ii) keep only those paths that enter exactly n vertices if the graph has n vertices (i.e., $n = |V|$).

(iii) keep only those paths that enter all of the n distinct vertices at least once.

Finally, if any paths remain, output "yes", or otherwise, output "no". All of the above three steps can be tested chemically, i.e., they all can be done using standard methods of molecular biology. More generally, we have the following nondeterministic DNA-based biological algorithm for the Hamiltonian Path Problem:

Algorithm 5.2.1 (DNA-Based Biological Algorithm for Hamiltonian Path Problem)

[1] (Generate random paths through the directed graph) Represent each vertex and edge of the random graph by a short nucleotide sequence as shown Figure 5.2.

[2] (Keep only those paths that begins with v_{in} and ends with v_{out}) This can be done by using polymerase chain reaction (PCR) amplification. After this step, only those molecules encoding paths that begin with v_{in} and end with v_{out} were amplified.

[3] (Keep only those paths that contain n vertices, where $n = |V|$) Run the product of [2] on an "agarose gel" and extract the DNA with the right number of bases through molecular weight. The product of this step is thus PCR-amplified and gel-purified.

[4] (Keep only those paths that enter all of the n distinct vertices at least once) The product of [3] is further affinity-purified by a series of steps known as "biotin-avidin magnetic beads", so that only those DNA that actually pass through each of the n vertices of the graph will be remained.

[5] (Output "yes" if any paths remain, or otherwise, output "no") Analysing the remaining DNA by using the "graduated PCR". This will give the desired output.

The above algorithm is essentially Adleman's, which solves the HPP in a totally brute force way: simply try all the possible paths of the given vertices in the graph. Lipton [84] later on extended Adleman's result to show that we can build a DNA-based biological computer that can solve any $\mathcal{NP}$ problem such as SAT directly.

5.2.2   Formal Models of DNA Biological Computation

The fundamental idea of DNA-based biological computation is that of a set of DNA strands. Since the set of DNA strands is usually kept in a test tube, the test tube is just a collection of pieces of DNA. In this section, we shall study some formal DNA-based biological computation models in terms of the test tube based on [4].

Definition 5.2.1 A *test tube* (or just tube for short) is a set of molecules of DNA (i.e., a multi-set of finite strings over the alphabet $\Sigma = \{A, C, G, T\}$). Given a tube, one can perform the following four elementary biological operations:

(i) **Separate** or **Extract**: Given a tube T and a string of symbols $S \in \Sigma$, produce two tubes $+(T, S)$ and $-(T, S)$, where $+(T, S)$ is all the molecules of DNA in T which contain the consecutive subsequence S and $-(T, S)$ is all of the molecules of DNA in T which do not contain the consecutive sequence S.

(ii) **Merge**: Given tubes T_1, T_2, produce the multi-set union $\cup(T_1, T_2)$:

$$\cup(T_1, T_2) = T_1 \cup T_2 \qquad (5.17)$$

(iii) **Detect**: Given a tube T, output "yes" if T contains at least one DNA molecule (sequence) and output "no" if it contains none.

(iv) **Amplify**: Given a tube T produce two tubes $T'(T)$ and $T''(T)$ such that

$$T = T'(T) = T''(T). \qquad (5.18)$$

Thus, we can replicate all the DNA molecules from the test tube.

These operations are then used to write "programs" which receives a tube as input and returns either "yes" or "no" or a set of tubes.

Example 5.2.1 Consider the following program:

(i) Input(T)

(ii) $T_1 = -(T, C)$

(iii) $T_2 = -(T_1, G)$

(iv) $T_3 = -(T_2, T)$

(v) Output(Detect(T_3))

The model defined above is an unrestricted one. We now present a restricted biological computation model:

Definition 5.2.2 A tube is a multi-set of aggregates over an alphabet Σ which is not necessarily $\{A, C, G, T\}$. (An aggregate is a subset of symbols over Σ). Given a tube, there are three operations:

(i) **Separate**: Given a tube T and a symbol $s \in \Sigma$, produce two tubes $+(T, s)$ and $-(T, s)$ where $+(T, s)$ is all the aggregates of T which contains the symbols s and $-(T, s)$ is all of the aggregates of T which do not contain the symbol s.

(ii) **Merge**: Given tube T_1, T_2, produce

$$\cup(T_1, T_2) = T_1 \cup T_2 \tag{5.19}$$

(iii) **Detect**: Given a tube T, output "yes" if T contains at least one aggregate, or output "no" if it contains none.

Example 5.2.2 (3-colourability problem) Given an n vertex graph G with edges $e_1, e_2, \cdots, e_z$, let

$$\Sigma = \{r_1, b_1, g_1, r_2, b_2, g_2, \cdots, r_n, b_n, g_n\}.$$

and consider the following restricted program on input

$$\begin{aligned} T = \{\alpha \mid \ & \alpha \subseteq \Sigma, \\ & \alpha = \{c_1, c_2, \cdots, c_n\}, \\ & [c_i = r_i \text{ or } c_i = b_i \text{ or } c_i = g_i], i = 1, 2, \cdots, n\} \end{aligned}$$

(i) Input(T).

(ii) for $k = 1$ to z. Let $e_k = \langle i, j \rangle$:

    (a) $T_{\text{red}} = +(T, r_i)$ and $T_{\text{blue or green}} = -(T, r_i)$.

    (b) $T_{\text{blue}} = +(T_{\text{blue or green}}, b_i)$ and $T_{\text{green}} = -(T_{\text{blue or green}}, b_i)$.

    (c) $T_{\text{red}}^{\text{good}} = -(T_{\text{red}}, r_j)$.

    (d) $T_{\text{blue}}^{\text{good}} = -(T_{\text{blue}}, b_j)$.

    (e) $T_{\text{green}}^{\text{good}} = -(T_{\text{green}}, g_j)$.

    (f) $T' = \cup(T_{\text{red}}^{\text{good}}, T_{\text{blue}}^{\text{good}})$.

    (g) $T = \cup(T_{\text{green}}^{\text{good}}, T')$.

(iii) Output(Detect(T)).

Theorem 5.2.1 (Lipton, 1994) Any SAT problem in n variables and m clauses can be solved with at most $\mathcal{O}(m+1)$ separations, $\mathcal{O}(m)$ merges and one detection.

The above theorem implies that biological computation can be used to solve all problems in $\mathcal{NP}$, although it does not mean all instances of $\mathcal{NP}$ can be solved in a feasible way.

5.3 Comparison of Quantum and DNA Biological Models

In this final chapter of the book, we have briefly introduced some basic ideas of the two non-classical computation models, namely the *quantum-mechanical* computation model and the DNA *biological* computation model. As a common goal, both models are designed to be used for solving intractable computation problems (see Figure 5.3). For example, quantum models are used to solve problems such as integer factorization and discrete logarithms, that are believed to be intractable in classical computation models, whereas biological models are used to solve problems such as directed Hamiltonian path problem (HPP) and Boolean formula satisfaction problem (SAT), which are proved to be $\mathcal{NP}$-complete.

Note that from a computability point of view, neither the quantum computation model nor the biological computation model has more computational power than the Turing machine. Thus we have an analogue of Church-Turing Thesis for quantum and biological computations:

> **Quantum and Biological Computation Thesis**: A number-theoretic function is computable or a decision problem is decidable by a quantum computer or by a biological computer if and only if it is computable or decidable by a Turing machine.

But from a complexity point of view, both the quantum computation model and the biological computation model do have some more computational power than the Turing machine. More specifically, we have the following complexity results about quantum and biological computations:

(i) Integer factorization and discrete logarithm problems are believed to be intractable in Turing machines; no efficient algorithms have been found for these two classical, number-theoretic problems, in fact, the best algorithms for these two problems have the worst-case complexity $\Theta\left((\log n)^2 (\log \log n)(\log \log \log n)\right)$. But however, both of these two problems can be solved in polynomial time by quantum computers [129].

(ii) The famous Boolean formula satisfaction problem (SAT) and directed Hamiltonian path problem (HPP) are proved to be $\mathcal{NP}$-complete, but these problems, and in fact any other $\mathcal{NP}$-complete problems can be solved in polynomial biological steps by biological computers (see [3] and [84]).

Undecidable Problems (or Noncomputable Functions)
e.g., The Turing Machine Halting Problem,
The Hilbert's Tenth Problem and the Busy Beaver Function

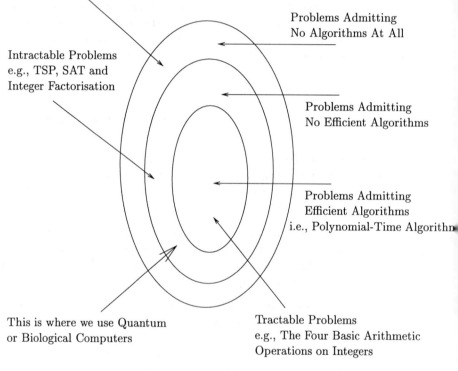

Intractable Problems
e.g., TSP, SAT and
Integer Factorisation

Problems Admitting
No Algorithms At All

Problems Admitting
No Efficient Algorithms

Problems Admitting
Efficient Algorithms
i.e., Polynomial-Time Algorithm

This is where we use Quantum
or Biological Computers

Tractable Problems
e.g., The Four Basic Arithmetic
Operations on Integers

Figure 5.3: The Usefulness of Quantum/Biological Computation

Although quantum and biological computations seem to be very promising, we at present do not know if we can actually build a quantum computer or a biological computer. For biological computers, the key issue is errors: the biological operations are not perfect, whereas the key issue in building a quantum computer is how to implement the quantum gates electronically and mechanically. Nevertheless, we hope that the new-coming technologies will overcome these difficulties, and a truly quantum or biological computer will be eventually built in the near future!

5.4   Comparison of Connectionist and DNA Biological Models

Recall from Chapter 2 that a connectionist machine is a 4-tuple algebraic system defined by

$$M = (n, \ N, \ B, \ A) \tag{5.20}$$

where

(i) $n \in N$ defines the size of the model,

(ii) $N = \{C_1, C_2, \cdots, C_n\}$ is a set of n formal neurons,

(iii) $B = (b_1, b_2, \cdots, b_n)$ is a real vector of dimension n, which serves as the *threshold* vector of the model.

(iv) $A = (a_{ij})$, $i, j, = 1, 2, \cdots, n$, is a matrix of real numbers whose entry a_{ij} represents the connection weight, with which C_i affects C_j.

One thing we would like to pint out is that connectionist models are suitable computation models for any computable functions discussed in Chapter 3. Given the state values $v(i, t)$, $i = 1, 2, 3, \cdots, n$, for some t, then $v(j, \ t + 1)$ is the next state value determined by

$$v(j, \ t + 1) = \begin{cases} 1, & \text{if } \sum_{i=1}^{n} a_{ij} v(i, t) > b_j, \\ 0, & \text{otherwise.} \end{cases} \tag{5.21}$$

Given a binary *input* w of length m with $m \leq n$

$$w = c_1 c_2 \cdots c_m, \tag{5.22}$$

then

$$v(i, t(w)), \quad i = 1, 2, \cdots, n \tag{5.23}$$

is called the *output*, where $t(w)$ is the computation time for input w, satisfying

$$v(i, t) = v(i, t + 1), \quad i = 1, 2, \cdots, n, \tag{5.24}$$

with t being the smallest value. The time complexity for this model can thus be defined by

$$t(n) = \max\{t(w) : \ |w| \leq n\}. \tag{5.25}$$

From a computability point of view, connectionist models have the same computation power as DNA molecular biological models or any other computation models [60], and the Church-Turing Thesis can extend to neural computation. Thus we have:

Connectionist/Neural Computation Thesis: A number-theoretic function is computable or a decision problem is decidable by a connectionist model if and only if it is computable or decidable by a Turing machine. Moreover, for any connectionist computation machine with inter-neuron defined by an $n \times n$ real matrix, there is a family of Boolean gates of size $\mathcal{O}(n^3 \log n)$ simulating the connectionist model with a time slow-down fact of only $\mathcal{O}(\log n)$.

The above thesis actually says that connectionist models do not possess any basic properties that are different from those of Turing machines or other parallel computation models. It should be noted that several authors (see e.g., [45] and [132]) do claim that connectionist models have more computation power than Turing machines, but their arguments seem not to be so convincible; see Shor's comments [130] on [132] for more information.

Note also that connectionist computation models are a kind of *brain-like* biological computation models. Although both DNA molecular biological models and connectionist models are *biological* in nature, they differ significantly in at least two aspects:

(i) DNA molecular biological models aim to solve $\mathcal{NP}$-complete problems (e.g., HPP and TSP) as discussed in Chapter 3,

(ii) connectionist models aim to achieve the learning/training abilities as the human brain.

The connectionist computation model was first studied by McCulloch and Pitts [90] in 1943. Since then it has attracted much attention from scientists and engineers in many areas such as physics, electronics, biology, computer science and mathematics, etc. In spite of their learning/training ability, connectionist machines have a serious problem with their computational complexity, i.e., the learning/training speed; say, for example, training even a very simple 2-layer, 3-node, n-input neural network is $\mathcal{NP}$-complete [14]. So at least from a complexity point of view, connectionist models are not very much useful in practice. How to speed-up the learning/training speed for connectionist machines, or how to design efficient (i.e., polynomial training time) neural networks would certainly be one of the most important research topics in the area.

Further Reading

Quantum and DNA-based biological computations are relatively new areas in computer science. Although few textbooks can be found on the market, many journal/conference papers and reports are available. We recommend the following papers for further reading: [3], [66], [83], and [84] for DNA-based biological computation, and [5], [12], [13], [16], [33], [40], [129], [131], and [134] for quantum computation. On the other hand, there are many well-written textbooks on neural computation on the market, we just randomly list two of them: [55] and [119].

Bibliography

1. L. M. Adleman, "Subexponential Algorithmic for the Discrete Logarithms Problem with Applications to Cryptography", *Proceedings of the 20th Annual IEEE Symposium on Foundations of Computer Science*, IEEE Press, 1979, pp 55–60.

2. L. M. Adleman, "Algorithmic Number Theory – The Complexity Contribution", *Proceedings of the 35th Annual IEEE Symposium on Foundations of Computer Science*, IEEE Press, 1994, pp 88–113.

3. L. M. Adleman, "Molecular Computation of Solutions to Combinatorial Problems", *Science*, **266**, 11 November 1994, pp 1021–1024.

4. L. M. Adleman, "On Constructing a Molecular Computer", In: *DNA Based Computers*, R. Lipton and E. Baum, editors, AMS, 1996, pp 1–21.

5. L. M. Adleman, J. DeMarrais and M. D. A. Huang, "Quantum Computability", *SIAM Journal on Computing*, **26**, 5(1997), pp 1524–1540.

6. L. M. Adleman, C. Pomerance, and R. S. Rumely, "On Distinguishing Prime Numbers from Composite Numbers", *Annals of Mathematics*, **117**, (1983), pp 173–206.

7. L. M. Adleman and M. D. A. Huang, *Primality Testing and Abelian Varieties over Finite Fields*, Lecture Notes in Mathematics 1512, Springer-Verlag, 1992.

8. A. V. Aho, J. E. Hopcroft and J. D. Ullman, *The Design and Analysis of Computer Algorithms*, Addison-Wesley, 1974.

9. A. O. L. Atkin and F. Morain, *Elliptic Curves and Primality Proving*, Mathematics of Computation, **61**, (1993), pp 29–68.

10. E. Bach and J. Shallit, *Algorithmic Number Theory*, Vol I: Efficient Algorithms, MIT Press, 1996.

11. R. J. Baillie and S. S. Wagstaff, Jr., "Lucas Pseudoprimes", *Mathematics of Computation*, **35**, (1980), pp 1391–1417.

12. C. H. Bennett, E. Bernstein, G. Brassard, and U. Vazirani, "Strengths and Weaknesses of Quantum Computing", *SIAM Journal on Computing*, **26**, 5(1997), pp 1510–1523.

13. E. Bernstein and U. Vazirani, "Quantum Complexity Theory", *SIAM Journal on Computing*, **26**, 5(1997), pp 1411–1473.

14. A. L. Blum and R. L. Rivest, "Training a 3-Node Neural Network is NP-Complete", *Neural Networks*, **5**, 1992, pp 117–127.

15. G. Brassard, *Modern Cryptology: A Tutorial*, Lecture Notes in Computer Science 325, Springer-Verlag, 1988.

16. G. Brassard, "A Quantum Jump in Computer Science", in: *Computer Science Today: Recent Trends and Development*, Lecture Notes in Computer Science 1000, Springer-Verlag, 1995, pp 1–14.

17. R. P. Brent, "Some Integer Factorization Algorithms using Elliptic Curves", *Australian Computer Science Communications*, **8**, 1986, pp 149–163.

18. R. P. Brent, "Primality Testing and Integer Factorization", *Proceedings of Australian Academy of Science Annual General Meeting Symposium on the Role of Mathematics in Science*, Canberra, 1991, pp 14–26.

19. R. P. Brent, "Uses of Randomness in Computation", Computer Sciences Laboratory, Australian National University, Canberra, 1994.

20. E. F. Brickell, D. M. Gordon and K. S. McCurley, "Fast Exponentiation with Precomputation" (Extended Abstract), *Proceedings of EUROCRYPT '92*, Lecture Notes in Computer Science 658, Springer-Verlag, 1992, pp 200–207.

21. J. G. Brookshear, *Theory of Computation: Formal Languages, Automata, and Complexity*, The Benjamin/Cummings, 1989.

22. A. W. Burks, *Essays on Cellular Automata*, University of Illinois Press, Urbana, 1970.

23. L. Childs, *A Concrete Introduction to Higher Algebra*, Undergraduate Texts in Mathematics, Springer-Verlag, 1979.

24. G. Cattaneo and C. Q. Vogliotti, "The Magic Rule Spaces of Neural-like Elementary Automata", *Theoretical Computer Science*, **178**, (1997), pp 77–102.

25. A. Church, "An Unsolved Problem of Elementary Number Theory", *The American Journal of Mathematics*, **58**, (1936), pp 345–363.

26. D. Cohen, *Introduction to Computer Theory*, 2nd Edition, John Wiley and Sons, 1995.

27. H. Cohen, *A Course in Computational Algebraic Number Theory*, Springer-Verlag, 1993.

28. A. Colmerauer, H. Kanoui, P. Roussel and R. Pasero, "Un Systeme de Communication Homme-Machine en Francais", Groupe de Recherche en Intelligence Artificielle, Université d'Aix-Marseille, 1973.

29. M. Davis (editor), *The Undecidable*, Raven Press, Hewlett and New York, 1965. (Collection of Papers by K. Godel, A. Church, A. Turing, J. Rosser, S. Kleene, and E. Post).

30. M. Davis, *Computability and Undecidability*, Dover Publications, New York, 1982.

31. M. D. Davis and E. J. Weyuker, *Computability, Complexity and Languages*, Academic Press, 1983.

32. P. J. Denning, J. K. Dennis and J. E. Qualitz, *Machines, Languages, and Computation*, Prentice-Hall, 1978.

33. D. Deutsch, "Quantum Theory, the Church-Turing Principle and the Universal Quantum Computers", Proceedings of the Royal Society of London, Series **A**, **400**, 1985, pp 96–117.

34. L. E. Dickson, *History of the Theory of Numbers*, Vol. I, Divisibility and Primality, G. E. Stechert & Co., New York, 1934.

35. W. Diffie and E. Hellman, "New directions in Cryptography", *IEEE Transactions on Information Theory*, **22**, 5(1976), pp 644–654.

36. P. F. Dierker and W. L. Voxman, *Discrete Mathematics*, Harcourt Brace Jovanovich Publishers, San Diego, 1986.

37. E. R. Dougherty and C. R. Giardina, *Mathematical Methods for Artificial Intelligence and Autonomous Systems*, Prentice-Hall, 1988.

38. S. S. Epp, *Discrete Mathematics with Applications*. PWS Publishing Company, Boston, 1995.

39. P. A. Fejer and D. A. Simovici, *Mathematical Foundations of Computer Science*, Springer-Verlag, 1991.

40. R. P. Feynman, *Feynman Lectures on Computation*, edited by A. J. G. Hey and R. W. Allen, Addison-Wesley, 1996.

41. W. Freeman and T. Jackson, The University of York, UK, *private communication*, 1997.

42. R. W. Floyd and R. Beigel, *The Languages of Machines*, Computer Science Press, 1994.

43. J. A. Gallian, "Error Detection Methods", *ACM Computing Surveys*, **28**, 3(1996), pp 503–517.

44. M. R. Garey and D. S. Johnson, *Computers and Intractability – A Guide to the Theory of NP-Completeness*, W. H. Freeman and Company, 1979.

45. M. Garzon, *Models of Massive Parallelism – Analysis of Cellular Automata and Neural Networks*, Springer-Verlag, 1995.

46. P. Giblin, *Primes and Programming – An Introduction to Number Theory with Computing*, Cambridge University Press, 1993.

47. S. Goldwasser and J. Killian, "Almost all Primes can be Quickly Certified", in: *Proceedings of the 18th ACM Symposium on Theory of Computing*, 1986, Berkeley, pp 316–329.

48. E. M. Gold, "Language Identification in the Limit", *Information and Control*, **10**, (1967), pp 447–474.

49. D. M. Gordon and K. S. McCurley, "Massively Parallel Computation of Discrete Logarithms", *Proceedings of Crypto '92*, Lecture Notes in Computer Science 740, Springer-Verlag, 1992, pp 312–323.

50. D. M. Gordon, "Discrete Logarithms in GF(P) using the Number Field Sieve", *SIAM Journal on Discrete Mathematics*, **6**, 1(1993), pp 124–138.

51. E. Gurari, *An Introduction to the Theory of Computation*, Computer Science Press, 1989.

52. G. H. Hardy and E. M. Wright, *An Introduction to Theory of Numbers*, Clarendon Press, Oxford, 5th Edition, 1979.

53. D. Harel, *Algorithmics: The Spirit of Computing*, Addison-Wesley, 1987.

54. M. A. Harrison, *Introduction to Formal Language Theory*, Addison-Wesley, 1978.

55. S. Haykin, *Neural Networks – A Comprehensive Foundation*, Macmillan College Publishing, 1994.

56. J. L Hein, *Discrete Mathematics*, Jones and Bartlett Publishers, Sudbury, MA, and London, 1996.

57. F. Hennie, *Introduction to Computability*, Addison-Wesley, 1977.

58. R. Herken, *The Universal Turing Machine – A half-Century Survey*, Oxford University Press, 1988.

59. J. W. Hong, *Computation: Computability, Similarity and Duality*, Longman, London, 1986.

60. J. W. Hong, "On Connectionist Models", *Communications of Pure and Applied Mathematics*, **XLI**, (1988), pp 1039–1050.

61. J. E. Hopcroft and J. D. Ullman, *Introduction to Automata Theory, Languages and Computation*, Addison-Wesley, Reading, MA, 1979.

62. R. M. Huizing, *An Implementation of the Number Field Sieve*, Note NM-R9511, Department of Numerical Mathematics, Centre for Mathematics and Computer Science, Amsterdam, 1995.

63. G. Jaeschke, "Reciprocal Hashing: A Method for Generating Minimal Perfect Hashing Functions", *Communications of the ACM*, **24**, 12(1981), pp 829–833.

64. D. S. Johnson, "A Catalog of Complexity Classes", In: Chapter 2 of [76].

65. B. S. Kaliski, "A Pseudo-Random Bit Generator Based on Elliptic Curve Logarithms", *Advances in Cryptography – CRYPTO '86 Proceedings*, Lecture Note in Computer Science 263, Springer-Verlag, 1986, pp 84–103.

66. L. Kari, "DNA computing: Arrival of Biological Mathematics", *The Mathematical Intelligencer*, **19**, 2(1997), pp 9–22.

67. K. M. Karp and V. Ramachandran, "Parallel Algorithms for Shared-Memory Machines", In: Chapter 17 of [76].

68. D. Kelley, *Automata and Formal Languages*, Prentice-Hall, 1995.

69. D. E. Knuth, *The Art of Computer Programming*, Vol II, Seminumerical Algorithms, Second Edition, Addison-Wesley, 1981.

70. N. Koblitz, *A Course in Number Theory and Cryptography*, 2nd Edition, Springer-Verlag, 1994.

71. S. Konyagin and C. Pomerance, "On Primes Recognizable in Deterministic Polynomial Time", In: *The Mathematics of Paul Erdős, I – Algorithms and Combinatorics 13*, Edited by R. L. Graham and J. Nesetril, Springer-Verlag, 1997, pp 176–198.

72. R. A. Kowalski, "Predicate logic as a Programming Language", *Information Processing 74*, Stockholm, North-Holland, 1974, pp 569–574.

73. D. C. Kozen, *The Design and Analysis of Algorithms*, Springer-Verlag, 1992.

74. H. Krishna, B. Krishna, K. Y. Lin, and J. D. Sun, *Computational Number Theory and Digital Signal Processing*, CRC Press, Boca Raton, Florida, 1994.

75. J. C. Lagarias, "Pseudorandom Number Generators", In: *Cryptology and Computational Number Theory*, Proceedings of Symposia in Applied Mathematics 42, American Mathematics Society, 1990, pp 115–143.

76. J. van Leeuwen (editor), *Handbook of Theoretical Computer Science*, Vol. A. Algorithms and complexity, Vol. B. Formal models and semantics, Elsevier Science Publishers, 1990.

77. R. S. Lehman, "Factoring Large Integers", *Mathematics of Computation*, **28**, 126(1974), pp 637–646.

78. H. W. Lenstra, Jr., "Factoring Integers with Elliptic Curves", *Annals of Mathematics*, **126**, (1987), pp 649–673.

79. A. K. Lenstra and H. W. Lenstra, Jr., *The Development of the Number Field Sieve*, Lecture Notes in Mathematics 1554, Springer-Verlag, 1993.

80. M. Li and P. M. B. Vitányi, "Inductive Reasoning and Kolmogorov Complexity", *Journal od Computer and System Science*, **44**, 2(1992), pp 343–384.

81. R. Lidl and G. Pilz, *Applied Abstract Algebra*, Springer-Verlag, 1985.

82. P. Linz, *An Introduction to Formal Languages and Automata*, D. C. Heath and Company, 1990.

83. R. J. Lipton, *Speeding Up Computations via Molecular Biology*, Department of Computer Science, Princeton University, NJ 08540, 9 December 1994.

84. R. J. Lipton, "DNA Solution of Hard Computational Problems", *Science*, **268**, 28 April 1995, pp 542–545.

85. J. W. Lloyd, *Foundations of Logic Programming*, Second Extended Edition, Springer-Verlag, 1993.

86. Z. Manna, *Mathematical Theory of Computation*, McGraw-Hill, 1974.

87. J. C. Martin, *Introduction to Languages and the Theory of Computation*, McGraw-Hill, 1990.

88. Y. V. Matiyasevich, *Hilbert's Tenth Problem*, MIT Press, 1993.

89. K. S. McCurley, "The Discrete Logarithm Problem", In: *Cryptology and Computational Number Theory*, Proceedings of Symposia in Applied Mathematics 42, American Mathematics Society, 1990, pp 49–74.

90. W. S. McCulloch and W. Pitts, "A Logical Calculus of the Ideas Immanent in Nervous Activity", *Bulletin of Mathematical Biophysics*, **5**, (1943), pp 115–133.

91. J. F. McKee, *Lecture Notes in Elliptic Curves and Factoring*, Dept of Mathematics, The University of Edinburgh, 1997.

92. A. R. Meijer, "Groups, Factoring, and Cryptography" *Mathematics Magazine*, **69**, 2(1996), pp 103–109.

93. A. Menezes, *Elliptic Curve Public Key Cryptosystems*, Kluwer Academic Publishers, 1993.

94. B. Meyer and and V. Müller, "A Public Key Cryptosystem Based on Elliptic Curves over $\mathbb{Z}/n\mathbb{Z}$ Equivalent to Factoring", *Advances in Cryptology – EUROCRYPT '96*, Lecture Notes in Computer Science 1070, Springer-Verlag, 1996, pp 49–59.

95. G. Miller, "Riemann's Hypothesis and Test for Primality", *Journal of Systems and Computer Science*, **13**, (1976), pp 300–317.

96. M. L. Minsky, *Computation: Finite and Infinite Machines*, Prentice-Hall, Engewood Cliffs, N. J., 1967.

97. R. N. Moll, M. A. Arbib and A. J. Kfoury, *An Introduction to Formal Language Theory*, Springer-Verlag, 1989.

98. P. L. Montgomery, "Speeding the Pollard's and Elliptic Curve Methods of Factorization", *Mathematics of Computation*, **48**, (1987), pp 243–264.

99. P. L. Montgomery, "A Survey of Modern Integer Factorization Algorithms", *CWI Quarterly*, **7**, 4(1994), pp 337–394.

100. F. Morain, *Courbes Elliptiques et Tests de Primalité*, L'Université Claude Bernard - Lyon I, 1990.

101. M. A. Morrison and J. Brillhart, "A Method of Factoring and the Factorization of F_7", *Mathematics of Computation*, **29**, (1975), pp 183–205.

102. I. Niven, H. S. Zuckerman and H. L. Montgomery, *An Introduction to the Theory of Numbers*, 5th Edition, John Wiley & Sons, 1991.

103. O. Ore, *Number Theory and its History*, Dover Publications, New York, 1988.

104. A. M. Odlyzko, "Discrete Logarithms in Finite Fields and their Cryptographic Significance", *Advances in Cryptography – Proceedings of EURO-CRYPT '84*, Lecture Notes in Computer Science 209, Springer-Verlag, 1984, pp 225–314.

105. V. Pan, *How to Multiply Matrices Faster*, Lecture Notes in Computer Science 179, Springer-Verlag, 1984.

106. R. G. E. Pinch, *Mathematics for Cryptography*, Queen's College, University of Cambridge, 1997.

107. P. Pleasants, The University of the South Pacific, Laucala Campus, Fiji, *private communication*, 1997.

108. S. C. Pohlig and M. Hellman, "An Improved Algorithms for Computing Logarithms over GF(p) and its Cryptographic Significance", *IEEE Transactions on Information Theory*, **IT-24**, (1978), pp 106–110.

109. J. M. Pollard, "A Monte Carlo Method for Factorization", *BIT*, **15**, (1975), pp 331–332.

110. J. M. Pollard, "Monte Carlo Methods for Index Computation (mod p)", *Mathematics of Computation*, **32**, (1980), pp 918–924.

111. C. Pomerance, J. L. Selfridge and S. S. Wagstaff, Jr., "The Pseudoprimes to $25 \cdot 10^9$", *Mathematics of Computation*, **35**, 151(1980), pp 1003–1026.

112. M. O. Rabin, "Complexity of Computations", In: *ACM Turing Award Lectures 1966–1985*, ACM Press, New York, pp 319–338.

113. M. O. Rabin, "Probabilistic Algorithms for Testing Primality", *Journal of Number Theory*, **12**, (1980), pp 128–138.

114. V. J. Rayward-Smith, *A First Course in Formal Language Theory*, (and also *A First Course in Computability*), Blackwell, Oxford, 1987.

115. P. Ribenboim, *The Little Book on Big Primes*, Springer-Verlag, New York, 1991.

116. P. Ribenboim, "Selling Primes", *Mathematics Magazine*, **68**, 3(1995), pp 175–182.

117. P. Ribenboim, *The New Book of Prime Numbers Records*, Springer-Verlag, 1996.

118. H. Riesel, *Prime Numbers and Computer Methods for Factorization*, Birkhäuser, Boston, 1987.

119. H. Ritter, et al., *Neural Computation and Self-Organising Maps – An Introduction*, Addison-Wesley, 1992.

120. R. L. Rivest, A. Shamir and L. Adleman, A Method for Obtaining Digital Signatures and Public-Key Cryptosystems, *Communications of the ACM*, **21**, 2(1978), pp 120–126.

121. J. A. Robinson, "Machine-Oriented Logic", *Journal of the ACM*, **12**, 1(1965), pp 23–41.

122. K. Rosen, *Elementary Number Theory and its Applications*, 3rd Edition, Addison-Wesley, 1993.

123. G. Rozenberg and A. Salomaa, *Cornerstones of Undecidability*, Prentice-Hall, 1994.

124. A. Salomaa, *Computation and Automata*, Cambridge University Press, 1985.

125. E. S. Santos, "Probabilistic Turing Machines and Computability", *Transactions of American Mathematical Society*, **8**, 1969, pp 701–710.

126. R. Schoof, "Elliptic Curves over Finite Fields and the Computation of Square Roots mod p", *Mathematics of Computation*, **44**, (1985), pp 483–494.

127. M. R. Schroeder, *Number Theory in Science and Communication*, 2nd Edition, Springer-Verlag, 1990.

128. A. Shamir, "Factoring Numbers in $\mathcal{O}(\log n)$ arithmetic Steps", *Information Processing Letters*, **8**, 1(1979), pp 28–31.

129. P. Shor, "Algorithms for Quantum Computation: Discrete Logarithms and Factoring", *Proceedings of 35th Annual Symposium on Foundations of Computer Science*, IEEE Computer Society Press, 1994, pp 124–134.

130. P. Shor, "Analog Computational Power" (Technical Comment), *Science*, **271**, Jan 1996, page 92.

131. P. Shor, "Polynomial-Time Algorithms for Prime Factorization and Discrete Logarithms on a Quantum Computer", *SIAM Journal on Computing*, **26**, 5(1997), pp 1484–1509.

132. H. T. Siegelmann, "Computation Beyond the Turing Limit", *Science*, **268**, 28 April 1995, pp 545–548.

133. R. D. Silverman, "Massively Distributed Computing and Factoring Large Integers", *Communications of the ACM*, **34**, 11(1991), pp 95–103.

134. D. R. Simon, "On the Power of Quantum Computation", *SIAM Journal on Computing*, **26**, 5(1997), pp 1474–1483.

135. D. Slowinski, "Searching for the 27th Mersenne Prime", *Journal of Recreational Mathematics*, **11**, 4(1978-79), pp 258–261.

136. M. Sipser, *Introduction to the Theory of Computation*, PWS Publishing Company, Boston, 1997.

137. R. Solovay and V. Strassen, "A Fast Monte-Carlo Test for Primality", *SIAM Journal on Computing*, **6**, (1977), pp 84–85.

138. T. A. Sudkamp, *Languages and Machines – An Introduction to Theory of Computer Science*, Addison-Wesley, 1988.

139. S-A. Tärnlund, "Horn Clause Computability", *BIT*, **17**, 1977, pp 215–226.

140. J. K. Truss, *Discrete Mathematics for Computer Scientists*, Addison-Wesley, 1991.

141. A. M. Turing, "On Computable Numbers, with an Application to the Entscheidungsproblem", *Proceedings of the London Mathematical Society*, **42**, (1936), pp 230–265. "A Correction", *ibid*, **43**, (1937), pp 544–546.

142. D. Welsh, *Codes and Cryptography*, Oxford University Press, 1989.

143. H. Wiener, "Cryptanalysis of Short RSA Secret Exponents", *IEEE Transactions on Information Theory*, **IT-36**, 3(1990), pp 553–558.

144. H. C. Williams, "Factoring on a Computer", *Mathematical Intelligencer*, **6**, 3(1984), pp 29–36.

145. H. Woll, "Reductions among Number Theoretic Problems", *Information and Computation*, **72**, (1987), pp 167–179.

146. S. Y. Yan, "Primality Testing of Large Numbers in Maple", *Computers & Mathematics with Applications*, **29**, 12(1995), pp 1–8.

147. S. Y. Yan, *Perfect, Amicable and Sociable Numbers – A Computational Approach*, World Scientific Publishing, 1996.

148. K. C. Zeng, C. H. Yang, D. Y. Wei and T. R. N. Rao, "Pseudorandom Bit Generators in Stream-Cipher Cryptography", *IEEE Computer*, 2(1991), pp 8–17.

Index